AF564711

ADVANCE STUDIES IN AGRICULTURE

ENCYCLOPEDIA OF AGRICULTURE I

ADVANCE STUDIES IN AGRICULTURE

By
Dr. Renuka Sharma

Published by:
Tilak Wasan
DISCOVERY PUBLISHING HOUSE PVT. LTD.
4383/4B, Ansari Road, Darya Ganj
New Delhi-110 002 (India)
Phone : +91-11-23279245; 23253475; 43596065
E-mail: discoverybooksindia@gmail.com
discoverypublishinghouse@gmail.com
namitwasan9@gmail.com
web : www.discoverypublishinggroup.com

First Edition: **2013**

Reprinted: **2019**

ISBN: 978-93-5056-020-4 (Set)

ISBN: 978-93-5056-021-1

Advance Studies in Agriculture

Printed at:
Infinity Imaging Systems
Delhi

Preface

The breakthrough in science that permitted genes to be identified and manipulated as molecular ushered in the agricultural technology era, which is now more than a decade old. The new tools of agricultural technology are changing the way scientists can address problems in the life sciences; agriculture is one area facing major changes as a result of this new technology. The unanticipated rapid ratio at which discoveries and their applications in technology have unfolded has stored the capacity of society – more specifically, our agricultural research and educational institutions to absorb and adjust to change. We are challenged by pressing decisions opportunities, and problems that we face now and will continue to face in the future. Competition from abroad impels us to devise and use new technologies that can improve the efficiency and quality of agricultural production. These concerns led to this study – an overview of how the agricultural research system is responding to the latest technology and how it might prepare for future opportunities. Agricultural technology is moving in many directions with positive results – crop improvement, production of transgenic plants, vaccine development and diagnostic methods are some impending applications – but the development of genetic engineering's tools can be found in almost every agricultural discipline. The exception is that, through lessions learned, adequate food will be made available to the whole world in the future. The present tile **"Endcyclopedia of Agriculture"** has been planned, written, and edited with the intention of being useful for the beginners, researchers and scientists involved in the field of agricultural genetic transformation.

I would like to express my sincere thanks to Dr. M.P. Arora, whose continuous inspiration and encouragement initiated me in bringing out this title.

I am specially indebted to my husband Mr. Rajeev Sharma for his enthusiastic support, constant helpfulness and good spirit during the writing of this title.

Quite frankly, this book would not have been written without the aid of my daughter Shreya, who co-operated patiently during periods of neglect.

Special and sincere thanks to my parents and in laws for their blessings and continuous encouragement in bringing out this title.

To make the work more comprehensive and informative, I have consulted many authoritative books, research journals, abstracts, monographs etc., so there can be no claim to originality except in the manner of treatment.

I also express thanks to my friends and colleagues whose continuous inspirations have initiated me to bring out this title.

I express my gratitude to Mr. Wasan and staff of M/s Discovery Publishing House Pvt. Ltd. for their whole hearted co-operation in the publication of this title.

I acknowledge the fact that the development and publication of this book would not have been possible without the assistance and encouragement of colleagues, research scholars and students. They are so numerous to mention. I thank all of them most warmly for helping to create the present title.

Any worthwhile criticism and suggestions for improvement would be thankfully acknowledged.

Renuka Sharma

CONTENTS

Chapter 1

INTRODUCTION

Biotechnology is a scientific discipline with *focus* on the *exploitation* of *metabolic* properties of living organisms for the production of valuable products of a very different structural and organizational level for the benefit of men. The products can be the *organisms* themselves (i.e., biomass or parts of the organismic body), products of cellular or *organismic* metabolism (i.e., enzymes, metabolites), or products formed from endogenous or exogenous substrates with the help of single *enzymes* or complex metabolic routes.

The organisms under question vary from microbes (*bacteria, fungi*) to animals and plants. In addition to intact organisms, isolated cells or enzyme *preparations* are employed in *biotechnology*. The possibility to submit the producing organisms or the cellular systems to technical and even industrial procedures has led to highly *productive* processes.

The products of biotechnology are of importance for *medicine, pharmaceutical sciences, agriculture, food production, chemistry,* and *numerous* other *disciplines*. Biotechnology receives the necessary scientific and technical information from a considerable number of disciplines. Cell biology, *morphology* of the employed organisms, *biochemistry, physiology, genetics,* and various technical fields are major sources.

In the last two decades, molecular biology and gene technology have *substantially* contributed to the *spectrum* of scientific disciplines *forming biotechnology*. As is always true for progress in natural sciences, it is especially true for biotechnology that more rapid development and gain of higher standards depend on the improvement of methods.

In the historical development of biotechnology, microbes have been used *preferentially*. They still offer an extremely rich potential for biotechnological application. Animal systems and their cells are also *valuable* systems, especially in view of the very costly products (i.e., *antibodies, vaccines*).

Although much later in the chronological process, plant biotechnology has made an impressive development in gaining basic and applicable knowledge as well as in establishing production processes. It is therefore justified to speak of an emerging field. Major steps will be discussed in this chapter.

A LONG HISTORY TO REACH A HIGH STANDARD

In each *ecosystem* plants and other *photosynthetically* active organisms are responsible for primary production, which provides the energetic and nutritional basis for all *subsequent* trophic levels. The extremely high ability of plants to adapt to all kinds of environmental conditions and *ecosystems* has led to an *extremely* wide and differentiated spectrum of plants.

Since ancient times higher plants have formed the main source of food for men, and therefore, concomitant with early phases of settlements and agriculture, men started to establish and improve crop plants. *Archeological* evidence has clearly shown how long well-known crop species (i.e., *maize, cereals, legumes*) have been grown, modified by selection, and thus improved in quality and yield.

Plant breeding is indeed an old art that has been *continuously* developed in efficiency and scope. Quite *typical* for quality breeding of, for instance, cereals is the long *procedure* required (sometimes *decades*) to reach particular genotypes and to cross in specific genes or traits.

An interesting achievement in breeding of wheat is characterized by the term *green revolution,* in which wheat genotypes from many different countries were used successfully on a very large scale to breed *high-yielding* and durable lines. For many countries such new varieties were a very great *improvement* for their agriculture.

Another important goal in breeding improved crop plants is the often achieved adaptation to unfavourable environmental conditions (i.e., *heat, drought, salt,* and other cues). Although good results have been obtained, such efforts will undoubtedly remain in the focus of future efforts. Better insight into the *physiology, biochemistry,* and *chemical* reactions as well as the gene regulation of the *endogenous adaptation* and defense *mechanisms* that plants can express will contribute to these objectives.

Gene technology will be an essential component in these efforts. Another characteristic feature of the long-term breeding of cereals, potatoes, or vegetables is the fact that during the long periods the shape and the outer appearance of the plants have changed so much that the original wild types were either lost or no longer easily identified as *starting material.* A typical example is corn.

Modern agricultural crop plants are also bred for very uniform physical appearance, time of flowering, and maturity so that harvest by machines in an industrial manner is possible (examples are cotton, maize, and cereals). It is a feature of our high-yielding agriculture that all possible mechanical techniques are being employed.

Very precious treasures for future agriculture and for plant biotechnology are the gene banks and the International Breeding Centers, where great numbers of genotypes of crop plants are multiplied and carefully preserved for long periods of time. Such *"pools of genes"* represent the basis for *sustainable* development and allow future *programs* for improved adaptation of plants to *human* needs.

Fortunately, the understanding has gained ground in recent years that in addition to crop plants all types of wild plants, in every ecosystem, must be preserved because of the genetic resources to be possibly *exploited* in the future. An interesting development in itself, with a long history and remarkable contributions to culture and art, is the *numerous* and sometimes highly *sophisticated ornamental* plants produced in many countries.

Beauty of colour and flower shape were the guidelines in their breeding and selection. Rather early in this development the value of mutagenetic reagents was learned, and these ornamentals also served to shape the term of a mutant. Recent biochemical studies with, for example, *snapdragon, tulip, chrysanthemum,* or *petunia* and their flavonoid constituents clearly presented evidence that the various flower colours can contribute to identifing *biosynthetic* pathways.

In connection with flower pigments, which are secondary metabolites, it should be remembered that numerous other secondary constitutents of very different chemical structures are valuable *pharmaceuticals.* In many countries knowledge of plants as sources of *drugs* has been cherished for long times. Modern *pharmacological* and chemical studies have helped in the *identification* of the relevant compounds.

Such investigations are still considered important objectives of plant biotechnology. In some cases extensive breeding programs have already achieved the selection and mass cultivation of high-yielding lines. In *modern pharmacy,* about 25% of drugs still contain active compounds from natural sources, which are primarily *isolated* from plants.

For a good number of years in the period from 1950 to 1980, plant biochemistry and plant biophysics concentrated on elucidation of the photosynthetic processes. The pathways of CO_2 *assimilation* as well as structure, energy transfer reactions, and *membrane* organization of *chloroplasts* and their *thylakoids* were objectives of primary interest.

Chloroplast organization and molecular function of this organelle can be regarded as well-understood fields in plant biochemistry and physiology. The last three decades of the 20th century were characterized by very comprehensive molecular *analyses* of chemical reactions, metabolic pathways, cellular organization, and adaptctive responses to *unfavourable* environmental conditions in *numerous* plant systems.

A very broad set of data has been accumulated so that plant biochemistry and closely related fields can now offer a good understanding of plants as multicellular organisms and highly *adaptative* systems. From a molecular point of view, the construction and the functioning of the different tissues and organs have become clear.

Numerous experimental techniques have contributed to this development and some are typical plant-specific methods (i.e., cell culture techniques) with a very broad scope of application. A *fascinating* field of modern plant biochemistry concerns the elucidation of the function and the molecular mechanisms of the various *photoreceptor* systems of higher plants. Red/far red receptors, blue light-absorbing *cryptochromes*, and *ultraviolet* (UV) light *photoreceptors* are essential components of plant development.

These systems translate a light signal into physiological responses via gene activation. Quite *remarkable, phosphorylated/ nphosphorylated* proteins are the essential components of the signal transduction system. Biotechnology will gain from this knowledge, and highly sensitive sensor systems could possibly be constructed. In the history of plant sciences and biotechnology, the recent development of molecular biology and the introduction of gene technology deserve emphasis.

Isolation, characterization, and functional determination of genes have become possible. Many plant genes were rather rapidly identified, and the number is increasing at enormous speed. Promoter analyses and identification of promoter binding proteins have decisively contributed to an understanding of the organization and function of plants as organisms consisting of multiple tissues and different organs.

The phenomena of *multigenes* and multiple enzymes in one protein family were further revealed. Many different techniques in molecular biology and gene technology turned out to be extremely valuable. Recognition of the biology of *Agrobacterium tumefaciens* and application of its transferred DNA (*T-DNA*) system represented giant leaps forward. In general, because of these modern gene technological methods, plant biotechnology has grown into a new *dimension* with *putative* future possibilities that can hardly be *overestimated.*

In the following sections of this chapter several recent and future aspects of biotechnological relevance will be discussed.

PLANT TISSUE AND CELL CULTURES—A VERY VERSATILE SYSTEM

The present status of plant biotechnology cannot be evaluated without appreciation of the many possibilities and the potential of organ, tissue and cell suspension cultures. Plants of wide taxonomic origin have been subjected to culture under *strictly aseptic* conditions. Completely chemically defined media supplemented with growth regulators and *phytohormones* are the basis for the exploitation of this technique. Depending on the explant and the culture conditions cells either preserve their state of biochemical and morphological *differentiation* or return to a status of *embryogenic, undifferentiated cells.*

The former situation can be used for organ cultures (e.g., pollen, anthers, flower buds, roots), whereas the latter leads to many callus and *suspension* types of cultures. For example, the cell culture technique has opened a facile route to haploid cells and plants, and such systems are of great importance for genetic and *breeding* studies.

Whenever a *heterogeneous* group of cells can be turned into a state of *practically* uniform cells, this much less complicated cellular system can then be exploited to study many problems. This has been performed with plant cell cultures for some 30 years now. Growth of cells in medium-size and large volumes has opened interesting applications for plant biotechnology. *Numerous physiological, biochemical, genetic,* and *morphological* results and data on cellular regulation stem from such *investigations.*

Various primary and secondary metabolic routes have been elucidated with the help of cell culture systems. The typical sequence in pathway identification was first product and intermediate characterization, then enzyme studies, and finally isolation of genes. Furthermore, application of gene technology in the field of transgenic plants depends to some extent on the tissue and cell culture techniques.

Plants are characterized by *totipotency,* which means that each cell possesses and can express the total genetic potential to form a fully fertile and complete plant body. This fact, highly remarkable from a cell biological point of view, is the genetic basis for important and widely used *applications* of the cell culture technique.

Differentiation of single cells or small aggregates of cells into embryos, tissues, and even plants allows the selection of interesting genotypes for several different fields of plant application. The *well-established* procedures for mass *regeneration* of valuable *specimens* of ornamental and crop plants constitute an important business section in agriculture and gardening. *Endangered* plant species can be saved from extinction so that valuable gene pools will not *disappear.*

Remarkable progress has been achieved in mass regeneration of trees from single plants or tissue pieces. This will undoubtedly be of further great benefit for forestry because several problems in tree *multiplication* can thus be *circumvented.* Furthermore, it should be mentioned that plant cell suspension cultures possess a great potential for *biotransformation* reactions in which *exogenously* applied substrates are converted in sometimes high yields.

Position and *stereospecific hydroxylations, oxidations, reductions,* and especially interesting *glucosylation* of very different substrates have been found. The plant cell culture technique has allowed the facile isolation of mutants from many plant species. The overwhelming importance of mutants for biochemical and genetic studies has been known for decades. Over the years, *mutations* from all areas of cellular *metabolism* have been selected and characterized.

A good deal of our basic knowledge of the functioning and regulation of organisms and cells and their organelles stems from work with mutants. The various techniques of plated, suspended, or feeder cell-supported cell systems and even protoplasts have found wide applications. The normal rate of mutation and also increased levels of mutated cells induced by physical (UV light, high-energy irradiation) or chemical mutagens (many such compounds are known) have been used.

The specific advantage of cell cultures for mutant selection is the possible isolation of single cells from a mass of unmutated ones. *Heterotrophic, photomixotrophic,* and *photoautotrophic cells* are available, and thus different areas of cell metabolism can be screened for mutations. In the cell culture field, regulatory mutants (i.e., excessive accumulation of products of primary and secondary metabolism including visible pigments), uptake mutants (i.e., the normal cellular transport systems of nutrients into cells are invalidated), and resistance mutants (i.e., pronounced cellular tolerance against toxic compounds such as *mycotoxins, pesticides, amino* acid *analogues,* or salt) have especially been *characterized.* Various *auxotrophic* mutants in the field of growth regulators have also been of *considerable* value.

As an example, a series of studies using photoautotrophic cell *suspension* cultures and the highly toxic herbicide metribuzin blocking electron transport in photosystem II will be cited. A series of single, double, or even triple mutants of the D1 protein coded by the *chloroplast psbA* gene were selected and thoroughly characterized. The various lines allowed interesting insights into the mechanism of *herbicide* interference with the D1 protein.

In a discussion of plant *mutants* resistant to herbicides, the impressive results on herbicide-resistant crop plants require mention. Many of the *mechanistic* and *metabolic* aspects of *herbicide* resistance were first elucidated with cell cultures.

Modern plant biotechnology has a wide choice of biochemical solutions for *herbicide* resistance by *inactivation* and *detoxification* reactions. Several major crop plant species (i.e., soybean, cotton, maize, rape) are presently cultivated to a large extent in the form of appropriately *manipulated genotypes*. This development is on the one hand regarded as a major advantage for *agriculture* but on the other hand as a subject of extensive and often very critical public debate.

FROM GENES TO PATHWAYS TO BIOTECHNOLOGICAL APPLICATION

A landmark in our understanding of the structure, the organization, and the functioning of multicellular organisms is described by the extensive *eukaryotic* genome *sequencing* projects in the last decade. The genomes of the yeast *Saccharomyces cerevisiae,* the nematode *Caenorhabditis elegans,* and the fruit fly *Drosophila melanogaster* clearly revealed the genetic basis of the similarities and the differences of diverse *multicellular* organisms.

This modest number should perhaps be compared with the 56 completed prokaryotic genome sequences (10 strains of archaea and 46 of bacteria) and the more than 200 in progress. In general, the number of genome sequencing projects is increasing rapidly.

The three eukaryotic genomes have a similar set of 10,000-15,000 different proteins, suggesting that this is the minimal complexity required by extremely diverse eukaryotes to execute development, essential metabolic pathways, and adequate responses to their environment. These available *eukaryotic genome* sequences thus also document basic lines of organismic evolution.

The recent completion and publication of the first complete genome sequence of a flowering plant, the brassica *Arabidopsis thaliana,* represents a further giant step forward. The genome of this model plant, dispersed over five chromosomes, documents for plant scientists a complete set of genes *controlling* developmental and growth patterns, primary and secondary *metabolism,* adaptative responses to *environmental cues,* and disease resistance.

This full genomic sequence provides a means for analyzing gene function that is also important for other plant species, including commercial and agricultural crops. Plant biotechnology greatly benefits from the *Arabidopsis* genome project. The large set of identified genes and also the hitherto functionally unknown, predicted genes form the basis for more sophisticated plant genetic analysis and plant improvement by construction of plants better adapted to human needs.

The complete *Arabidopsis* genome appears to harbor 25,498 genes, of which 17,833 can presently be classified as predicted from careful sequence *comparisons* with genes from other organisms. Again, functional classification comprises altogether nine areas of metabolism as known from the *aforementioned* other *eukaryotic* genome sequence projects.

Thus 7665 genes (roughly 30%) remain to be functionally identified. In order to outline the amount of research still to be fulfilled with the *Arabidopsis* genome, it should be mentioned that altogether only some 10% of the genes have been characterized experimentally. Although a detailed description of the *A. thaliana* genome cannot be given in this chapter, a few plant-specific aspects will be presented because they appear to be of importance for future plant *biotechnological* application, i.e., selection of specific lines, genetic modification, or transformation at sites of characteristic *plant-specific* properties.

A considerable number of the nuclear gene products (approximately 14%) are predicted to be targeted to the chloroplasts as indicated by *appropriate* signal peptide sequences. Such a value indicates the massive influx of nuclear-coded proteins into plastids. *Protein kinases* and the proteins containing a disease resistance protein marker as well as domains characteristic of *pathogen* recognition molecules are quite abundant in the *Arabidopsis* genome.

The essential elements are *domains* (*intracellular* proteins with an amino terminal leucine zipper domain, a nucleotide binding site typical of small G proteins and *leucine-rich* repeats) that were already known from the *Arabidopsis RPS2* and *RPM1* genes as well as from other plant *R* genes (R, plant disease resistance genes).

These findings in the *Arabidopsis* genome as well as all other molecular data on plant mechanisms for recognizing and responding to pathogens indicate that pathways transducing signals in response to pathogens and various other environmental factors are more essential elements in plants than in other eukaryotes. Uptake, distribution, and compartmentalization of organic and *inorganic nutrients*; energy and signal transduction; and *channeling* of *metabolites* and end products are very essential elements in a plant's life.

Membrane transport systems are especially decisive for a sessile organism composed of many different organs and tissues such as higher plants. Therefore, the *comparatively* large number of predicted membrane transport systems in the *Arabidopsis* genome appears understandable. Furthermore, it is not surprising that these transport systems are the well-known plant proton-coupled *membrane potentials* (in contrast to the animal and the fungal *sodium-coupled* systems).

Proteins with sequence homologies to channel proteins and peptide transporters are further prominent components in the *Arabidopsis* genome. The importance of peptide transporters is further emphasized by the great number of *Arabidopsis* genes encoding Ser/Thr protein kinases; thus, plant signaling pathways are presumably performed by the *peptidepeptide phosphate* mechanism. Future biotechnological applications of these documented plant *transporters* are, for example, construction of more facile cellular sequestration processes for biologically active or *toxic xenobiotics* (i.e., pesticides, parasitic toxins) into *vacuoles*.

Arabidopsis has over three times more transcription factors than identified in the *genomes* of the other eukaryotes. This great number should best be seen in connection with the expanded number of genes (also found in *Arabidopsis*) encoding proteins functioning in inducible metabolic pathways controlling defense and environmental *interaction*. Such routes are *characteristic* features of higher plants.

Increased numbers of transcription factors are logically required to integrate gene function in response to the vast range of environmental factors that plants can perceive. Needless to say, these genes and their products represent very important tools for future biotechnology.

Finally, one aspect of the complex *Arabidopsis* genome analyses referring to signal transduction will be mentioned. The very high number of *mitogen-activated* protein (MAP) kinases in combination with the high number of PP2C protein phosphatases and biochemical evidence from inducible signal transduction studies support the assumption that plants operate *signaling* pathways with MAP kinase *cascade moduls*.

As mentioned before, the presence of genes encoding enzymes for pathways that are unique to *vascular* plants is of great importance for biotechnology. Thus, several hundred genes with probable roles in the synthesis and modification of cell wall *polymers* clearly emphasize the *decisive* role of cell walls in the life of plants.

Cellulose synthetases and related enzymes involved in polysaccharide formation, polygalacturonases, pectate lyases, pectin esterases, β-1,3-glucanases, and numerous groups of polysaccharide *hydrolases* were among the most prominent enzymes indicated by the gene sequences of *Arabidopsis*. Again, this knowledge offers a wide range of experimental tools for either structural modification of cell walls in the living plant or in vitro studies with suitable substrates and isolated enzymes.

Furthermore, the considerable number of genes encoding peroxidases and diphenol oxidases (*laccases*) points at the importance of oxidative processes most likely in connection with lignin, suberin, and other polymers. *Decisive* reactions are thought to be cell wall stiffening and *modification* processes including cross-linking reactions of cell wall proteins. In connection with cell wall-located proteins, the large group of glycine-rich proteins (GRPs) may be used to show that plant molecular biology and *biotechnology* are quite often confronted with very complex *metabolic* systems.

The plant GRPs possess a remarkable sequence homology with numerous animal proteins that are well known for their pronounced adhesive properties (GRPs from shells), extreme mechanical flexibility (*spider silk*), or ability to resist high mechanical *pressure* (human skin *collagen*). These properties result from the specific amino acid sequence and certain repetitive *motifs*.

In the case of plant GRPs, where again interesting and valuable properties for biotechnological application can be predicted from the sequences and functions, the characteristic elements are many glycine-rich motifs (i.e., GGGx or GGxxxGGx with x *tanding* for *tyrosine*, *histidine*, or *serine*).

The gene sequences allow differentiation between two large groups of GRPs. Proteins with an N-terminal signal peptide are designed for apoplastic transport and cell wall *localization*. Protection of cells during antimicrobial defense and increased cell wall resistance toward enzymic digestion by *microbial enzymes* are logical functions.

GRPs without N-terminal signal peptides are, in contrast, characterized by *RNA-binding motifs*, *zinc finger* domains or regions with oleosin character, i.e., proteins that *stabilize oil* droplets ("*oleosomes*") in the cytoplasm. In the last point, the ability of GRPs to form conformations with *hydrophobic* surfaces can be seen.

The real complexity in the GRP field results from the very different cues leading to their induction. *Phytohormones*, *water stress*, *cold*, *wounding*, *light*, *nodulation*, and *pathogen* attack have been demonstrated. Cytosolic compartments or matrix structures such as *xylem*, protoxylem, cell walls, epidermal cells, anthers, or root tips are the alternative expression sites. The *putative* function always appears to be to impregnate sensitive *compartments* with *hydrophobic* seals.

A complete understanding of this complexity requires, in addition to the genes, identification of the various transcription factors and regulatory genes in order to open the *GRP field* for biotechnological application. As a further illustration of surprising data obtained from the *Arabidopsis* genome project, the great number of genes encoding cytochrome P450 *oxygenases* will be mentioned.

The P450 oxygenases represent a superfamily of heme-containing proteins that catalyze various types of *hydroxylation* reactions using NADPH and 0_2. In plants these membranebound (endoplasmic reticulum) enzymes are known to be involved in *pathways* leading to various secondary metabolites as well as routes to plant growth *regulators*.

Although of great importance in plant metabolism, the various plant P450s are poorly understood with only a very limited number characterized to any extent. In this context the very high number (-286) of *Arabidopsis* P450 genes must be seen in contrast to the 94 genes in *Drosophila,* the 73 genes in *C. elegans,* and just 3 genes in *S. cerevisiae.*

Intensive analyses of plant P450 oxygenases will represent a major task in future years. Biotechnology will greatly benefit from such studies because hydroxylation-oxygenation pathways are already known as routes to valuable compounds. Further aspects of P450 will be discussed in connection with flavonoid and phytoalexin formation as well as xenobiochemical metabolism.

The great number of sequenced genes, especially in cases of *isoenzymes* encoded by *multigene* families, leads to an important question: Under which conditions of growth, tissue, and organ development or changing *environmental* conditions are such genes (*selectively*) transcribed? Knowledge of selective gene activation and changes therein under normal or adverse conditions is of great importance for our understanding of the complexity of multicellular *organisms* and for *plant biotechnology.*

A *fascinating* new technique (*DNA microarray technology*) allows the determination of RNA expression profiles on the genome level with many hundred genes at the same time. Samples of *sequenced* genes or characteristic gene fragments are immobilized as *microspots* on membranes or glass slides. These *arrays* are treated with mRNA preparations from the plant material under *investigation.*

The process of specific DNA-mRNA hybridization can be followed or *automatically* recorded by various techniques of light emission or colour formation. It is easy to predict that in the future plant sciences will benefit from *DNA microarray* technology to the same extent as already shown for medical and *pharmaceutical* applications.

Among other features, plants are characterized by their overwhelming number of structurally highly diverse secondary metabolites. These compounds (examples are *alkaloids, terpenoids, flavonoids,* and many other classes) are not essential for growth, energy conversion, and other primary metabolic pathways. They are, however, essential for interaction of the plant with its environment and other organisms; they are said to determine the "*fitness*" of a plant.

Elucidation of many of their *biosynthetic* pathways, characterization of the enzymes involved, and cloning of the genes have been performed over many years. Detailed knowledge of the organ- or tissue-specific localization and integration of these compounds in developmental processes has been accumulated. Furthermore, many secondary metabolites of numerous different structural classes are well known for their biological (i.e., roles as *attractants, repellents,* defense compounds of plants to interact with other *organisms*) and physiological (organoleptic and other sensory properties and UV protection) characteristics as well as their *medicinal* and *pharmaceutical* value.

Pharmaceuticals from plants still form a large portion of drugs in human therapy. Secondary metabolites will undoubtedly continue to be of great importance. Detecting, isolating, and producing biologically or *pharmaceutically* active secondary plant *metabolites* are *high-priority* objectives in many laboratories around the world. Such studies greatly benefit from the *tremendous* progress in analytical processes for valuable product recognition.

The search for valuable plant secondary metabolites can make use of the large number of plants that have so far not been analyzed to any extent. Another interesting aspect of the search for new secondary products is the fact that most likely all plants have a much greater genetic *potential* for the formation and *accumulation* of such products than actually expressed during normal growth conditions.

Under stress (i.e., heat, drought, cold, high salt concentrations) and other difficult environmental conditions (i.e., UV irradiation, high light intensity) plants tend to form a much wider spectrum of secondary metabolites. Thus, numerous compounds (e.g., alkaloids, quinones, phenolics, lignans) are found as stress-related metabolites.

Especially in response to pathogen (i.e., *bacterial, fungal, viral*) infection, a wide range of antimicrobial compounds called phytoalexins are inducibly formed de novo around infection sites. The large number of such phytoalexins indicates the reservoir of genetic information for *secondary* product formation that will be *activated* under particular *circumstances.*

The importance of phytoalexins as efficient antimicrobial defense compounds is elegantly demonstrated by the transfer of genes encoding key biosynthetic enzymes into plants that do normally not produce these compounds. The ability to synthesize the groundnut stilbene phytoalexin *resveratrol* has been expressed in tobacco, which resulted in much improved resistance of the *transgenic* plant toward established tobacco *fungal parasites.* This strategy to alter the spectrum of secondary metabolites in a plant by directing the flow of *constitutive* precursors into new products represents a valuable approach for modern plant biotechnology. Other examples, especially in the field of flavonoids and isoflavonoids, are feasible and are under *investigation.* With regard to *pharmaceutical* products, the value of *transgenic* plants has repeatedly been *demonstrated.*

A challenging field for plant biotechnology is the *anthocyanin pigments* in flowers. The introduction of additional *hydroxyl* functions in ring B by transfer of genes coding specific cytochrome P450 monooxygenases opens the possibility to create flowers with deeper (red-blue) colour shades. *Prerequisites* are correct vacuolar pH conditions and copigmentation. Plants kept under adverse conditions accumulate not only *phytoalexins* but also normal secondary *metabolites,* various of their biosynthetic *intermediates,* and many new compounds in sometimes high *concentrations.*

In this context, a valuable technique for biotechnological application is connected with cell suspension cultures of the experimental plants in which secondary product *accumulation* and phytoalexin formation are stimulated or induced by treatment of cultures with microbial elicitors. These signal compounds of very different chemical structure (oligo- or polysaccharides of microbial cell wall structures, peptides or proteins of pathogens, as well as regulator compounds as *glutathione, jasmonic acid, salicylic acid*, or *heavy metal ions* as *abiotic stress* compounds) all tend to interfere with cell metabolism via signal transduction cascades to induce stress and defense responses.

A large spectrum of secondary metabolites has been shown to accumulate. Because heterotrophic, photomixotrophic and photoautotrophic cell cultures can be used, the experimental possibilities are quite variable and wide. Elicitation of cell cultures has also been determined as a simple but efficient technique to detect new *cytochrome* P450 *monooxygenases* that are not expressed under *normal* conditions.

The findings on new secondary products formed de novo under particular conditions support the interesting data from the *Arabidopsis* genome project showing that this plant as judged by sets of *unexpected* genes possesses (at least the genetic) potential to form secondary metabolites not yet isolated from *A. thaliana*.

In conclusion, the search for secondary products can use both new plants, not yet analyzed, and plants with known sets of these products because the genetic *potential* has not yet been fully *exploited*.

THE PLANT CELL ORGANELLES CONTAINING GENETIC INFORMATION

Plants possess three cellular compartments containing *genetic* information, namely the nucleus, the plastids, and the *mitochondria*. The genomes of these three *compartments* differ greatly in size and thus in number of heritable traits. The *nucleus* (size of the haploid genome ~1.2×10^8 to 2.4×10^9 bp; ~20,000-40,000 genes) possesses a linear *genome* distributed over several *chromosomes* that normally occur as *diploid* sets of genes with the DNA *material* highly complexed with proteins.

Identification and cloning of *nuclear* genes, their elimination or silencing, and introduction of foreign genes have almost become a routine procedure in *numerous* plant species. The highly *sophisticated* and efficient techniques of modern molecular *biology* that allow substantial *modifications* of nuclear *genomes* will be of utmost importance for plant biotechnology.

The mitochondria of plants carry circular genomes 200-2000 kb in length, differ in the number of genes (~50-70), and even vary considerably between species and sometimes within one plant. Transformation of mitochondrial genomes is in its infancy. The plastids harbor a *circular double-stranded* DNA molecule of 120160 kb with about 130 genes. This genome has been found in all cellular types of *plastids* (i.e., *proplastids, photosynthetically active chloroplasts, chromoplasts*, and *amyloplasts*), and quite remarkably each chloroplast may contain up to 100 identical copies of the plastid *genome*.

Given the fact that each leaf cell may possess as many as 100 chloroplasts, an exceptionally high degree of ploidy (up to approximately 10,000 plastid genomes) is the result for each cell. Successful attempts to engineer the chloroplast genome have so far been restricted to very few systems (i.e., *Chlamydomonas reinhardtii, Nicotiana tabacum, Arabidopsis thaliana)*, but routine procedures with other crop plants suitable for biotechnological application are slowly emerging.

Recent data on the stable genetic transformation of tomato (Ly*copersicum esculentum)* plastids and expression of a foreign protein in fruit represent a major step forward in technology. A key step in the chloroplast transformation experiments was the use of a specified region in the chloroplast genome as a component of transformation vectors in order to target transgenes by homologous recombination.

The transplastomic tomato plants finally obtained were shown to transfer the foreign gene to the next generation via uniparentally maternal transmission. This work also represents a significant breakthrough with regard to biotechnology because of the great advantages of transplastomic plants over conventional transgenic plants generated by transformation of the nuclear genome.

Some advantages can be summarized as follows. Due to the polyploidy of the plastid genome, high levels of transgene expression and foreign protein *accumulation* (up to 40% of total cellular protein) can be expected. Because the chloroplast DNA lacks a compact *chromatin* structure, position effects of gene integration are most likely not involved.

As mentioned earlier, transgene integration by homologous recombination provides an efficient integration system. Finally, as shown for the transplastomic tomato plants, most higher plants follow a strict uniparentally maternal inheritance pattern of chloroplasts, i.e., *absence* of pollen transmission of *transgenes*. Thus, the often criticized spread of transgenes from plants *generated* by nuclear transformation experiments can be avoided. This aspect will undoutedly be of major ecological importance.

It is easy to predict that the availability of transplastomic plants offers a wide range of biotechnological applications. The new technology can be offered for the introduction of new biosynthetic pathways, resistance management of crop plants, and the use of plants as factories for *biopharmaceuticals, proteins, enzymes,* or *peptides.*

METABOLISM OF XENOBIOCHEMICALS

Higher plants are often confronted with a wide range of exogenous organic compounds, of either natural or anthropogenic origin. Products in the latter category (especially prominent are herbicides, insecticides, and various other groups of *pesticides*) are *intentionally* applied to *agricultural* plants and thus are also introduced into the general biosphere.

As expected from the very reason for their application, these environmental chemicals differ greatly in their biological activity or *toxicity* toward different plant species; this variability ranges from highly toxic to nontoxic because the plants' responses vary from very sensitive to highly resistant. This difference in itself allows important *conclusions* for biotechnology when the decisive mechanistic reason has been deciphered.

Great progress has been made in our understanding of the metabolism of these *environmental* chemicals in crop plants. In contrast to *previous* belief, plants have developed a pronounced potential for the metabolism of foreign compounds. Metabolism proceeds such that after uptake of *foreign* products, structural modifications (phase I: generation of functional groups such as —OH, —NH, or —SH) are introduced that finally allow transfer of hydrophilic metabolites (phase II: conjugation metabolism, formation of polar, water-soluble products by addition of *glucosyl* or *amino acyl* residues) into vacuolar long-term storage or peroxidative *polymerization* (phase III) of xenobiotic derivatives into polymeric structures such as lignin or cell wall-localized *polyphenolic matrices*.

Because plants cannot excrete organic waste or end products outside the plant body (as animals normally do), metabolic excretion aims at vacuoles or long-term durable polymers. A great variety of very different chemical structures can thus be changed to harmless metabolites. Complete degradation of the carbon skeleton of foreign products to CO_2 and water is very rare in plants.

For plant biotechnology aiming at the generation of (crop) plants with a higher level of resistance toward xenobiochemicals, two enzyme systems are of special interest. Decisive hydroxylation reactions of phase I are catalyzed by cytochrome P450 monooxygenases. Numerous dealkylation, epoxidation, and hydroxylation reactions (at aromatic, heterocyclic, alicyclic, or aliphatic substrates) are the key introductory steps that convert toxic compounds into much less toxic or nontoxic metabolites. In addition to xenobiotic metabolism, P450 enzymes are involved in numerous reactions of primary (phytohormones) and secondary (e.g., flavonoid pigments, many phenylpropanoid compounds, terpenoids, alkaloids, phytoalexins) metabolism.

The importance of cytochrome P450 oxygenases in plant metabolism can hardly be overestimated. The great number of P450 genes detected in the *Arabidopsis* genome (see earlier) adequately supports this statement. Furthermore, the well-characterized mammalian P450 enzymes and their documented decisive role in detoxification of drugs and other exogenous compounds have stimulated research in this field.

Therefore, based on the knowledge that numerous xenobiochemicals are converted by P450 enzymes, clear identification of the relevant enzymes, determination of their substrate specificities, analysis of gene regulation (constitutive expression versus inducible formation), and cloning of the genes are now preferential objectives. Because cloning of P450 genes is often easier than isolation of the membrane-bound proteins and their bio-chemical characterization, numerous known gene sequences await functional identification.

It has clearly been shown that in *Helianthus tuberosum* a P450 enzyme was highly induced by exogenous chemicals (a phenomenon well known from animal systems). Upon heterologous expression in yeast, the enzyme converted a wide range of xenobiotics and herbicides to nonphytotoxic compounds. For plant biotechnology such genes are potential tools for the control of herbicide tolerance as well as soil and groundwater bioremediation.

In accordance with this statement, a cytochrome P450 monooxygenase cDNA selected from a soybean P450 cDNA library was also shown to catalyze the oxidative metabolism of a range of herbicides and to enhance tolerance to such compounds in transgenic tobacco. The preceding data would never have been obtained without the application of molecular biological techniques. Such procedures are of great importance for biotechnology in the search for other specific genes and their functional characterization.

With the great number of genes obtained from the genome sequencing projects (e.g., *Arabidopsis)* or from the facile cloning of P450 genes, techniques for gene selection and functional determination become more and more of interest. In this context, new and elegant applications of the well-known technique to identify and characterize genes by constructing knockout mutants should be mentioned.

Using T -DNA of A. *tumefaciens* as an insertional mutagen and PCR techniques with primers directed at the wanted gene(s), large collections of transformed *Arabidopsis* lines (or other plants if they can readily be transformed) have been made available for screening studies. In essence, any gene can thus be identified and the mutant plant analyzed for the resulting phenotype.

Highly lipophilic xenobiotics, especially those carrying conjugated double bonds, halogen substituents (Cl, Br) at aromatic or aliphatic structures, or nitro and nitroso groups are metabolized in plants by glutathione S-transferases (GSTs).

This highly complex set of isozymes is involved in the metabolism of endogenous substrates (protection against oxidative stress in respiratory and photosynthesis pathways, carrier systems for vacuolar transport of anthocyanin pigments and xenobiochemicals) as well as exogenous compounds (detoxification of herbicides and other foreign products, especially by nucleophilic attack of the S atom and displacement of the halogen or nitro substituent).

The resulting peptide derivative may be processed further but will eventually be stored in vacuoles. The GSTs are homo- or heterodimers with the various subunits either expressed constitutively or formed inducibly upon treatment of plants with suitable substrates.

Each distinct subunit is encoded by a different gene. Multiple homo- and heterodimers exist, and the isoenzymes show distinct but only partly overlapping substrate specificities. Intensive studies with maize, wheat, and soybean have shown that the constitutive expression or the manipulated overexpression of certain GST subunits represents a tool for promoting tolerance of crop plants toward specific agrochemicals.

The data collected so far on plant metabolism of xenobiochemicals and other foreign compounds clearly indicate that powerful techniques exist that provide interesting applications for plant improvement.

CROP PLANTS AND RENEWABLE RESOURCES

With the *Arabidopsis* genome in hand, plant scientists are now eagerly looking for sequence data for crop plants such as rice and maize. In these cases the scientific challenge of genome sequencing is much bigger because these plants have genomes 4 to 25 times larger than the *Arabidopsis* genome. This results from the tendency of many plants to carry duplicate or multiple copies of large sections of DNA.

In view of the economic importance of rice and maize as staple food for more than half of the world's population, the results of such projects will undoubtedly form the basis for better knowledge of the genetics of these plants. These efforts will eventually also lead to continued progress in improving the productivity and the quality of these crop plants.

Thus, a challenging and fascinating chapter of plant biotechnology will be opened in a few years. In general, the productivity of modem agricultural crop plants has been increased manyfold over the last decades. Adaptation of the various genotypes either to the often complex factors of the environment (i.e., soil, climate, temperature, water supply), to the specific prevailing agricultural conditions, or to pests and pathogens has been achieved very successfully at sometimes impressive speed.

Furthermore, the different demands of markets and consumers with regard to product quality and fields of product application have been leading guides in the breeding programs. These programs were conducted by conventional techniques of crossing and selection, but more recently molecular biological procedures [e.g., restriction fragment length polymorphism (RFLP)] have also been introduced. In general, in addition to yield and quality, modem agricultural crop plants have been optimized for high consumption of fertilizers and water.

This last aspect will have to be at least partly reversed because future agricultural practice in many countries will be confronted by a reduced water supply. Plants with appropriate mechanisms for low water management are a challenging scientific task in the future. A few lines of foreseeable development in plant breeding and construction are certain. Plant breeding will more and more apply molecular biological and gene technological methods.

The data from genome sequencing programs will be essential prerequisites. The diversification of lines within a given species will increase because of the diverse demands for product quality and product application. The overall productivity of our crop plants has to be greatly increased in order to feed the rapidly growing population. A very interesting and scientifically important step into this modern field has been taken by the recent release of "Golden Rice."

This transgenic rice supplies provitamin A and iron and is expected to reduce major micronutrient deficiencies in substantial populations where rice is the major diet. Iron deficiency (a health problem in many women) is compensated by several transgenes leading to better iron uptake and hydrolysis of phytate. Vitamin A (required to prevent eye problems and blindness) is provided by substantial levels of β-carotene accumulating in the rice grains due to four transgenes to allow carotinoid formation.

The wide field of renewable resources represents a further challenge for plant biotechnology and modern *agriculture*. Petrol oil and many *mineral oil*-derived chemicals as well as coal are to be replaced by plant biomass or plant-derived raw materials, various chemicals, biopolymers, and all sorts of high-molecular or low-*molecular* products formed by and *isolated* from plants. Such plant production requires little if any *exhaustable* energy resources.

Potato lines with structurally modified starch (changes in amylose/amylopectin ratios), rape transgenic genotypes *accumulating* seed oil with other than the normal C16 and C18 fatty acids, or crop plants mainly storing *fructans* instead of sucrose in their roots are well-established suitable examples. From rape-derived "*bio-diesel*" as petrol for cars to highly sophisticated organic chemicals from suitably constructed plant lines, the design of new "*industrial plants*" opens wide possibilities for plant biotechnology on a *practically unlimited* scale.

CONCLUSIONS

Plant biotechnology has developed into a scientific discipline with substantial value in itself. In addition to the microbial and the animal systems, plants and their cells can be used with great benefit for biotechnological questions. This application will undoubtedly continue and most likely will increase in importance.

This is especially mandatory because plants are the major and most important source of our nutrition. It is easy to predict that the use of transgenic plants will more and more become routine and a matter of course. The development that started a number of years ago is of so much value that there will be no way and no need to go back.

All the biotechnological efforts have to be seen in the context of the pressure that the rapidly growing *population* exerts on the production of food and all materials that can be produced with plants.

Chapter 2

SIGNAL TRANSDUCTION ELEMENTS

Optimal growth and differentiation of plants require coordinated regulation of cellular and intercellular processes and their continuous adaptation to the variable environment. Efficient mechanisms evolved during plant evolution for sensing the environment, for translation of these data into biological information, for transfer of this information within the organism, and for initiation of appropriate reactions.

These processes of biological signal transduction comprise the perception of endogenous and environmental signals, the generation of endogenous cellular and systemic signals, and their transmission to the appropriate response targets. Receptor proteins of the plasma membrane or the cytoplasm specifically bind and thereby recognize the signals and either alone or in concert with other proteins initiate cellular signaling processes.

Intracellular mediators form the basis of complex cellular signaling networks that are responsible for signal transmission, integration, and evaluation and response activation. Although plants lack an equivalent of the circulating bloodstream, the responses can include the production and secretion of systemic signals that are transported throughout the plant and are recognized by receptors of target cells. In order to allow optimal adaptation to the environment, the signals perceived by individual receptors need to be integrated and, most important, evaluated according to their importance.

Therefore, linear signal transduction pathways are the exception, if they exist at all. Rather, complex signaling networks with points of signal convergence and divergence support cross talk between the signaling pathways initiated by individual input signals. Thereby, an overall assessment is possible that guarantees an appropriate response to the general environmental situation. This complexity of interconnected signaling networks and the limited knowledge of molecular details of plant signal transduction mechanisms have so far not supported the employment of genes encoding signal transduction components in plant biotechnology.

However, some promising results have been reported, primarily using the end points of signal pathways, receptors and transcription factors, respectively. These will be summarized in this chapter in the form of a speculative outlook on future possibilities to modulate developmental processes in plants.

HORMONE LEVELS

Alterations in the level of individual or in the balance of different hormones usually result in pleiotropic phenotypes of little interest for application. However, if these changes were spatially and temporally tightly regulated, meaningful effects might be obtained. Seed dormancy delays germination despite favourable conditions and, thereby, allows seed survival in soil for long periods of time. Although an advantage for weeds, this is undesirable for crop plants. The plant hormone abscisic acid (ABA) is synthesized during seed development and is involved in the induction of primary seed dormancy. ABA biosynthesis or sensitivity mutants displayed reduced seed dormancy.

Consequently, constitutive expression of the ABA biosynthetic gene *ABA2,* encoding zeaxanthin epoxidase, in *Nicotiana plumpaginifolia* resulted in increased ABA levels and delayed germination, whereas antisense suppression of this gene resulted in reduction of ABA seed levels and rapid germination. Constitutive expression of the *AtGA2ox1* gene from *Arabidopsis thaliana* encoding the key regulatory gene for gibberellin (GA) biosynthesis, GA 20-oxidase, in hybrid aspen resulted in increased levels of several gibberellins in internodes and leaves.

The transgenic trees showed an improved growth rate, produced larger biomass, and had more numerous and longer xylem fibers than wild-type plants. As one of only a few negative effects, poor root initiation was observed when the transgenic plants were transferred to soil. Methyl jasmonate (MeJA) is a naturally occurring volatiie derivative of the plant hormone jasmonic acid (JA) that also stimulates many typical plant responses to JA when applied exogenously.

The plant enzyme S-adenosyl-L-methionine:jasmonic acid methyltransferase (JMT) catalyzes the formation of MeJA from JA. The gene is differentially expressed in *A. thaliana* during development and with environmental stimuli. Transgenic *A. thaliana* plants constitutively expressing the *JMT* gene from the same plant show increased MeJA but unaltered JA levels. Although visually not distinguishable from wild-type plants, the transgenic plants display elevated transcript levels of JA-responsive genes without any stimulus.

Furthermore, their degree of resistance against *Botrytis cinerea* was found to be significantly increased when compared with untransformed *A. thaliana* plants. The signal molecule salicylic acid (SA) plays a central role in pathogen defense of plants but is also involved in other regulatory processes, such as cell growth, stomatal closure, flower induction, and heat production.

Transformation of tobacco plants with bacterial genes encoding the enzymes isochorismate synthase (ICS) and isochorismate pyruvate lyase (IPL), under control of a constitutive promoter, resulted in strongly increased levels of SA and SA glucoside in healthy plants when the enzymes were targeted to the chloroplast. The transgenic plants displayed a normal phenotype but constitutively expressed pathogenesis-related genes and showed elevated resistance against tobacco mosaic virus and the fungal pathogen *Oidium lycopersicon.*

These few examples of rather crude modulation of hormone levels demonstrate the large potential of such an approach for molecular engineering, particularly if more sophisticated regulatory tools become available.

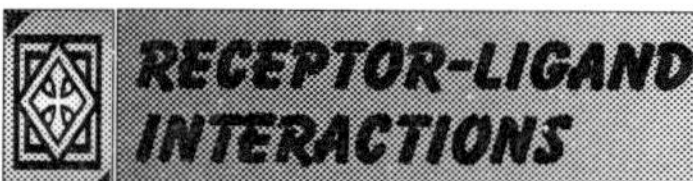

RECEPTOR-LIGAND INTERACTIONS

The idea of employing ligand-receptor pairs in transgenic plants has been pursued most intensively with the goal of generating plants with durable disease resistance, although no commercial product has yet been generated by this approach. Plants successfully resist the attack of most potential pathogens in their environment through an efficient nonself recognition system that is similar to the innate immunity system of vertebrates and insects.

Elicitors, originating from the pathogen or released from the plant cell wall during pathogen attack, are specifically recognized by receptors of the plant plasma membrane and thereby initiate a multicomponent defense response. Although several such elicitors have been purified and shown to bind specifically to binding sites of the plasma membrane, only two genes encoding components of the corresponding plant receptors have been cloned so far.

One encodes the 75-kDa plasma membrane-associated binding protein of the hepta-, β-glucan elicitor from *Phytophthora sojae* that occurs in various Fabaceae. Transgenic tomato plants expressing the hepta-β-glucan-binding protein from soybean display high-affinity binding sites for this elicitor.

It is unknown, however, whether these plants display increased resistance toward pathogens harboring the hepta-β-glucan in their cell walls. A gene encoding a 129-kDa receptor-like kinase apparently involved in the recognition of bacterial flagellin was isolated from A. *thaliana.* Point mutations in different regions of this gene resulted in loss of ligand recognition and elicitor responsiveness.

Although receptor function of the corresponding plasma membrane protein needs to be demonstrated, it represents an essential component in binding of a 22-amino-acid fragment of flagellin and may play an important role in the recognition of bacterial pathogens by plants. Overexpression of such receptors involved in nonhost recognition has the potential to increase basal pathogen resistance in many different crop plants. In addition to the basal nonhost resistance, plants have developed a complex host defense system relying on receptor-mediated recognition of pathogen avirulence (Avr) gene products by plant resistance *(R)* gene products.

The existence of corresponding pairs of Avr and *R* genes results in an incompatible interaction between the pathogen and its host plant; i.e., disease development is efficiently stopped by the rapid activation of a multicomponent plant defense response. Although evidence for direct ligand/receptor-type interaction of Avr and *R* gene products is lacking in most cases, both may be components of larger signal perception complexes, whose formation is required for an incompatible interaction.

The *R* genes introduced into crop plants by conventional breeding techniques did function in the expected way, but, with a few exceptions, the resulting resistance was found to lack durability in the field. Pathogens could rapidly break *R* gene-mediated resistance because conventional breeding allowed only the generation of *R* gene monocultures.

The availability of a growing number of different cloned R genes directed against specific pathogen races would allow generating transgenic R gene polycultures of crop plants. Using a population of a given crop plant species consisting of individual plants expressing specific *R* gene patterns would significantly reduce the speed of adaptation of the pathogen and thereby result in an overall reduction of disease development.

Interestingly, increasing the level of a specific *R* gene product by overexpression can activate defense responses in the absence of pathogens and thereby result in broad resistance of the transgenic plants as shown for the *Pto* gene in tomato. A different strategy has been suggested by de Wit involving the coexpression of pairs of Avr and *R* genes in one plant. This approach requires constitutive expression of a plant R gene and tightly regulated coexpression of a pathogen-derived Avr gene under control of a broadly pathogen-responsive promoter.

Transformation of tomato plants carrying the *Cf9* resistance gene against *Cladosporium fulvum* with the *Avr9* gene from this fungal pathogen under control of a pathogen-responsive promoter rendered the transgenic plants resistant to a broad spectrum of pathogens. Although the application of this approach appears to be limited by the lack of functional expression of *Cf9* and *Avr9* genes in certain crop species, different pairs of matching *R* and *Avr* genes may function in different crop species. Interestingly, this strategy to generate broad-spectrum disease resistance appears not to be limited to *R/Avr* gene-mediated plant-pathogen recognition.

All plant pathogenic *Phytophthora* species analyzed so far secrete elicitins, homologous proteinaceous elicitors that induce an efficient resistance response in tobacco and a few other plant species. Tobacco plasma membranes harbor high-affinity binding sites for elicitins that appear to function as elicitin receptors. *Phytophthora* species pathogenic on tobacco, such as the causal agent of black shank disease, *Phytophthora parasitica* var. *nicotiana,* do not produce elicitins and thereby escape recognition .

Expression of the elicitin cryptogein from *Phytophthora cryptogea* under control of a strictly pathogen-responsive promoter in tobacco resulted in broad-spectrum resistance of the transgenic plants against *P. parasitica* var. *nicotiana, Thielaviopsis basicola, Erysiphe cichoracearum,* and *Botrytis cinerea.*

GTP-BINDING PROTEINS

In mammals, heterotrimeric G protein complexes link a large number of heptahelical transmembrane G protein-coupled receptors (GPCRs) to cellular signaling networks and thereby regulate a multitude of cellular processes. The importance of this signaling mechanism is reflected by the presence of more than 1000 different GPCR-encoding genes and several Ga, Go and Gy isoform genes in mammalian genomes.

Although in plants the existence of only a limited number of GPCRs, G(3 and Gy, and only a single Ga is predicted from the *A. thaliana* genome sequence, heterotrimeric G proteins appear to play a central role in plant hormone signaling. Knockout mutants of the only Ga isoform of *A. thaliana (GPAI)* have reduced cell division in their aerial tissues and lack ABA inhibition of guard cell inward K^+ channels as well as pH-independent ABA activation of anion channels.

Consequently, stomatal opening in these knockout mutants was found to be insensitive to ABA inhibition and, therefore, the water loss rate was greater in mutant than in wild-type plants. Because inducible overexpression of *GPA1* in *Arabidopsis* conferred inducible cell division to transgenic tissue, targeted modulation of the status of Ga in stoma cells may provide a new tool to control water balance. A large group of small GTP-binding proteins in plants are involved in a broad spectrum of cellular signaling processes.

In rice, three genes *(OsRacl-3)* were identified encoding proteins with 60% identity to human Rac proteins. Expression of a constitutively active derivative of *OsRacl* in rice resulted in enhanced production of reactive oxygen species (ROS) and phytoalexins, stimulation of programmed cell death, and increased resistance against bacterial and fungal pathogens.

Loss-of-function experiments with a dominant-negative *OsRacl* derivative showed the expected suppression of elicitor-stimulated ROS production and pathogen-induced programmed cell death in transgenic rice. These findings suggest that processes regulated via small GTP-binding proteins can be modulated in transgenic plants and cause modification of distinct physiological processes.

CALCIUM SIGNALING

Release of Ca^{2+} from internal stores and influx from the apoplastic space resulting in transient increases or oscillation of cytosolic Ca^{2+} levels are involved in many signaling networks in plants. The duration, intensity, and spatial distribution of Ca^{2+} transients appear to encode signalresponse specificity. Direct modulation of cytosolic Ca^{2+} levels might therefore be a difficult task.

However, downstream targets of Ca^{2-} regulation may represent future tools for engineering plants with improved stress tolerance. Two calcium-dependent protein kinases (CDPKs) from tobacco (NtCDPK2 and 3) have been shown to be involved in defense and hypoosmotic stress signaling. Both CDPKs are phosphorylated and thereby activated in a Ca^{2+}-dependent manner and trigger an oxidative burst and programmed cell death.

Ectopic expression of a related CDPK from *A. thaliana* in tomato protoplasts stimulated NADPH oxidase activity and an oxidative burst. Tightly regulated expression of constitutively active CDPK derivatives might offer a tool to engineer plants with enhanced flooding and disease resistance. Calmodulin is a well-known target of calcium in multiple signal transduction pathways.

Transgenic tobacco cells expressing a dominant-acting calmodulin derivative responded with a stronger oxidative burst and enhanced NADPH levels to treatments with various elicitors, infection with avirulent bacteria, and osmotic and mechanical stress compared with wildtype cells. Two specific calmodulin isoforms, SCaM4 and 5, and their transcripts were found to accumulate in cultured soybean cells upon elicitor treatment.

These two calmodulin isoforms are most divergent from other isoforms described from plants and animals. Furthermore, their ability to activate calmodulin-dependent enzymes in vitro differed greatly from that of the other calmodulin isoforms of soybean.

Transgenic tobacco plants constitutively expressing SCaM4 or SCaM5 exhibited spontaneous lesion formation, constitutive expression of defense-related genes, and increased resistance to virulent viral, bacterial, and oomycete pathogens. These findings suggest that specific calmodulin isoforms represent promising tools for engineering stress-tolerant plants.

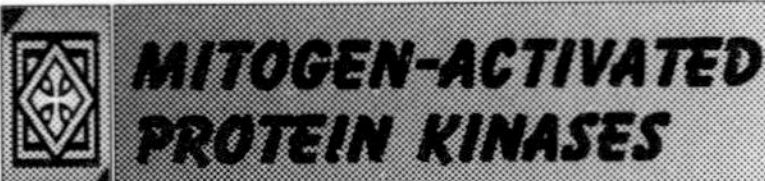

MITOGEN-ACTIVATED PROTEIN KINASES

Mitogen-activated protein kinase (MAPK) cascades are universal signaling modules of eukaryotic cells that fulfill essential regulatory functions primarily in transduction of extracellular signals to cellular and nuclear responses. Each cascade consists of a module of at least three protein kinases. The most downstream kinase, MAPK, is activated by dual phosphorylation of a typical threonine-X-tyrosine motif catalyzed by a MAPK kinase (MAPKK). MAPKKs are themselves activated by serine/threonine phosphorylation by a MAPKK kinase (MAPKKK).

The activation mechanism of plant MAPKKKs and their linkage to the corresponding receptor are not known. Also, no substrate of a plant MAPK has been identified. According to sequence similarities, the *Arabidopsis* genome contains up to 23 MAPKs, 9 MAPKKs, and at least 25 MAPKKKs. Therefore, plants have more MAPKs than any other eukaryotic organism.

The presence of more MAPKs than MAPKKs furthermore shows that MAPKKs cannot phosphorylate only a single MAPK. Plant MAPK cascades have been found to be involved in hormonal responses, cell cycle regulation, abiotic stress signaling, and pathogen defense, where they exhibit positive as well as negative regulatory functions.

Phenotypes of recently identified MAPK cascade mutants and transgenic plants expressing constitutively active MAPKKK or MAPKK derivatives suggest that it might be possible to employ these signaling elements in order to generate transgenic plants with improved stress tolerance. The *mpk4* mutant of *A. thaliana* lacks a functional MAPK, AtMPK4. Besides being a severe dwarf, this mutant contains high levels of salicylate, is jasmonate insensitive, and displays enhanced resistance to bacterial and oomycete pathogens.

Because active AtMPK4 is required to complement the *mpk4* mutant, this MAPK apparently suppresses salicylate accumulation in wild-type plants and thereby negatively regulates activation of salicylate-dependent defense responses. Although complete lack of AtMPK4 results in this pleiotropic dwarf phenotype, tightly regulated inactivation of this MAPK would possibly increase disease resistance without deleterious effects for the corresponding plant.

Another MAPK cascade mutant of *A. thaliana, edrl,* carries a mutation in a putative MAPKKK. This mutant has a normal phenotype and does not constitutively activate defense responses. However, defense responses are more rapidly activated upon pathogen attack in *edr 1* than in the wild-type plant, conferring enhanced resistance against powdery mildew to the mutant.

These results suggest that EDR 1 negatively regulates the activation of defense reactions and that its inactivation results in enhanced disease resistance. MAPK cascades also positively regulate disease resistance. In tobacco, the MAPKs SIPK and WIPK are rapidly activated upon treatment with pathogen-derived elicitors and infection with avirulent pathogens.

Transient expression of a constitutively active derivative of the tobacco MAPKK, NtMek2, in tobacco leaves resulted in activation of endogenous SIPK and WIPK and furthermore stimulated typical defense reactions, such as programmed cell death and activation of defense-related genes. Transient inducible expression of SIPK in the same experimental system led to increased protein kinase amounts, enhanced enzyme activity caused by phosphorylation, and activation of multiple defense reactions, demonstrating that SIPK alone is sufficient to stimulate the defense response. Transgenic *Arabidopsis* plants expressing constitutively active derivatives of the MAPKKs, AtMEK4 and 5, under control of an inducible promoter displayed programmed cell death preceded by activation of endogenous MAPKs and ROS production.

The *Arabidopsis* orthologue of SIPK, AtMPK6, was also found to be activated by the flagellin-derived oligopeptide elicitor flg22. Expression of components of the corresponding MAPK cascade confers pathogen resistance to leaves of the transgenic *Arabidopsis* plants. Together, these findings suggest that spatially and temporally regulated modulation of specific MAPK cascades in transgenic plants can be employed to engineer disease resistance. Reactive oxygen species such as superoxide anion radical (O_2^-), hydroxyl radical (OH), and hydrogen peroxide (H_2O_2) are generated by plant cells in response to diverse abiotic and biotic stresses.

Although the sensing mechanisms of ROS by plant cells are unknown, it has recently been demonstrated that $H,0_2$ activates MAPK cascades in *A. thaliana*. Specifically, the MAPKKK ANP1 and the two MAPKs AtMPK3 and 6 were found to be activated by H_2O_2 treatment. Interestingly, transient expression of a constitutively active version of ANPI in *Arabidopsis* protoplasts initiated activation of AtMPK3 and 6 in the absence of H_2O_2.

Most important, when a constitutively active derivative of the tobacco orthologue of ANP1, NPK1, was expressed in transgenic tobacco, these phenotypically normal plants exhibited enhanced tolerance to various stress conditions, such as freezing, heat shock and salt stress. Thus, modulation of specific MAPKKK activities clearly has the potential to engineer broad stress tolerance in plants.

TRANSCRIPTION FACTORS

Transcription factors represent the final signaling elements that determine the gene expression pattern of a cell. Analysis of the *A. thaliana* genome using a consistent conservative threshold revealed more than 1700 putative transcription factors. Among the 29 classes of transcription factors, 16 appear to be unique to plants. Functional analysis by knockout techniques was found to be hampered by the frequent occurrence of functionally redundant genes.

In contrast, overexpression of transcription factors in several cases indicated functional links and possible usefulness in molecular engineering. The tomato *Pti5* gene encodes a pathogen-responsive transcription factor of the ethylene response element-binding protein (AP2/EREBP) class that interacts with the *R* gene product Pto.

The *Pti5* gene is believed to be specifically involved in pathogen defense because its transcript accumulates in response to infection but not to abiotic stresses or hormone treatments. Overexpression of PtiS in a tomato cultivar lacking the corresponding *R* gene, *Pto,* conferred enhanced resistance against the virulent bacterial pathogen *Pseudomonas syringae pv. tomato* to the normal looking transgenic plants.

Interestingly, the plants did not constitutively express defense-related genes but their activation was drastically accelerated upon infection in comparison with wild-type plants. Overexpression of the *Pti4* gene, also encoding a transcription factor interacting with Pto, in *A. thaliana* resulted only in slightly enhanced resistance to bacterial pathogens. The AP2/EREBP-type transcription factor gene, *Tsil,* was identified as a salt-responsive transcript in tobacco. Transgenic tobacco plants overexpressing *Tsi* exhibited increased salt tolerance. In addition, these plants showed constitutively elevated transcript levels of defense-related genes and enhanced resistance to the virulent bacterial pathogen, *Pseudomonas syringae pv. tabaci.* The *NPRI* (also designated *NIMI* and *SAIL)* gene of *A. thaliana* is an important regulator of inducible pathogen defense.

It encodes a protein with ankyrin repeats and homology to the human transcription factor IKB. Despite these homologies and the requirement of its nuclear localization for function, it probably does not act directly as transcription factor but acts upon differential interaction with bZIP transcription factors of the TGA family. Overexpression of *NPRI* renders *Arabidopsis* plants resistant against virulent bacterial and oomycete pathogens without altering their normal phenotype.

Most interestingly, ectopic expression of the *Arabidopsis NPRI* gene in rice also results in increased resistance of the transgenic plants against the virulent bacterial rice pathogen *Xanthomonas oryzae pv. oryzae.* The transgenic plants displayed a normal phenotype like the wild-type plants. These results make *NPRI* an excellent candidate tool to engineer broad-spectrum disease resistance in monocotyledonous and dicotyledonous plants. Overexpression of transcription factors has also been described as a useful approach in engineering abiotic stress tolerance. The dehydration response of *A. thaliana* is regulated in a complex manner via at least four independent pathways, two of which are ABA dependent and two ABA independent.

The dehydration-responsive cis-element (DRE) is essential for regulation of dehydration- and cold-responsive gene expression via one of the ABA-independent signaling pathways. The *Arabidopsis* transcription factors, DREBIA and DREB2A, bind to the DRE element and thereby activate transcription of the corresponding gene.

Constitutive overexpression of DREBIA in *Arabidopsis* plants caused dwarfed phenotypes, expression of stress response genes under nonstressed conditions, and improved tolerance to drought, salt stress, and cold temperatures. However, expression of DREBIA under control of the stress-responsive promoter, *rd29A,* did not significantly affect growth of transgenic *Arabidopsis* plants but resulted in even greater drought, salt, and freezing tolerance than constitutive expression.

The *Arabidopsis* transcriptional activator CBFI (C-repeat/DRE binding factor 1) binds to the C-repeat/DRE cis-acting element that regulates coldand drought-responsive gene expression. Constitutive overexpression of *CBF1* in *A. thaliana* activated cold-responsive genes at normal temperatures and enhanced the freezing tolerance of the transgenic plants without obvious effects on their growth and development.

A cold-responsive zinc finger protein from soybean, SCOF-1, appears not to act directly as a transcription factor but rather associates with a soybean G-box binding bZIP transcription factor, SGBF-1, in the nucleus and thereby dramatically enhances its binding affinity to the ABA responsive *cis*acting element (ABRE) of cold-responsive genes.

Tobacco plants constitutively overexpressing SCOF-1 had normal phenotypes and did not exhibit altered cold sensitivity. However, cold-stressed SCOF-1-transgenic plants recovered significantly faster than wild-type plants under normal growth conditions. SCOF-1-expressing *Arabidopsis* plants displayed constitutive activation of cold-responsive genes and enhanced freezing tolerance.

The synthesis of heat stress proteins (HSPs) protects plants against damage by high temperatures. Heat stress transcription factors (HSFs) regulate the heat-induced transcription of *HSP* genes. Constitutive overexpression of *HSF3* but not *HSF4* stimulated HSP synthesis in *Arabidopsis* plants without altering their normal phenotype.

Most important, the HSF3-transgenic plants displayed increased basal thermotolerance. The *Alfin 1* gene, originally cloned as a salt-responsive gene from alfalfa, was found to encode a root-specific zinc finger-type transcription factor. Constitutive expression of the *Alfin 1* gene in transgenic alfalfa led to increased root growth under normal and saline conditions and enhanced salt tolerance without having adverse effects on plant shoot growth.

Surprisingly, transcriptional regulators of complex developmental processes, such as flower development, may also be employed in this rather simple approach for genetic improvement of crop species. The *Arabidopsis* genes *LEAFY (LFY)* and *APETALA 1 (API)* function as meristem identity genes and promote flower initiation when constitutively expressed in *Arabidopsis*. Constitutive expression of either one of these genes in transgenic citrus resulted in the production of fertile flowers and fruits in the first year and thereby in a significant reduction of the generation time of the transgenic trees. These few examples of molecular engineering via transcriptional regulators nicely demonstrate the enormous potential of this strategy for crop plant improvement.

CONCLUSION

Plants apparently employ signal transduction elements similar to those of animals. However, most of them are present in larger numbers and they are often linked in a different and variable way. Most if not all plant signaling pathways are organized in networks with points of convergence and divergence allowing cross talk between pathways and evaluation of the importance of simultaneously perceived signals.

Significant progress has been made during the last decade in the analyses of all levels of signal transduction. Array and proteomic technologies together with the knowledge of total genome sequences will further complete our understanding of complex signaling processes and allow better targeted approaches than those described.

However, simple overexpression studies with model systems have already demonstrated that diverse signaling elements can be employed successfully in molecular engineering of plants. On the other hand, these experiments have also shown that modulation of a specific signaling pathway often interferes with other not related pathways and thereby negatively affects physiological processes.

In order to fine-tune expression of transgenes, spatially and temporally tightly regulated promoters are required. The increasing number of functionally characterized cis-acting elements might soon allow the construction of artificial promoters that better fulfill this purpose.

In summary, molecular engineering of signal transduction pathways has the potential to modulate all processes of plant growth and development including their responsiveness to biotic and abiotic stresses. In order to use this potential, a detailed understanding of the plant signaling components and their linkages within complex networks is required.

3 Chapter

MOLECULAR DEVELOPMENT

Plant cell walls are made up of the world's most *abundant* and most *durable* organic materials. Small wonder that they have been used by man since the dawn of time. Cell wall-derived materials played crucial roles in the cultural evolution of man, and they continue to be *integrated* yet often *unappreciated* ingredients of modern everyday life. Examples are wood as an energy source, used as construction material, or made into pulp and paper; *cotton, linen, hemp, ramie,* and *sisal* woven into tissues and/or made into strings and ropes; *alginic acid* and *pectins* used as gelling materials; and *dietary* fibers required for healthy *nutrition*.

Cell walls are essential in material sciences as well as in food and feed technology. Cell walls evolved to fulfill a range of important tasks in plants, both structural and functional, explaining their *versatility* when exploited by man. It is clear, however, that plant cell walls never evolved to fulfill the roles we use them for, so that there is a great potential for *optimizing cell* wall components or *architecture* for our purposes.

This optimization can be done and has been done using traditional breeding *procedures* leading to materials such as long-staple cotton varieties. Modern methods of genetic *engineering* will now open up new opportunities that we are only beginning to realize. *Cellulose*, the stress-bearing fiber component of plant cell walls, is estimated to be produced at a rate of 10^{10} tons per year, and the existing mass of *approximately* 10^{11} tons far exceeds that of any other molecule of organic origin.

The title "second most abundant organic molecule on earth" is given either to chitin, the fiber-forming component of fungal cell walls, or to lignin, the load-bearing matrix component of many secondary plant cell walls. The different *hemicelluloses* found in plant cell walls together form about as much *biomass* as cellulose, and *pectins* make up another sizable portion of organic plant material. Even structural proteins such as *extensin*, which are only a minor component of plant cell walls, have been considered to be among the most abundant proteins on earth.

In spite of the enormous importance of cell walls and of our ever increasing knowledge of the components that make up plant cell walls, *surprisingly* little is known about the *biosynthetic machinery* responsible for *polymerizing* the cell wall constituents, and even less is known about their integration and assembly into the architecture of the complex *organelle* that is a cell wall.

Almost terra incognita is the turnover and modification of plant cell wall components during plant cell division, growth, differentiation, senescence, and death. Although the paucity of information available is a serious drawback in genetically *optimizing* plant cell walls for biotechnological uses, it offers a large and promising field for future studies of both *fundamental* and applied relevance. A decade ago, the situation appeared to be far more advanced for *lignin* biosynthesis.

The enzymes necessary to produce the three different *monolignols* and to polymerize them into the three-dimensional network of the lignin polymer all appeared to be known, and most of the corresponding genes had been cloned from different plant species.

Consequently, the genetic engineering of designer lignin seemed an easy task, and studies were initiated to overexpress or silence these genes and to analyze the lignin of the *transgenic* plants.

To our surprise, however, these molecular genetic experiments revealed that the process of monolignol synthesis is far more complex than previously thought-and today we realize that we know less about lignin *biosynthesis* than we thought we knew. *Genetic* engineering of lignin remains a challenging and promising task.

This is the subject of other chapter of this book, and lignin will therefore not be considered in more detail here. Drawing up possible *biotechnological* uses of plant cell walls first requires an understanding of the roles cell walls play in the life of a plant. These roles have driven the evolution of the components and architecture of the cell wall, and we will have to understand which components fulfill the different roles and how they do their job cooperatively in the *complex* cell wall system.

From this knowledge we can define strategies to improve plant *performance*, and we can deduce possible ex plants uses for the different cell wall components. If *optimization* is to be achieved by molecular genetic engineering, we have to characterize the enzymatic machinery that builds, assembles, and modifies the cell wall components.

The genes coding for these enzymes will eventually be the target for *manipulation* of plant cell wall components or architecture with the goals of *improving* plant performance and of obtaining higher quality products specifically optimized for uses in material sciences or food and feed technology. This chapter aims at giving a-necessarily *nonexhaustive*-overview of the many roles cell walls fulfill in plants and the many possible ex plants uses for plant cell walls.

It will try to highlight both the enormous *opportunities* and challenges for the genetic engineering of plant cell walls. Necessarily, this *treatise* will have to be somewhat cursory, and the reader will, whenever possible, be referred to recent more focused and more detailed reviews.

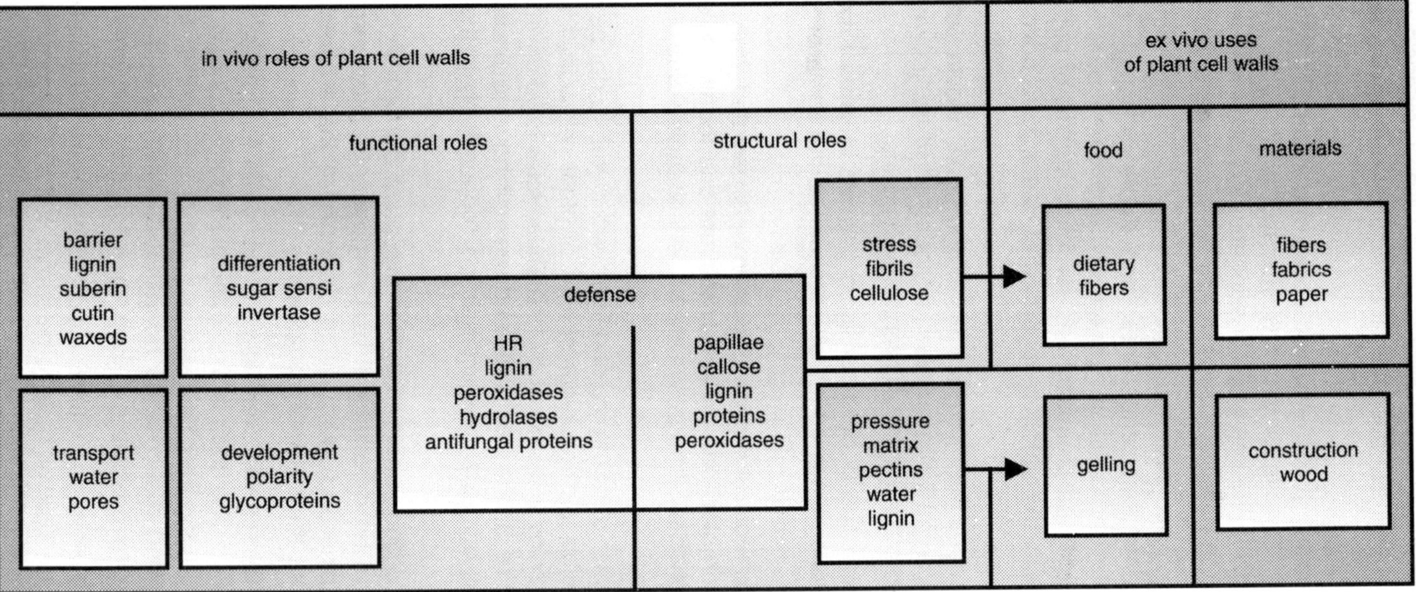

Figure 3.1: In vivo roles and ex vivo uses of plant cell walls and their components. Cell walls fulfill many roles-both structural and functional-in the life of a plant, and different components are responsible for the different roles. Biotechnology may aim at improving these in vivo roles according to human interest, e.g., increasing resistance of crop plants against pathogens (HR, hypersensitive reaction). Ex vivo, different cell wall components can be used according to their physical properties and, again, biotechnology may aim at improving these properties.

FUNCTIONS OF PLANT CELL WALLS IN PLANT

Cell walls have long been considered as dead *extracellular* material *possessing* solely static structural roles, namely to counteract the turgor pressure and thus give strength and form to plant cells, tissues, organs, and organisms. This view has changed dramatically with the insight modem *analytical* techniques and molecular probes have allowed into the fine structure and apparent highly complex spatial and temporal regulation of cell wall components and *architecture*.

It was concluded that the cell wall must have important functional roles, e.g., in cell-cell communication, transport of metabolites, differentiation, and development. To indicate this new appreciation of the cell wall as a *dynamic* and *regulatory extracellular organelle*, some authors prefer to call it an extracellular matrix. The growing realization that this matrix forms an integral and essential part of the plant cell may eventually even lead us to abandon the term extracellular in favour of, e.g., *pericellular*.

In spite of these considerations, plant cell walls do play important static structural roles. First, they have to bear the *turgor* pressure, which results in enormous *tangentially* oriented stress forces in the cell wall plane. This stress is taken up by the fibrillar component in the cell wall, namely the cellulose *microfibrils*, which are interconnected and held in place by *hemicelluloses*.

Second, the cell wall has to take up the pressure exerted on a plant cell by the surrounding tissue. This pressure is taken up by the matrix in which the fibrils are embedded, made up of polysaccharides containing uronic acids, namely pectins and glucuronoarabinoxylans, or-more *precisely*-by the water molecules held in the cell walls by these *negatively* charged polymers.

In extreme cases, this water is replaced by the incorporation of lignin, a heavily cross-linked three-dimensional *polyphenolic* network able to withstand very high pressures. As the *hydrophobic* lignin replaces the water in the cell wall, lignification also *renders* cell walls *impermeable* to water. Other ways to achieve the same goal-impermeability to water-are the *incrustation* with suberin or the deposition of cutin along with cuticular and *epicuticular* waxes.

These layers form essential barriers to the free transport of water and solutes, such as nutrients and assimilates. Suberin in the *Casparian* strand of the endodermis effectively seals the central *cylinder* of the root from the surrounding cortex tissue, allowing control over the uptake of nutrients and water from the soil.

Lignin in the tracheary elements of the xylem allows water and solute transport from the roots to the aboveground plant parts by stabilizing the vessel cell walls against the strong internal negative pressure of the transpiration stream and by preventing losses to the *surrounding* tissue.

Cutin and the waxes protect the aboveground plant organs from desiccation by restricting the transpiration to the stomata with their *regulated* diffusion resistance. Cell walls restrict and dictate cell size and form. Consequently, growth is possible only when the cell walls partially yield to *turgor* pressure. In effect, plant growth is growth of the plant cell walls.

Similarly, the shape of a plant is formed by the shape of the plant cell walls. The processes taking place in the cell wall during growth have been the subject of intense research over many decades. Although we are still far from understanding this highly complex and tightly regulated process, it appears today that the *hemicelluloses* interconnecting cellulose fibrils play a *crucial* role in this process.

Enzymes loosening and retightening these connections, such as *expansins* and *xyloglucanendotransglyc-osylases*, are critically involved in cell wall growth. Shape of a plant cell, on the other hand, appears to be dictated by the orientation of the cellulose fibrils that is most likely brought about by the movement of the cellulose synthase complex in the plasma membrane, which may or may not be guided and driven by the *cytoskeleton*. In contrast to the preceding *morphological* roles of plant cell walls, our understanding of the functional roles of cell walls in the *physiology* of a plant still remains rather *patchy*.

It is clear that cell walls both separate and connect neighboring cells in a tissue, and it can be anticipated that information and *metabolites* travel from cell to cell both *symplastically*, i.e., via *plasmodesmata*, and *apoplastically*, i.e., across the cell wall. One example of *apoplastic* transport of metabolites is *sucrose* on its way from source mesophyll cells to the phloem and from the *phloem* to sink cells in the root or fruit tissues.

Sucrose in the cell wall may be hydrolyzed into its constituent hexoses by apoplastic invertase, and the glucose produced can act as a local signal in sugar *sensing* influencing the *metabolic* state of the plant cells. Another example of information-bearing molecules traveling in the apoplast is auxin, which is thought to be moved through the plant by polar secretion into the cell wall and uptake from the cell wall by adjacent cells. Although the water present in the cell wall provides the space for *intercellular* travel of molecules, transport in the cell wall is not *unrestricted*.

We have already seen that special *elaborations* of the cell wall, such as suberin in the Casparian band, can completely stop flow of water and solutes. In addition, the pore size in the cell wall determines the diffusion rate of larger molecules, such as proteins. The size exclusion limit of plant cell walls has been estimated for globular proteins at between 10 and 30 (and sometimes up to 100) kDa. Different types of cross-links between *pectins* appear to be responsible for the *porosity* of the plant cell walls.

Phenolic cross-links possibly involving tyrosine residues of cell wall (*glyco*)proteins have been postulated to restrict porosity further. It remains to be seen whether plant cells are able to change the pore size of their surrounding cell walls actively in reaction to internal or external *stimuli* such as pathogen attack. Immunological approaches have shown that the wall surrounding a plant cell is far from being a uniform structure.

Owing to the availability of a set of two complementary antibodies, the distribution of methyl esterified versus non-methyl *esterified* pectins has been analyzed in many plant tissues, and distinct distributions both over the thickness and in the plane of plant cell walls have been found. Even more spectacular are the distributions of certain proteins that form minor constituents of the cell walls.

These epitopes are sometimes restricted to a very small number of cells and their presence may indicate or dictate determination of cell fate, e.g., to become metaxylem cells. Also, cell polarity may be determined by markers immobilized in the cell wall, as evidenced by a unique localized cell wall domain stabilizing the spatial orientation of the fucus *zygote*.

The role of cell walls in the development of plant cells, tissues, organs, and organisms is likely to be a most deserving field for future *research*. Cell walls are also of paramount importance in the defense of plants against pests and pathogens and in their tolerance to stress. Forming the outer shell of plant cells and, *consequently*, of total plants, cell wallsand most notably outer epidermal cell walls-are the first barrier put up by the plant against external biotic and *abiotic* threats.

Cell walls are the single most important *protection* of plants against the *myriads* of *microorganisms* that otherwise would be potential plant pathogens. Whereas successful pathogens must have evolved means to degrade and/or *penetrate* standard plant cell walls, plants have counteracted this attack by a multitude of pre- or *postinfection* cell wall modifications, such as callose deposition to form a periplasmic papilla, production of active oxygen species in an odixative burst for the peroxidative cross-linking of *phenolic acids* and proteins, or deposition of suberin or lignin, which as a true polymer is extremely difficult to degrade *enzymatically*.

Active oxygen species produced in the cell wall have also been supposed to act as second messengers inducing the *hypersensitive* reaction, a form of programmed cell death involved in plant disease defense that in some cases may be brought about by the intracellular performance of a process usually confined to the cell wall, namely radical coupling of *monolignols* to the lignin polymer. The cuticle including the *epicuticular* wax layers does not only protect plants from *desiccation*.

The *microroughness* of the plant surface brought about by the *wax crystals* creates the *Lotus* effect ensuring that dust particles and fungal spores are easily washed away by runoff raindrops. The extreme hydrophobicity of the epicuticular wax prevents the formation of a continuous water layer so that motile bacteria cannot easily reach *stomates*.

Wax protrusions may protectively cover stomatal openings, effectively reducing *stomatal* transpiration and preventing *recognition* of stomates by fungal pathogens searching for easy ingress points. Cell walls form the arena where *potential* microbial pathogens deploy their pathogenicity factors, often a plethora of plant cell wall hydrolyzing enzymes, and where host plant cells mount their defenses, e.g., in the form of *pathogenesis-related* (pr-) proteins such as inhibitors of the microbial hydrolases or hydrolytic enzymes attacking the microbial cell walls.

Not all of the microbes interacting with a plant are potential foes. Both the rhizosphere and the phylloplane are complex ecosystems *colonized* by a multitude of mutualistic or commensal microbes. The interactions of these microorganisms with the plant are surface interactions and, as such, interactions of the cell walls of both *partners*, at least initially.

However, little is known about the molecular details of these interactions, except for arbuscular *mycorrhiza* and nitrogen-fixing rhizobia interacting with plant root cells. The fungi and bacteria involved in these interactions form symbiosome structures within the host cells, similar to the haustorial complexes of biotrophic *pathogenic* fungi such as the rusts and mildews.

Cell wall *penetration* must be accomplished by these symbiotic microorganisms in a manner preventing the elicitation of the active resistance *mechanisms* just described for the defense against pathogenic *microorganisms*. Mutualists may achieve this by *actively* suppressing plant defense responses such as callose *deposition* or by inducing cell wall autolytic processes within the plant cell *reminiscent* of phragmosome formation during cell division.

Similarly, successful biotrophic pathogens must *continuously* suppress resistance reactions by the penetrated host cells, and they may achieve this by generating plant cell wall *fragments* acting as endogenous suppressors of disease resistance *elicitation*. Cell walls do not only form the stage of interaction between plant cells and microbial symbionts and pathogens. Cell walls of the male pollen tube and the female pistil cells are also the site where *self-recognition* in the selfincompatibility system prevents *self-pollination*.

This system has been elucidated on the molecular level in *Brassica*, where a receptor kinase in the plasma membrane of the stigma epidermal cells cooperates with a stigma cell wall glycoprotein in the recognition of a cysteine-rich pollen cell wall protein. The outcome of this molecular interaction is most likely a multiple response certainly including cell wall modifications that *eventually* arrest further pollen tube growth. The examples described here are recent corroborations of the much older *oligosaccharin* concept postulating oligosaccharide fragments from plant cell walls as hormone-like signal molecules influencing plant metabolism and development. Intermediate-sized products of fungal endopolygalacturonase digestion of plant pectins act as endogenous elicitors of active defense *reactions* in plants.

Small products of pectin digestion act as endogenous suppressors of elicitor-induced plant resistance reactions. The biological activity of the endogenous suppressors that are beneficial to the fungal plant pathogens but not to the host plant *suggests* that these may mimic as yet unknown endogenous plant signals involved in determining differentiation events in plant cells. Indeed, oligomeric fragments of pectin and xyloglucan have been implicated as tissue *hormones* regulating plant growth and development.

Similarly, the nodulation factors of rhizobia initiating *meristem* formation may mimic endogenous plant signal molecules. It has been speculated that the diverse family of plant *arabinogalactan* proteins may form a cell wall-located reservoir of such signal molecules involved in local *differentiation* and development.

COMPONENTS AND ARCHITECTURE OF PLANT CELL WALLS

In reviewing the *structural* and functional roles of plant cell walls, we have already named most of their constituents, and we have indicated which of the different *components* contribute to the specific functions. In this section, we will briefly *summarize* our current knowledge of the structure of these components and our understanding of how they interact to assemble into the complex and dynamic three-dimensional *pericellular* matrix.

Over the past decades, a number of ever more refined models of plant cell wall architecture have been proposed, mostly based on detailed investigations of the primary cell walls of *suspension-cultured* cells. This prototype cell wall of dicot plants is opposed by the somewhat different cell wall of somethe commelinoid-monocot plants, most *notably* the grasses and cereals.

The general *architectural* plan of both types of cell walls, however, is *identical*: stress-bearing fibrils are embedded in a pressure-bearing matrix. In both types of plant cell walls, the fibrillar component is represented by cellulose *microfibrils*, accounting for about one third of the cell wall dry weight. In the dicot cell wall, each cellulose microfibril is thought to be completely wrapped in an envelope of *xyloglucan* chains mediating the surface interaction between the *fibrillar* component and the surrounding matrix.

Xyloglucan molecules are thought to span the interfibrillar space, their ends *hydrogen* bonding to different cellulose microfibrils. The cellulose and xyloglucan molecules thus form a three-dimensional scaffolding network in the dicot cell wall. In commelinoid monocot cell walls, xyloglucan is mostly replaced by *glucuronoarabinoxylan* chains, which are similarly hydrogen bonded to several cellulose *microfibrils*, thus leading to a *cellulose-xylan* network.

The cellulose-xyloglucan network of the dicot cell wall is embedded in the pectic *matrix*, which makes up about one third of the dry weight of the cell wall. The pectins are believed to form a second, independent network in the cell wall that may not be *covalently* linked to the *cellulosexyloglucan* network.

The intermolecular cross-links knitting pectic *polysaccharides* together are most likely Ca^{2+} bridges between *nonesterified* stretches of homogalacturonan, hydrophobic interactions between methyl esterified stretches of *homogalacturonan*, and, possibly, boron diesters between rhamnogalacturonan II stretches.

Commelinoid monocot cell walls contain the same pectic polymers as dicot walls but at much lower contents (below 10%). Their role as a water-binding matrix component is *substituted* by the major hemicellulose, *glucuronoarabinoxylan*, which alone can account for about half of the dry weight of the commelinoid *monocot* wall.

It can be expected that the same types of interpectic cross-links exist in the commelinoid monocot cell wall as discussed for the dicot wall. Many of the arabinose residues of the glucuronoarabinoxylan carry ferulic acid or *p-coumaric* acid esters, and these may *dimerize* oxidatively to form diphenolic acid cross-links possibly substituting for missing interpectic links.

Thus, a pectin and/or a glucuronoarabinoxylan-phenolic acid network may exist in *commelinoid* monocot cell walls. Both types of cell wall contain *structural* glycoproteins-roughly 1020% of the dry weight of the dicot wall compared with about 2-10% in the commelinoid monocot wall-which may form a third independent network in the primary plant cell wall. The major *protein* component of dicot walls appears to be the *hydroxyproline*-rich glycoprotein (HRGP) *extensin*, but proline-rich proteins (PRPs) and glycine-rich proteins (GRPs) may be involved as well.

The only cross-links proposed so far to hold the protein network together are *intermolecular* isodityrosine bridges, but their existence still awaits *experimental* support. The classical cell wall models depicting primary plant cell walls as a largely covalently cross-linked system of the constituent *polysaccharides* and proteins have been *extended* and improved to yield a composite cell wall model with a more flexible and dynamic *architecture*.

According to our current view, primary cell walls of both *dicot* and monocot plants are built of the three independent interwoven networks already described-the cellulose-glycan network, the pectin network, and the *proteinaceous* network -held together by noncovalent links. Primary cell walls are capable of enormous rates of largely planar, two-dimensional growth, believed to be *restricted* by the cellulose-glycan network.

In dicot cell walls, controlled activity of *expansins* and *xyloglucanendotransglycosylases* is supposed to allow controlled *yielding* of this network, so that the orientation of the cellulose microfibrils determines the direction of cell elongation. In contrast, some of the xyloglucan-poor monocot cell walls may, during phases of elongation growth, produce an *additional* polysaccharide component-a mixedlinkage α-1,3-β-1,4-glucan-that may hydrogen bond to cellulose and may *transiently* replace the xylan, which appears to be turned over rapidly during growth.

In contrast to the artificial situation of plant cells growing in liquid suspension where all cells are *permanently* in a similar state of differentiation, *preferentially* rapid spherical *growth*, cells of many different types, and in many different states of differentiation, interact with each other to form a highly complex tissue in the *intact* plant.

Cultured plant cells are *surrounded* only by a primary cell wall. In a plant tissue, all cells still in the process of growing and even many fully *differentiated* cells, such as leaf mesophyll cells, also contain only a primary and no secondary cell wall. However, many *specialized* plant cells produce highly sophisticated secondary cell walls, e.g., *collenchyma* and *sclerenchyma* cells, tracheids, and pollen cells.

Moreover, all the different cells that make up a tissue are "glued" together by the middle lamella, a structure that may not be present in cultured cells, at least if they grow as a very fine *suspension* of *individual* cells. Furthermore, it appears that the primary wall of a cell in a tissue is much less *uniform* than that of a cell grown in suspension culture.

Immunocytological electron microscopic studies using antibodies to different cell wall components revealed microdomains in the *seemingly* uniform primary cell wall *exhibiting* striking differences in their molecular composition. Antibodies cross-reacting with *non-methyl* esterified epitopes of pectin stained the middle *lamella, especially* in cell *corners*, but also lined the inner surface of the primary cell walls along the plasma membrane.

Antibodies cross-reacting with highly methyl esterified pectic components, in *contrast*, evenly bound across the whole width of the primary cell wall. Cell wall angles where three or more different cells meet are *distinct* from cell wall stretches in contact with a single neighboring cell, and these again differ from areas that face the *intercellular* space. The mature primary cell wall of about 80 nm *thickness* consists of only about three layers of cellulose *microfibrils*.

Each layer, thus, resides in a different *nanoenvironment*, the outer layer facing the *pectic* middle lamella, the middle layer *surrounded* by the other two, and the inner layer facing the plasma membrane or, later, the secondary cell wall layers. Virtually nothing is known about the *interaction* of the primary cell wall with these *neighboring* layers.

These novel micro- and nanoscopic approaches will *certainly* shed new light on our *understanding* of the plant cell wall, but they can be *expected* to pose more *questions* initially than they will answer-e.g., concerning *assembly* and regulation of this *enormous* complexity outside the *cytoplasm*. To make things even more complex, many cells that fulfill special functions within a plant tissue exhibit *striking* and often highly *localized* modifications of the standard primary plant cell wall described so far.

Epidermal cells have to play multiple roles in growth restriction and stress protection, *necessitating modifications* of the outer primary cell walls. These are thicker than the other epidermal cell walls, and the cutin monomers and cuticular and *epicuticular* waxes produced in epidermal cells are transported to the outer surface of this outer periclinal wall to reach the plant surface, where they are laid down in layers and where the cutin is *polymerized* in situ.

One of the most prominent examples of secondary cell walls is certainly the intricate elaboration of cell wall thickenings in tracheary elements of the xylem, where additional cellulose fibers and *embedding* matrix polymers are laid down only on *certain* parts of the cell walls. *Eventually,* lignin is *incorporated* into the walls of these cells.

The *polymerization* of the monoligno s in muro leads to a growing lignin polymer infiltrating the primary cell wall, displacing the water in the process. Eventually, when the polymerization process is complete, the *lignin* polymer serves to cement in place all the other components of the cell wall. The wall is then no longer *permeable* to water, and the mature *tracheid* dies.

The ensuing loss of turgor pressure would result in a shrinking of the dead cell and most likely a collapse due to the negative pressure (tension) of the *transpiration* stream, were it not for the lignin *impregnating* and stiffening the cell wall of the tracheid. *Clearly,* the wall of an individual cell is a highly differentiated and coordinated organelle with domains differing in structure and-most likely -also in function. In addition, we can expect a highly dynamic temporal *differentiation* to be superimposed on this complex *spatial* differentiation.

Cells are born in meristematic zones where cell division occurs, they undergo elongation and differentiation, they may eventually senesce, and they will finally die. The wall of a cell most likely changes *continuously* during these processes, but today little is known of what these *changes* are and what they may entail.

BIOSYNTHESIS AND ASSEMBLY OF PLANT CELL WALLS

The *daunting* complexity of the many *constituents* that make up plant cell walls is *rivaled* by an equally *impressive* complexity of genes potentially involved in their synthesis, as revealed by the recent completion of the *Arabidopsis* genome sequencing project. However, little is known about the actual biochemistry of cell wall component biosynthesis, and virtually nothing is known about the processes involved in the *assembly* of these *components* into the three-dimensional complex plant cell wall.

In the plant cell cycle, new cell walls are typically produced during mitosis when the internuclear *anaphase* spindle is transformed into the *phragmoplast.* Within this cytoskeletal apparatus, the new cell plate is assembled from cell wall material synthesized and delivered by the fusing Golgi-derived vesicles. No *reliable* information is available on the *biochemical* nature of the cell wall *material* initially laid down in the outward-expanding cell plate.

Circumstantial evidence suggests that callose, hemicelluloses, and pectins are involved, and the finding of a membrane-bound 8-1,4-glucanase in the cell plate may even point to cellulose already being *synthesized* during the birth of the new cell wall. For decades, the search for cellulose synthase has resembled the quest for the Holy Grail of plant biochemistry.

The rosette structure seen in the electron microscope at the end of newly deposited cellulose microfibrils has long been suspected to harbor cellulose synthase, but biochemical evidence was and still is lacking. The *identification* of the first plant gene with sequence similarity to bacterial cellulose synthase only a few years ago has been followed by a deluge of related *sequences* forming the cellulose synthase *superfamily.*

At least 12 presumed cellulose synthase genes and around 30 members of the six classes of cellulose *synthase-like* genes have been identified in the *Arabidopsis* genome. They all belong to the class of multipass *transmembrane* proteins with a large, *presumably* cytoplasmic catalytic domain. An antibody raised against a recombinant cellulose synthase confirmed the presence of this *enzyme* in the rosette structure, which is now thought to be composed of six complexes, each with six catalytic subunits.

It is still unknown whether one or two catalytic sites are involved in chain *elongation* and whether the chain grows by the addition of one or two glucose units at a time. It is generally assumed, though, that the rosette structure allows transmembrane synthesis of an entire cellulose *microfibril* containing a few dozen parallel cellulose *chains* through the plasma membrane into the cell wall.

The *apoplastic* self-assembly of the fibril, thus, may be intimately coupled to the biosynthesis of the cellulose molecules. The *surface* of the nascent microfibril might be immediately covered by hemicelluloses present on the outside of the plasma membrane, *ensuring* the individual *identity* of the nascent fibril by preventing its fusion with concomitantly synthesized neighboring microfibrils. The different cellulose synthase genes appear to be expressed in different cell types, depending on their state of *differentiation* and development.

Different isoenyzmes appear to be involved in the synthesis of cellulose microfibrils for primary and secondary cell walls that *differ* both in the length of the *individual* cellulose chains and in the number of chains in individual microfibrils. Interestingly, the cellulose *synthases* may operate in pairs in both cases, possibly *forming* dimers in the rosette structure.

A mutant unable to assemble *rosette* structures but *unimpaired* in cellulose *biosynthesis* forms noncrystalline cellulose aggregates instead of crystalline microfibrils. Unlike the plasma membrane localization of cellulose synthesis, noncellulosic *polysaccharides* are generally believed to be *synthesized* in the *Golgi* apparatus. This is in agreement with delivery of *hemicelluloses* and pectins and/or their synthesizing enzymes with Golgi vesicles to the cell plate during mitosis.

As for cellulose synthase, the *glycosyltransferases* involved have long *resisted biochemical approaches* of purification and *characterization*. The first two glycosyltransferases involved in hemicellulose side chain attachment and elongation have only recently been isolated and characterized. Both enzymes are membrane *anchored* by a single transmembrane domain near the *N-terminus*, with the *catalytic* domain most likely *oriented* toward the interior of the Golgi *vesicles*.

This raises the interesting question of how the *sugar-nucleotide substrates* enter the vesicles. No biochemical data are yet available on the biosynthesis of the hemicellulose *backbones*, but the multitude of genes encoding cellulose *synthase*-like glycosyltransferases leaves plenty of room for enzymes involved in the *generation* of linear β-glycosidically linked *glycans*.

Presumably, these cellulose synthase-like *polymerases* are located in the Golgi vesicle membranes. Their *orientation*, however, would lead to polysaccharide synthesis in the cytoplasm, as these enzymes are not *believed* to channel their products through the membrane. The *linear* glucan and *xylan* polymers would have to be imported into the Golgi lumen, where the glycosyltransferases just *described* could attach and *elongate* side chains to build xyloglucans and glucuronoarabinoxylans.

The distribution of the different side chains along the linear backbone of the hemicelluloses is thought to control their roles in the overall architecture of the cell wall. However, nothing is known about the *regulation* of this *distribution*. Possibly even less is known about pectin biosynthesis. Having a backbone of a-glycosidically linked *galacturonic* acids, interspersed or not with α-glycosidically linked *rhamnose residues*, these cannot be synthesized by any of the cellulose synthase-like family-2 glycosyltransferases that produce β-linked *glycans* from α-linked UDP-sugar donors.

Homogalacturonan biosynthesis has been *achieved* in vitro at very low *rates*, and pectin *methyltransferase* activity was *associated* with the Golgi *preparations* used. *Consequently*, the pectin is presumed to be *delivered* into the cell wall in a highly methyl esterified state. Apoplastic pectin methyl esterases are then believed to convert these immature pectin polymers to the diverse *partially* methyl esterified *pectins* found in the different domains of mature plant cell walls.

The presence of up to a *dozen isoenzymes* of pectin methyl esterase may be related to the many roles pectins have been *suggested* to play in the physiology of plant tissues. Nothing is known *concerning* the biosynthesis of the linear backbone of rhamnogalacturonan I consisting of the *α-glycosidically* linked repeat disaccharide galacturonic acid -rhamnose. The *rhamnogalacturonan* I backbone appears to be contiguous with the backbone of *homogalacturonan*, posing the interesting question of a possible switch in enzyme activity during biosynthesis.

Moreover, those homogalacturonan regions close to rhamnogalacturonan appear to be acetylated but not methylesterified another unresolved riddle of regulation. *Microsomal* membranes have been shown in vitro to *catalyze* the attachment and elongation of galactan side chains to the rhamnogalacturonan I backbone, but no activities have been shown to account for arabinan and *arabinogalactan* side chains.

Perhaps the most complex and most *puzzling* plant cell wall polysaccharide is rhamnogalacturonan II, *possessing* a linear homogalacturonan backbone decorated with an apparently strictly controlled sequence of four highly complex side chains. Unsurprisingly, our knowlegde of its biosynthesis is nil. Much easier to understand is the biosynthesis of cell wall proteins.

These are conventionally polymerized on *ribosomes* of the rough endoplasmic reticulum and are delivered to the cell wall via vesicle trafficking through the Golgi apparatus. Nonstructural cell wall proteins, such as invertase or *antimicrobial* hydrolytic enzymes, are typically glycoproteins, and the typical plant oligosaccharide side chains are transferred to the polypeptides via the usual dolichol pathway. Secretion into the cell wall appears to be the default *pathway* for *proteins* entering the secretory pathway; no further *sorting* signals appear to be required. The situation is less clear for the *structural* cell wall proteins.

Often, these do not carry the typical oligomeric N-linked glycosidic side chains mentioned earlier. Instead or in addition, they may carry O-glycosidically linked oligomeric or polymeric glycan moieties. Moreover, *hydroxyproline*-rich glycoproteins are *characterized* by an additional posttranslational *modification*, the hydroxylation of some but not all proline residues.

Again, these modifications appear to be carried out in the *endoplasmic* reticulum and in the Golgi apparatus. Golgi vesicles contain large amounts of plant proteoglycan-members of the large family of arabinogalactan proteins (AGPs). Their *relatively* high concentration in the *vesicles* is in contrast to the low abundance of these proteoglycans in the cell wall. AGPs may, therefore, be interesting *candidates* for chaperone-like assembly assistants preventing premature selfassembly of cell wall matrix material in the Golgi vesicles and/or ensuring correct assembly of the cell wall components. Although our knowledge of the synthesis of most cell wall polymers is sketchy at best, we seem to know a little more about the biosynthesis of the monomeric precursors of these polymers.

All nucleotide sugars required for polysaccharide *biosynthesis* are produced from UDP- and GDP-glucose, by the action of epimerases, dehydratases, oxidoreductases, and carboxylases-or they are *generated* in salvage pathways from cell wall degradation products by the action of *kinases* and nucleotide pyrophosphorylases.

The nucleotide sugars are synthesized in the *cytosol*, as are the phenylpropenol monomers of lignin. In contrast, cutin and wax monomers are synthesized in the endoplasmic reticulum, and the hydrophobic, water-insoluble components are most likely transported through the outer epidermal cell wall to the surface of the plant via *lipid* transfer proteins.

The only *unusual* amino acid found in structural plant cell wall proteins is hydroxyproline, but hydroxylation occurs posttranslationally so that no *specific* amino acid *biosyntheses* are involved in cell wall generation.

BIOTECHNOLOGICAL APPROACHES TO OPTIMIZE CELL WALL PERFORMANCE IN PLANTS

The oldest known fossils are cells with a wall. Cell walls had over 3.5 billion years of time to evolve, so that they can be expected to be fairly optimized for the many roles they play in a plant's life. It may appear totally presumptuous and vain to try to further optimize this admirable pericellular organelle. However, as with all biological structures and systems, the cell wall components and architecture are optimized to *fulfill* diverse and sometimes *contradictory* roles concomitantly, so that the result represents an optimal compromise.

Consequently, the outcome varies depending on the environmental situation of the species, individual plant, and individual *cellgiving* rise to the described complexity and diversity of plant cell walls. There is, hence, some room for gentle adjustments of the balance between the different optimization goals so as to better adapt crop plants to the man-made environment of agriculture (or of a fermenter) and to better suit human needs. However, today our *possibilities* are greatly limited by the paucity of *information* available about the biosynthesis, turnover, and modification of the components and by our lack of information on the assembly and regulation of the complex plant cell wall.

This relative *ignorance* of the targets far outweighs the limitations posed by the still less than optimal genetic engineering tools for plant transformation. For the time being, our chances of success are most promising if the aspect to be *optimized* can clearly be *ascribed* to a single cell wall component.

Ideally, this component should be a nonstructural protein that is synthesized, modified, and delivered to the cell wall by the conventional secretory pathway. Today's agriculture is characterized by vast homogeneous fields of genetically uniform crop plants. The high-yield cultivars grown are bred for fast growth and maximum yield. Often, the contents of secondary plant metabolites have been *reduced* in favour of product quality.

Typically, disease resistance has not been a primary breeding goal, and the *inherent* resistance traits of the wild progenitors have largely been lost in our crop species. There is a clear need to breed old and new resistance traits back into modern crop cultivars, and one way to achieve this is by genetic engineering. We have seen that cell walls play crucial roles in plant defense against microbial *pathogens* so that here are promising targets.

The most easily handled are cell wall-located pathogenesis-related (pr-) proteins. These include *antimicrobial* defensins, thionines, proteases, peroxidases, β-1,3-glucanases, and chitinases. Transgenic plants overexpressing a single pr-protein usually do not exhibit significantly increased disease resistance, but *combinations*, e.g., of *fungal* cell wall hydrolyzing β-1,3-glucanases and chitinases, are more efficient.

The hyphal cell walls of some phytopathogenic fungi *growing* endophytically in the host plant tissues have been shown to contain *chitosan* rather than chitin, so that *chitosanases* may be more effective than chitinases in protecting plants from disease. Systematic *approaches* to analyze cooperative, *synergistic* actions of the different classes of pr-proteins are currently under way. A number of further strategies toward transgene-mediated disease resistance are currently being pursued, and most of them involve the plant cell wall.

The most efficient resistance reaction of many plants against a variety of pathogens is the hypersensitive response, which is often induced as a result of the molecular interaction of the direct or indirect products of plant *resistance* genes and microbial *avirulence* genes. The former are believed to act as plasma membrane-bound receptors for the latter, which are therefore *termed* elicitors.

Expression and cell wall secretion of a microbial avirulence gene under the control of a strictly pathogen-induced promoter in a plant carrying the corresponding resistance gene has been shown to lead to pathogen-induced *triggering* of an artificial hypersensitive response. *Similarly*, the tightly regulated expression of glucose oxidase can lead to the generation of an artificial *oxidative* burst, which in turn activates a hypersensitive-like reaction.

As this reaction involves the activation of a cell death program, tight regulation of the process-both in time and in space, and under a broad regime of environmental conditions-is required for this approach. Of the incredible *biodiversity* of plant species, only a very few have been domesticated to provide the mainstay of the human food supply. Valuable traits from other plant species that cannot be inbred into our crop species can now be introduced by genetic *engineering*.

Examples of important traits are the production of pharmacologically active secondary plant metabolites and mechanisms of stress tolerance, such as salt tolerance and drought tolerance. The main problem with these complex traits is that they are brought about by the consorted action of multiple genes and are therefore difficult to transfer. We have seen that the outer epidermal cell walls are crucial for drought tolerance, mainly by virtue of the cuticle.

The classical *eceriferum* mutants of barley implicated around 100 genes as involved in the biosynthesis and assembly of the cuticle. Currently, tagged *eceriferum* mutants of barley and *Arabidopsis* are being selected to shed light on the molecular machinery involved. Eventually, we will have to transfer and broaden the knowledge gained from these model plants to include, e.g., drought-resistant desert plants.

Only then can we hope to learn on a molecular basis how these plants cope with their hostile environment, and we may be able to profit from this insight by generating drought-resistant crop plants that may flourish in the growing semiarid areas. Similar approaches are being undertaken for other stress factors. Cell walls can be anticipated to be prominently involved in many of these stress resistance mechanisms. Comparing modern crop plants with their wild relatives reveals the striking changes in morphology that sometimes thousands of years of breeding have generated and that are witness to the enormous flexibility in plant organization.

Today's maize cultivars producing corn cobs with up to 24 rows of perhaps 50 kernels each were bred from wild progenitors having cobs the size of a pencil *eraser*. Sugar beets with a diameter of up to 20 cm and a sugar content of up to 20% were bred from small beets with only trace amounts of *sucrose* within less than 150 years.

Plant morphology is dictated by the cell walls: walls shape the forms of plant cells and organs; walls restrict the growth of plant cells and organs; walls are responsible for the *strength* of plant cells and *organs*; and cell wall enzymes and oligosaccharin wall fragments are involved in the development of different types of plant cells and organs.

In this case, our basic handicap for biotechnological manipulations is the sparse knowledge we have of how cell walls influence these different aspects and the realization that these are multifactorial and *multigenic* traits. However, first results have been obtained, and if these have not yet led to commercially useful crop plants, they do provide proof of principle for the general approach. The changes forced upon the crop plants by conventional breeding all involved redistribution of *photoassimilates* toward the sink organs of human interest: shorter stems in favour of more grains, less aboveground growth in favour of more and *thicker* tubers, etc.

A century ago, the harvest index of wheat straw/grain was 70/30; it is 40/60 today. The *regulation* of sourcesink ratios is achieved by modifications in the relative phloem transport rates from the different source tissues toward the different sink tissues of a plant. Both phloem loading and phloem unloading, which are thought to be the main regulatory steps in phloem transport in most plants, involve an apoplastic step. Cell wall-located invertases are likely to be involved in the process of sugar sensing, which determines both source and sink activities.

Overexpressing a fungal invertase in the apoplast of potato leaves or tubers led to dramatic phenotypes, including modified root-to-shoot ratios, changed number and weight of tubers, and altered *morphology* of leaves and tubers. This is another example of genetic engineering of plants via *manipulation* of cell walls where the target is a *nonstructural* cell wall protein. Cell walls are essential for food *texture* and fruit juice turbidity, and modifications in cell wall composition and architecture are among the most prominent changes during fruit maturation.

Interfering with these changes by *genetic* engineering promises to increase shelf life and to improve the manufacturing processes in food technology. Perhaps the best known example of "novel food" is the Flavr-Savr tomato *expressing* an antisense gene for endopolygalacturonase. Decreasing the activity of this cell wall enzyme prevents maceration of the tomato tissue so that red-ripe fruits can be transported over long *distances* and stored for prolonged times.

However, the processes of fruit maturation are still far from being understood, and cause-consequence *relationships* are mostly unknown. The partial breakdown of cell wall components during maturation is most likely an integral part of a sequence of events, and disturbing one link of the chain will have unpredictable consequences for all *downstream* events. Flavr-Savr tomatoes have been reported to look but not to taste like red-ripe tomatoes.

BIOTECHNOLOGY APPROACHES OT OPTIMIZE CELL WALL COMPONENTS FOR USES EX PLANTS

Plant cell walls are the most abundant *renewable* resource available to man, and they have accompanied mankind from the beginning. Wood, *entirely* consisting of plant cell walls, is used in the form of timber for *construction* purposes and as a raw material to produce pulp and paper. Cell wall fibers have provided man with strings to be woven into tissues and drilled into ropes, and they are increasingly manufactured into insulation materials. Plant fibers are also healthy ingredients of man's diet and may actually have *antitumor* activities, and other cell wall *components* are used as gelling materials in food technology.

Many other uses of plant cell walls have been proposed even though they may not yet have been verified. *Optimizing* plant cell wall components for these ex plants uses appears a much more easily *achievable* goal in the short range, as cell walls are not naturally optimized for these *applications.*

However, as attempts at genetically engineering a reduced lignin content in wood for the pulp and paper industries have amply proved, *surprises* can result from this strategy, too. Genetic *engineering* of lignin is extensively described in other chapter of this book and will, therefore, not be treated in detail here. But we should *reflect* briefly on the lessons to be learned from these approaches as they relate to the general subject of biotechnology of plant cell walls.

Most of the initial experiments aimed at reducing the lignin content by expressing *antisense* genes for lignin biosynthetic enzymes, with the goal of improving rumen digestibility of forage grasses and easing the delignification step during paper production. *Invariably,* lignin content was almost *unchanged* in these transgenic plants, but lignin composition often changed to some *extent.*

Clearly, the cells that survived the switching off of specific steps in lignin biosynthesis did so by making use of the inherent flexibility in cell wall architecture. If some monolignol was no longer available due to a block in its biosynthesis, other, often unusual, monomers were incorporated into the lignin of the transgenics, and the *resulting altered lignins* were *apparently* able to take over at least *partially* the roles of normal lignin.

Similar flexibility can be expected in the case of other cell wall components as well, as exemplified by suspension-cultured cells growing in the presence of cellulose biosynthesis inhibitors or fucosyltransferase-lacking mutants of *Arabidopsis. Consequently,* the *results* of even very pinpointed genetic *engineering* experiments can hardly be predicted with certainty. Although we will not ponder further on genetic engineering of lignin *biosynthesis* per se, it is important to realize that lignin always forms part of complex cell walls, and the *composite* material is not uniform but able to react to environmental influences with compositional and, most likely, architectural changes.

If woody plants are *misaligned* from the vertical axis, so-called reaction wood is formed to realign the stem. Xylem cells involved in the transformation undergo massive changes in their cell wall composition. In *gymnosperms*, compression wood is formed in the areas of the stem that are subject to increased pressure, and in angiosperms, tension wood is formed in the areas of the stem subjected to increased stress.

Compression wood is increased in lignin content and decreased in cellulose content, while *tension* wood contains more cellulose and less lignin. It might be conceivable, then, to generate wood that is specifically adapted to certain construction needs. Such a *manipulation* would have to consider both the *polysaccharides* and the polyphenolics present in the cell wall. Fibers of plant cell walls form the basis of millennia-old techniques such as warping and weaving. Yarns, *strings*, ropes, fabrics, and tissues were and still are to some extent produced from plant cell wall-based fibers.

Increasingly, natural fibers are also used in the *manufacturing* of insulation materials for *construction* purposes. Cotton is the most widely used plant fiber today. Conventional *breeding* has already produced cotton varieties with cellulose fibrils of different *lengths* and colours. When we understand the roles of the different cellulose synthase isoenzymes in producing cellulose microfibrils of different lengths and diameters, we might be able to generate an even greater variety of cotton-based raw materials.

Flax, hemp, sisal, and ramie yield fibers used on a large scale, and the material properties of these fibers differ from those of cotton fibers owing to the different cell wall polysaccharides *accompanying* the cellulose microfibrils in the cell walls of these plants. Breeding and biotechnology with these fiber plants are far less *advanced* than with cotton, so that there is a great *potential* for future improvements. Many other plants such as palm trees and agava plants are used locally for fiber production. These plants are typically integrated in a complex system of human uses not *restricted* to fiber production, so that there is less room for *optimization* in a single direction.

Pectins are widely used as gelling agents in the food industries. However, only pectins of a very few plant species have been shown to be well suited for the purpose, as other *pectins* available in large *quantities* at competitive prices proved inadequate. Large volumes of pectin-rich waste pulp are produced by the potato starch industries, but the pectin present is less suitable for the food industries partly due to its high content of *rhamnogalacturonan* side chains, particularly neutral *galactans*, and its low degree of methyl esterification.

Expressing a fungal galactanase in potato tubers led to the formation of galactan-poor pectins with increased solubility characteristics, but the transgenic plants showed no *obvious* altered phenotype. It remains to be seen whether this recombinant pectin is better suited for industrial uses and whether the genetic engineering approach is superior to conventional *postharvest* enzymic treatments.

But clearly, expressing hydrolases in the cell walls is successful and interesting because it is presumably a very versatile means of *manipulating* plant cell wall composition and architecture in plants for uses ex plants. Many other biotechnological uses for plant cell walls and their components have been described or proposed.

In terms of market size, medicinal *applications* are among the most attractive. Cell wall components-in the form of nondigestible oligosaccharides or as dietary fibers are increasingly suspected to be responsible for some of the pronounced health effects of vegetables and fruits. Dietary fibers, and particularly *pectins,* have been implicated in protection against cancer, coronary heart disease, and vascular *diseases.* No causal relationships have yet been established, but functional foods are already on the horizon.

Cell wall components may also be useful in *environmental* protection. For instance, the cation chelating properties of pectins have been used for heavy metal decontamination of wastewaters. *Increased* knowledge of biosynthesis and assembly of cell wall constituents may eventually allow us to draw up totally new designer polysaccharides not known from any natural source. As an example, let us consider again the biosynthesis of *cellulose* and other *p*-glycosidically linked hemicellulose backbones.

All of these linear polysaccharides consist of disaccharide repeat units. The two *glycosyl* units involved are most likely attached to the growing polysaccharide chain by two *independent* catalytic sitespresent in either one or two cellulose synthase or cellulose synthase-like proteins. If this concept proves to be right, and when different catalytic domains with specificities for different nucleotide sugar substrates become known and available, domain swapping may lead to the creation of novel polysaccharides.

Even though these may not properly and cooperatively function in a primary cell wall, it may still be possible to produce them, e.g., in cotton fibers-and it may thus become possible to tailor novel designer polysaccharides for specific ex plants uses.

PROSPECTS FOR THE FUTURE OF PLANT CELL WALL BIOTECHNOLOGY

We have seen that the major limitation for biotechnology with plant cell walls today is our limited knowledge about the biosynthesis of the cell wall components and their assembly into the complex pericellular matrix. The genomic approaches initiated recently promise rapid progress in identifying the genes potentially involved in these processes, at least for some model plants such as *Arabidopsis,* maize, rice, barley, poplar, and pine.

Proteomic approaches will complement these studies, and they will help in identifying the enzymes involved in the biosynthesis and assembly of *specific* cell wall domains elaborated under certain sets of *environmental conditions.* As seen with the cellulose *synthase genes* and, even more clearly, with the six classes of cellulose synthase-like genes, *extensive functional genomic* experiments will be needed to assign *specific* functions to the genes identified, and these will keep us occupied for years to come.

Functional genomics will allow us to analyze which enzymes are involved in biosynthesis and assembly of complex cell walls, but they will not shed light on how these enzymes do their jobs. Methods for functional proteomics may have to be developed to investigate the molecular *interactions* between proteins, polysaccharides, and the other cell wall *constituents* to understand how the cell wall is assembled and how it functions.

This knowlegde, of course, is a *prerequisite* for targeted genetic engineering of plant cell walls for biotechnological uses. Functional proteomics will have to make use of the newly developed technical possibilities offered by the nanosciences. *Nanoanalytical* tools allow the visualization and the determination of surface properties of individual molecules. They also permit direct measurement of *intermolecular* forces holding together the noncovalent networks that make up plant cell walls. Interestingly, nanosciences not only offer analytical tools but also are currently developing *manipulating* tools with which to handle and influence individual molecules. These tools can be increasingly used in vivo, and this will allow us to learn about the discrete roles that individual *molecules* play in the complex fabric of the cell wall.

Eventually, cell walls may serve as ideal models for complex, versatile, and very stable composite materials, and the molecular machinery for assembly or the principles of self-assembly of the different *components* to build the complex cell wall may well serve as a model for biomimetic processes to be used in the construction of nanomachines or *nanomaterials* using nonnatural or semisynthetic molecules. It should be clear from the preceding *discussion* that we are only beginning to see the surface of the *incredible biodiversity* inherent in plant cell walls-as it is in almost every aspect of living systems.

We have seen the flexibility and dynamics in cell wall *architecture* of a single plant, and we have pointed to the differences in cell wall composition and architecture between dicot plants and the commelinoid monocot plants. However, this treatise has focused on the cell walls of the *angiosperms* while those of the *gymnosperms* have not been *mentioned.*

But to be comprehensive, we would even have to include cell walls of the nonflowering ferns and mosses and the green, red, and brown algae. Fungal cells are also surrounded by cell walls that function according to the same *principle* as plant cell wallsstress-bearing fibers in a *pressure-bearing* matrix-but this is realized using quite different polysaccharides and proteins, and the diversity within the *fungal* kingdom is enormous.

Furthermore, almost all prokaryotic cells possess cell walls, but these are built in a completely different way, such as the murein sacculus, which is one giant *covalently* linked molecule. We know little about most of these cell walls-and this is an exciting prospect for future researchers who are not intimidated by the enormous *complexity* of the pericellular matrices. Clearly, plant cell walls are an incredibly rich renewable resource for sustainable biotechnologies.

However, we still understand little of the causeand-effect relationships in most of the physiological processes involving plant cell wall components. Manipulating single components of such a complex structure has almost invariably yielded results that were different from what was expected. We should seize the *opportunities* molecular genetic tools offer, but we should take our time to analyze the effects of genetic engineering of cell wall components, assembly, and architecture.

There should be no haste to introduce transgenic plants and their products to the environment and to the market prematurely. Only well-planned, well-done, and well-analyzed transgenic plants with clear *environmental* and consumer benefits will stand a chance to overcome reasonable and irrational fears of the consumers.

4

Chapter

GENETIC ENHANCEMENT

The oils of plant origin have important edible and non-edible uses in human life. Various oil yielding crop species have been *domesticated* to produce high oil yielding seeds for edible or industrial purposes. The oilseed bearing crops include perennial trees like coconut and palm and annuals like *groundnut, soybean, sunflower* and *oleiferous brassicas*.

The oilseed brassicas include *B. carinata, B. nigra* and rapeseed-mustard (collective term for *B. napus, B. juncea* and *B. campestris).* Among various oilseed crops mentioned above oilseed brassicas command a substantial market proportion. Rapeseed mustard with an average world production of 3.6×10^7 million ton and an acreage of 24.7 million Ha, ranks third at global level preceded by soybean and cotton seed. A considerable proportion of the world production of oilseed brassicas has been contributed by developing countries, in particular China and India.

In India, among the nine annual oilseed crops grown in the country, oilseed brassicas collectively rank second in terms of production and acreage next only to groundnut. Among the various *Brassica* species, *B. juncea* occupies the maximum acreage followed by *B. campestris* especially in the north gangetic plain. *B. napus* is a relatively new introduction to India but is gaining rapid popularity in Punjab and Himachal Pradesh.

Thus, in India too, the oilseed brassicas are a major contributor to vegetable oil production and a considerable amount of research effort is directed towards their genetic enhancement. On account of its economic importance, oilseed brassica attracted the early plant breeders and geneticists, and pioneering research was undertaken for elucidating its cytogenetic composition. The cytogenetic relationship between various oilseed brassicas has been represented by U and is popularly referred to as the U's triangle. Brassicas comprise three diploid species namely *B. nigra, B. oleracea and B. campestris* and three allotetraploids namely *B. carinata, B. juncea and B. napus* which have arisen out of interspecific hybridization between the diploid species followed by spontaneous chromosomes doubling.

Evidence in support of this relationship has been accumulated from cytological, biochemcial and molecular investigations. The oilseed brassicas show variable pollination behaviour in terms of existence of selfcompatible and incompatible forms. *B. campestris* is cultivated in the form of three ecotypes namely brown sarson, yellow sarson and toria.

These ecotypes comprise both—self-compatible and incompatible plant types. *B. napus, B. carinata and B. juncea* are predominantly selfpollinated whereas *B. nigra and B. oleracea* are often cross-pollinated species . Therefore, a variety of breeding methodologies ranging from inbred development through pure line selection to hybrid cultivar development have been utilized for genetic enhancement of oilseed brassicas. The choice of the breeding methodology, irrespective of the breeding goal, largely depends upon the predominance of self- or cross-pollination, and the availability of naturally occurring genetic variations. Nevertheless, the recognition of breeding goals are essential before a strategy to attain the same is spelt out.

NEED FOR GENETIC ENHANCEMENT IN OILSEED BRASSICAS

The major breeding objectives for genetic enhancement of oilseed brassica are aimed at enhancing the commercial value of the end product either quantitatively or qualitatively. On one hand breeding objectives like development of high yielding varieties and generation of varieties having resistance to abiotic stress (salt, drought and frost) and biotic stress (insects and diseases) target production in terms of quantitative enhancement.

On the other hand, improvement in oil content, and quality of oil and meal target the qualitative improvement of the produce. In recent years greater emphasis has been laid upon nutritional quality enhancement of oilseed brassicas with the aim of providing better nutrition and value addition. Various methods are being routinely used for generation of genetic variability and genetic enhancement.

These range from conventional tools, such as selection from available germplasm and hybridization for transfer of desired genes (either *in vivo* or *in vitro* through embryo rescue and somatic hybridization) to the development of transgenics and use of molecular markers for selection. Among the biotechnological tools used for genetic improvement doubled haploids have emerged as an exciting tool for brassica breeders since this technology has a versatile ability to blend with and expedite the existing approaches.

DOUBLED HAPLOIDS: THE CONCEPT AND ITS UTILITY

Doubled haploids are plants produced by spontaneous or artificial doubling of the chromosomes of haploid plants. Such a plant is valuable because the chromosomes that are created by artificial/spontaneous doubling are exact copies of the chromosomes that were present in the haploid plant-justifying the term doubled haploid.

Doubled haploids offer a major advantage by attainment of homozygosity in a single step thus significantly reducing the breeding cycle along with its use in conjugation with other biotechnological methods for expediting the crop improvement programs. The major advantages of the doubled haploids are briefly summarized as follows.

Attainment of Homozygosity

According to conventional breeding approaches, homozygosity may be achieved by repeated selfing and rigorous selection for several generations. Normally this exercise requires about 10 to 12 years for varietal development programs, and is most efficient in self-pollinated crops that do not show inbreeding depression.

Doubled haploids greatly reduce the time required for obtaining homozygous plants if an efficient haploid generation protocol is available. Moreover, attaining homozygosity for recessive and quantitatively controlled traits is an even greater mammoth task because of the involvement of many loci and masking of recessive allels in heterozygous state.

Doubled haploids may be generated in a single step thus fixing the genotypic combinations in a single generation. This technique also considerably reduces the time required for homozygous line development in cross-pollinated crops and may be especially useful in parent development for hybrid production. Thus doubled haploids essentially compress the breeding cycle by accelerating the development of homozygous lines.

Utilization of Gametic Gene Combinations

Doubled haploids offer the unique advantage of utilization of the haploid phase for selection. This fact assumes greater importance for selection in case of induced mutations, which are generally recessive in nature or other quantitative traits, controlled by recessive allels.

Since haploids would express recessive genes, transgressive segregants for recessive traits can efficiently be recovered through diploidization of chromosomes. This concept also offers the advantage of significantly smaller population size required to find the least likely recombinant in case of quantitative traits, since in a doubled haploid population selection is actually effective on gametic gene combinations. Therefore, doubled haploid populations have been used to study the inheritance of important quantitative traits.

Versatile Compatibility with Other Approaches

In addition to the above stated advantages of the doubled haploids, they can also be profitably utilized for mutation breeding, disease resistance, biotechnological gene transfer etc. Some of the applications of doubled haploids in conjugation with other breeding approaches are reviewed below:

Mutation breeding

Mutations are immediately expressed in haploid and doubled haploid plants, hence these are very lucrative targets for mutation research. In *B. napus* imidazoline herbicide resistance has been introduced using doubled haploid technique in conjugation with chemical mutagenesis.

These resistant lines have been evaluated in field trials. Besides this, chemical as well as physical mutagens have been utilized to develop resistance to *Phoma lingam* and *Alternaria brassicola* in *B. napus* by treating cultured microspores.

In vitro mutagenesis at haploid level also led to the development of high oleic acid, thinner seed coat, high oil and protein, and low fibre content in *B. napus* lines and modified erucic acid content in *B. carinata.*

Disease resistance

Microspore cultures are one of the most excellent targets for *in vitro* selection for disease resistance, provided that the disease defence system is active at such an early stage of plant development. Gametoclonal variation exhibited by haploids generated through anther/microspore culture, along with host specific and non-host specific toxins as medium supplements, could be used for *in vitro* selection of resistant genotypes.

Biotechnological gene transfer

Microspores or anther form a good explant source for gene transfer systems such as PEG, electroporation, microinjection and biolistic methods. Microspore derived embyros have been used as recipient cell system for *Agrobacterium-mediated* gene transfer in *B. napus.*

Molecular breeding

DNA based procedures such as RFLP/AFLP analysis are being increasingly employed in plant breeding due to their enormous range of application. DNA markers provide unprecedented refinement in genetic analysis through the construction of nearly saturated genetic maps. This provides the breeder with a highly efficient marker aided selection tool.

Double haploids being truly homozygous for all loci are now being routinely used for genetic mapping of brassicas since they reduce the time required for making RFLP maps and for generating polymorphic mapping populations. Doubled haploid production have been utilized to study gene linkages and interactions.

Breeding for desired oil profile

Doubled haploid technique has also proved useful in the development of mustard cultivars expressing specific oil profile, such as modified erucic acid, reduced linolenic acid, increased linoleic, palmitic and oleic acid content for specific purposes. Hence double haploids provide a powerful breeding tool for obtaining designer mustard varieties having specific fatty acid profile.

DEVELOPMENT OF HAPLOIDS AND DOUBLED HAPLOIDS

Availability of haploids is a prerequisite for doubled haploid production. Spontaneously occurring haploid plants having half the normal number of chromosomes were discovered in 1920s, but utilization of haploid plants was not a practical technique until methods for the controlled production of haploid plants were developed.

Apart from spontaneous occurrence, which is rare and is confined to a few species, haploid plants may be produced by interspecific/intergeneric crossing followed by selective chromosome elimination, for example, as in barley and wheat × maize. Developments in tissue culture techniques in the early 1960's opened the doorway for haploid plant production by *in vitro* culturing of unfertilized ovules (gynogenesis) or from the mature/immature pollen grains (androgenesis).

Although doubled haploids have been obtained via gynogenesis, the majority of published successes have resulted from anther and microspore culture. A breakthrough in this direction was achieved when Guha and Maheshwari for the first time demonstrated that anthers of *Datura innoxia* cultured *in vitro* produce embryos that originate from immature pollen grains or microspores.

However, the difficulties associated with anther culture are low frequencies of haploids, difficulty in distinguishing spontaneous doubled haploids from diploids which regenerate from somatic tissue, and the considerable time and labour which may be needed to generate the desired doubled haploid population required for successful utilization in breeding program.

Developments in isolated microspore culture for several crop species attracted the interest of brassica breeders for generation of doubled haploids. Oilseed brassicas being responsive to cell and tissue culture techniques have been extensively researched upon for both anther and isolated microspore culture.

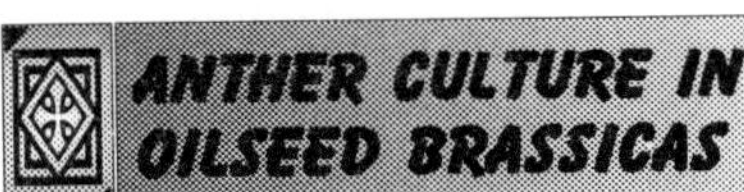

ANTHER CULTURE IN OILSEED BRASSICAS

Successful anther culture in brassicas was first reported by Canadian scientists in early 1970s. Keller and Armstrong reported development of embryoids from cultured anthers of *B. napus.* Following this several reports on *B. napus* anther culture were published. Similar to these, anther culture has also been reported in *B. nigra, B. campestris* and in *B. juncea.*

These reports on anther culture of various oilseed brassicas primarily elucidate the possibility of doubled haploid production. However, the major bottleneck of this technique lies in low frequency of embryogenesis resulting in the realization of a very few embryos. Several factors may be responsible for this low embryo yield.

It has been proposed that anther wall generated toxins may inhibit microspore embryo development inside cultured anthers and the congested conditions inside the anther may limit the nutrient supply to growing microspores. Much of these drawbacks of anther culture have been effectively overcome by the development of isolated microspore culture technique.

MICROSPORE CULTURE IN OILSEED BRASSICAS

As in the case of anther culture, the first report of successful isolated pollen culture was also reported in *B. napus.* This, in fact, was the first report of isolated pollen grain or microspore embryogenesis in plant species other than those belonging to the family *Solanacae.* Following this many reports have elucidated the potential of isolated microspore culture in oilseed brassicas for developing haploid embryos.

Among the various oilseed brassicas, *B. napus* is primarily the most researched upon species for doubled haploid production using isolated microspore culture. Many laboratories have developed specialized protocols for isolated microspore culture of *B. napus*. Several reports on isolated microspore culture have also been published in *B. carinata, B. nigra, B. campestris and B. juncea*. However, apart from *B. napus*, limited success has been achieved in other species towards cultivar development, due to the lack of efficient microspore embryogenesis.

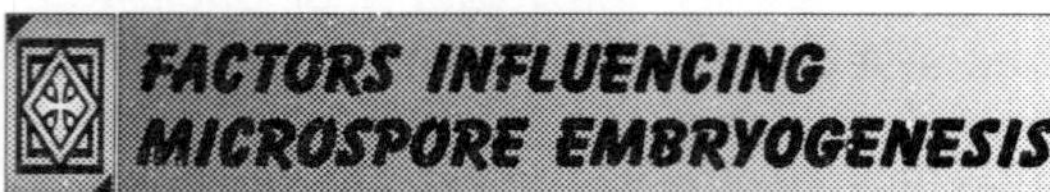

FACTORS INFLUENCING MICROSPORE EMBRYOGENESIS

The ability to induce totipotency in anther cultures/isolated microspore cultures is greatly influenced by several factors. These include genetic and exogenic factors that may have profound implications on microspore development *in vitro*. Various factors that influence microspore embryogenesis in oilseed brassicas are:

Genotype

The genotype of the donor plants have been reported to have a profound effect on the microspore embryogenic response. Genotypic variations in haploid embryo development have been observed in several brassica species. Genotypic variability for microspore embryogenesis response in isolated microspore culture has been reported in *B. campestris, B. juncea* and in most reports sited for *B. napus* above.

Recently microspore embryogenic ability has been studied as a stably inherited trait using the diallele mating system in *B. napus* by Zhang and Takahata. The study elucidates that both additive and dominant effects are significant for microspore embryogenesis as a genetically controlled trait.

Similar dominant gene action has been reported by Cloutier et al. reiterating the strong genotype influence on microspore embryogenic ability. Ajisaka et al. have further reported two putative chromomosome regions associated with microspore embryogenic ability in *B. campestris*.

Donor Plant's Growth Condition

The donor plant's growth condition has a marked effect on the physiological processes of the plant thus it invariably influences the microspore embryogenic ability. Proper light, temperature, humidity and nutrients are all necessary to develop healthy plants. Varying growth conditions of donor plants have been tested for their influence on microspore embryogenesis, ranging from plants grown under field conditions to plants grown under artificial or growth room conditions.

In *B. campestris*, plants grown under cold temperature conditions, 10/5°C day/night cycle showed enhanced microspore embryogenic capability. However, plants grown under relatively higher temperature regime 28/15°C day/night cycle upto bolting and then transferred to low temperature conditions as mentioned above, have also been reported to show enhanced microspore embryogenic response.

Low temperature regime for donor plants has been reported to show a positive correlation with increased microspore embryogenesis in *B. napus*. However, in this species too, donor plants grown till bolting at higher temperature regime (25–28°C day/ 12–15°C night) and then shifted to a low temperature (10/5°C day/night cycle) have been reported to show higher microspore embryogenic ability.

Limited reports are available on the influence of donor plant's growth condition on microspore embryogenesis in *B. juncea*. A relatively higher temperature regime (20–21°C day/15–18°C night) has been reported to be congenial for microspore embryogenic response in isolated microspore cultures of *B. juncea*. In view of this it seems likely that the plants may be grown under normal *in vivo* temperature conditions for healthy and vigorous vegetative growth, however, a shift to lower temperature regime is essential for enhancing microspore embryogenic response.

Microspore Development Stage

The microspore development stage is a prime important factor that influences the microspore's ability to turn totipotent. This is primarily due to the fact that microspores would only respond to embryo formation at a developmental stage when they are not committed to develop into pollen grains. Moreover, in oilseed brassicas the microspore development is asynchronous and microspores of different developmental stages may be observed in a developing anther.

Therefore, selection of buds that have maximum proportion of embryogenic microspores is essential for efficient microspore embryo yield. The microspore development may be divided into three basic stages viz. the tetrad stage (when the microspore mother cell splits into four haploid cells), the uninucleate stage (when the uninucleate microspore prepares for the nuclear division to form the vegetative and generative nuclei) and the binucleate stage (when the microspore contains a generative and a vegetative nucleus).

Each of the abovementioned microspore development stage has been extensively researched upon in *B. napus* to determine the exact stage at which the microspore is not under a differentiation pressure or is not committed towards pollen development. It has been established that there is an optimum development stage (embryogenic window) that corresponds to the late uninucleate to early binucleate stage of development, during which large number of microspores could undergo embryogenesis.

Further to this it has been proposed that non-embryogenic microspores produce inhibitory substances that suppress embryo development in the embryogenic microspores. This may be because of the rupturing of non-embryogenic binucleate microspores thus reducing the embryogenic frequency and altering the morphology of embryos. Replacement of culture media after microspore isolation helps in reducing the autotoxins thus allowing normal embryo development.

Similar influence of microspore developmental stage on microspore embryogenesis has been reported in *B. campestris*, *B. carinata*, and *B. juncea*. These studies have established that the late uninucleate stage is most responsive to embryogenesis and that selection of buds with majority of late uninucleate microspores increases the frequency of embryo formation in isolated microspore culture in oilseed brassicas.

Microspore Density

The density of isolated microspores in the culture media is another essential factor responsible for normal development of embryos. Microspore culture density ranging from 1 to 10,000 cells/ ml has been reported for different species. Varying density of microspores have been studied by various scientists to get the optimum embryo yield such as 1×10^4 microspores/ml, 2×10^4/ml, 3–4×10^4/ml, 8×10^4/ml and 10×10^4/ml. However, the most favourable density has been found to range between 5 and 8×10^4 cells/ml for *B. napus and* 1 and 4×10^4 cells/ml for *B. juncea.*

Media Composition

The basic media composition for microspore culture protocols in various oilseed brassicas have mostly been the same over the years. Initial reports of anther culture in oilseed brassicas emphasized the role of increased sucrose concentration in culture media. Subsequent to this, it was established that high sucrose concentration (10 to 13%) is essential for microspore embryogenesis in anther as well as isolated microspore culture in *B. napus*.

Keller et al. demonstrated that L-serine was an important constituent of the anther culture media in *B. napus*. Later, Lichter demonstrated that basal Nitsch and Nitsch medium supplemented with glutamine, glutathione and L-serine was most congenial for anther culture in the same species. Off late Nitsch and Nitsch medium modified by Lichter, commonly known as NLN medium, has become the most frequently used media for isolated microspore culture in oilseed brassicas.

Growth Additives

Activated charcoal has been reported to be beneficial for embryo growth and normal development in both anther and microspore culture of *B. napus*. However, its role in triggering microspore embryogenesis has not been reported, rather it has been reported to be beneficial for normal growth of induced embryos. Growth regulators have been used in both isolated microspore as well as anther culture of oilseed brassicas.

Initial reports on microspore culture emphasised the role of 6-benzyl amino purine and 1-naphthalene acetic acid for induction of microspore embryogenesis. However, off late it has been reported that growth regulators are not essential for inducing microspore embryogenesis. Several scientists have also propounded the role of colchicine as an embryogenesis inducing agent.

However, owing to high toxicity, potential hazard and great care required in handling colchicine, this concept has not been utilized extensively. Till date the major role of colchicine has been limited for doubling chromosome number of haploid plants. Apart from the abovementioned factors, post culture incubation conditions are also reported to have profound effect on induction and development of microspore derived embryos.

For most reports cited above an initial heat shock of 30–32°C for 3–10 days has been reported to be essential for microspore embryogenesis. Further to induction of microspore embryos, determination of ploidy of the produced embryos and colchiploidy for chromosome doubling are essential part of any successful doubled haploids production protocol. Determination of ploidy is usually done either through root cytology or using flow cytometry.

Colchiploidy may be carried out by axillary bud treatment or by dipping the plantlet's roots in colchicine. Having overcome the successful induction of the embryogenesis, the doubling of the chromosomes and regeneration of doubled haploid plants is not a severe bottleneck for utilizing this technique for various crop improvement programs.

CONCLUSIONS

The oleiferous brassica species have been extensively utilized as an important source of edible oil. Oilseed brassicas stand third in world's oilseeds production and acreage, whereas in India, it ranks second next to groundnut. A considerable amount of work has been undertaken for genetic enhancement of oilseed brassicas to increase its commercial value through various conventional as well as modern approaches. Among the biotechnological techniques used, double haploids have emerged as a promising tool.

In addition to compressing the breeding cycle by accelerating the development of homozygous plants, it has versatile compatibility with other approaches like mutation breeding, transgenics, molecular breeding, etc. However, for successful utilization of doubled haploids in crop improvement programs, generation of substantial doubled haploid population is essential, which in turn is possible with the availability of an efficient haploid production protocol.

Haploids have been produced by selective chromosome elimination, gynogenesis or androgenesis. Among the three, androgenesis, i.e. anther/microspore culture has been the most successfully utilized approach. Both anther and microspore culture have been researched upon in oilseed brassicas, however, extensive work has been done elucidating the potential of the isolated microspore culture technique for developing haploid embryos.

A number of genetic and exogenic factors influence the efficiency of microspore embryogenesis. This article has presented some of the important factors affecting microspore development and their implications on establishing an efficient microspore culture protocol. The doubled haploids are already being extensively utilized for the quality and agronomic improvement in *B. napus* but its practical utilization in other economically important oilseed brassicas is yet to be realized.

The work in this direction is in progress at several institutions opening many a new vistas for utilization of this versatile technique for genetic enhancement and value addition in the oilseed brassicas.

Chapter 5

OBTAINING NATURAL DRUGS

Commiphora wightii (Arnott.) Bhandari, commonly known as 'Indian bdellium', or 'guggul', is an important medicinal plant of the herbal heritage of India. For centuries, guggul has been used extensively by Ayurvedic physicians to treat a variety of afflictions, including arthritis, inflammation, bone-fractures, obesity and disorders of lipid metabolism.

It provides 'guggul', an oleogumresin whose medicinal and curative properties are mentioned in the classic Ayurvedic medical text, the Sushruta Samhita 3000 years ago. The plant has become endangered because of over exploitation for its gum-resin, associated with slow growth of the plant, poor seed set and excessive tapping for gum-resin, which causes mortality of the plant. Gum-resin yields guggulsterones effective against high blood cholesterol and lipids.

DISTRIBUTION

Commiphora is widely distributed in tropical regions of Africa, Madagascar and Asia. It is generally distributed in arid regions and is particularly widespread on the Indian side of Thar Desert. In the Indian subcontinent *Commiphora* species occur in Pakistan, Baluchistan and India. Of the total 185 species, only three (*C. wightii, C. stocksii and C. berryi)* have been found in India. *C. wightii* occurs in Rajasthan, Gujarat and Maharashtra.

BIOLOGY

Commiphora wightii (Arnott.) Bhandari (syn. *C. mukul, C. roxburghii, Balsamodendron mukul*) belongs to the family Burseraceae. A characteristic feature of the family is the presence of resinducts in the

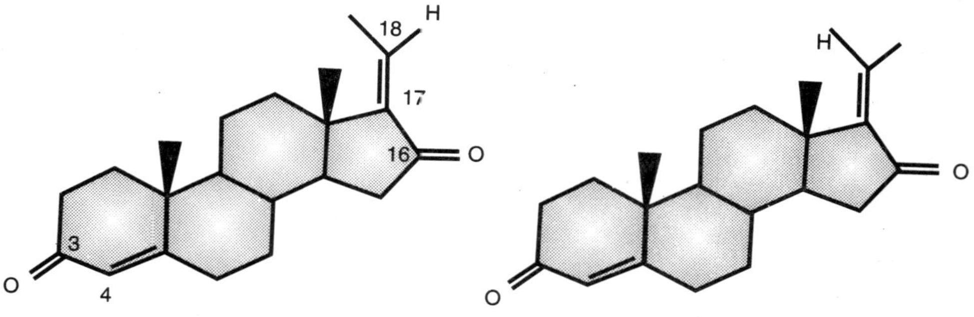

E

Z

Structure of E- and Z-guggulsterones

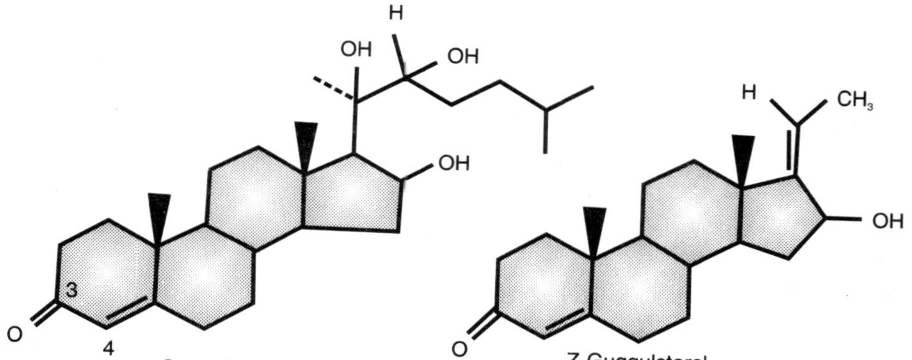

Guggulsterol-1

Z-Guggulsterol

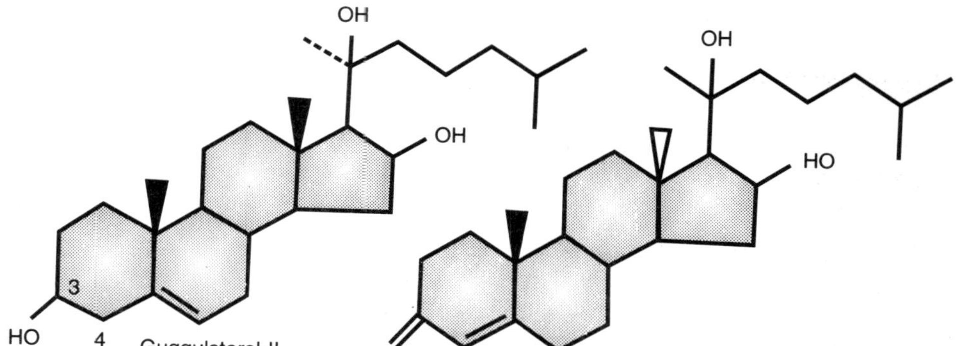

Guggulsterol-II

Guggulsterol-III

Figure 5.1: Structure of guggulsterones and guggulsterols isolated from C. wightii.

parenchymatous bark. The plant is a shrub reaching 3 m in height with crooked, knotty branches ending in sharp spines.

The papery bark peels in flakes from the older parts of the stem, whereas younger parts is pubescent and grandular leaves are trifoliate. The flowers are sessile and single or in groups of 2-3. The fruit, 6-8 mm in diameter is a drupe, which becomes red on ripening. Fruit yield and seed set is low (about 16% in Aravalli ranges).

In drier parts it is even lower. The chromosome number of C. *wightii* is *2n = 26.* Recently, Gupta et al. reported apomictic seed development associated with polyembryony in guggul. Female plants set seeds irrespective of the presence or absence of pollen. Hand pollination experiments and embryological studies have confirmed the occurrence of non-pseudogamous apomixis, nucellar polyembryony and autonomous endosperm formation.

It was inferred that apomixis may have a significant role in the speciation of tropical trees. Apomixis may be favoured by natural selection if the population densities are low and distance between individual trees is greater than the permissible cross-pollination range. Multiple sapling formation by the germination of polyembryonic seeds of *C. wightii* has also been observed thereby confirming the multiple embryo formation in seeds.

In another study, Gupta et al. described the cause of low seed set in *C. wightii* on the basis of pollen-stigma interaction in the non-pseudogamous apomictic plants. They observed that although pollen grain germinated on stigma, pistil did not support pollen tube growth perhaps due to changed orientation of the cells of transmitting tissue and absence of proteins in the intercellular matrix.

This results in poor seed set. *C. wightii* is an excellent fuel wood and burns even wet due to the presence of resin in the stem. The plant is cut mercilessly by villagers for cooking the food and used with other wet woods to facilitate burning. Due to abovementioned inherent biological and social problems the plant has become an endangered species.

An interesting biological property of ecological significance of resin has also been reported. Essential oil constituents of resin have been shown to enhance sexual maturation of immature adults of the desert locust.

CHEMISTRY OF GUM-RESIN

The presence of guggulsterones differentiates *C. wightii* from 184 other *Commiphora* species. Gum-resin obtained from *Boswellia serrata,* another tree from the family Burseraceae and common in the same regions, is also known locally as salai guggul or white guggul, but *B. serrata* does not contain guggulsterones.

It is used as an anti-inflammatory drug. Phytochemical investigation of guggul gum-resin has been carried out by the group of Dev. Guggul (oleogum-resin) of *C. wightii* is a mixture of 38.5% resins, 32.3 % gum, 1.45% volatile oil, 19.5% minerals, 3.2% organic foreign matter and 3.6% other impurities. During the separation of various products from the complex mixture, the neutral fraction was reported to contain ketonic compounds (5.13%).

It is this ketonic fraction that contains biologically important active principles of C_{21} or C_{27} steroid,

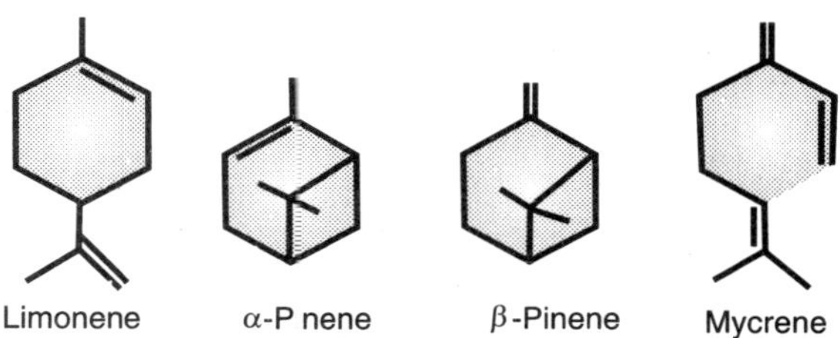

Figure 5.2: Structure of various monoterpenes isolated from C. wightii.

viz. Z-guggulsterol (0.01%), guggulsterol–VI (0.02%), Z-guggulsterone (1.6%), E-guggulsterone (0.4%), guggulsterol-III (0.03%), guggulsterol-I (0.8%), guggulsterol-IV, guggulsterol-V (Fig. 1) and some defence related secretory ketones.

Oleogum-resin is a complex mixture and needs stepwise separation. Purification involves separating soluble (45%) and insoluble (55%) components with the aid of ethylacetate, alcohol or petroleum-ether. Insoluble fraction is associated with toxic effects while the soluble fraction contains the guggulsterones and other constituents that are thought to impart the hypolipidemic and anti-inflammatory effects.

A method of high performance liquid chromatographic (HPLC) separation was proposed. Recently a HPLC method for quantitative determination of E- and Zguggulsterones in *C. mukul* resin, serum and diet supplement has been developed. In addition to these steroids, the gum-resin of *C. wightii* contains diterpenoids (combrene-A and mukulol), steroids derived from pregnane and cholestane, and various carbohydrate derivatives.

Upon steam distillation, the gum-resin furnishes an aromatic essential oil. The oil contains the monoterpenes—myrcene, camphorene, polymyrcene and caryophyllene. The aerial parts of *C. wightii* contain β-sitosterol, myricyl alcohol and amino acids. The flowers are rich in flavonoids, most notably quercetin. From the resin of C. *tenuis* growing in Ethiopia, 37 mono- and sesqui-terpenes were detected and identified by GLC and GLC-MS.

The main components of the monoterpenoid fraction were a-pinene (60.8%), β-pinene (8.8%), sabinene (6.3%), α-thujene (8.9%), limonene (5.5%), 3-carene (3.7%), β-myrcene (1.8%) and β-elemene (1.1%) constituting 97% of the oil. Identified sesquiterpenoid components constituted approximately 1.6% of the oil. Oleanolic acid acetate and three other triterpenes were also identified in this oil obtained by wounding the plant.

MEDICINAL PROPERTIES AND PHARMACOLOGY

Many Indian medicinal plants have come under scientific scrutiny since the middle of the nineteenth century. *Commiphora wightii* is one such plant from which a modern medicine for hyperlipidemia has been prepared based on ancient information. In ancient times, guggul was used primarily as treatment for inflammatory conditions, including arthritis.

The development of gum guggul as a potent hypolipidemic agent was first reported by Satyavati working at Banaras Hindu University, Varanasi, leading to the discovery of new anti-cholesterol drug from a plant source. This work was initiated on the basis of information gathered from ancient concept of the pathogenesis of atherosclerosis and obesity described in Sushruta Samhita. In Ayurveda, guggul is highly valued for the treatment of several ailments including rheumatoid arthritis, lipid disorder and obesity.

Currently several formulations of Ayurveda for arthritis, joint pain, sciatica and other ailments contain guggul. Several reports conclusively established scientifically with modern pharmacological tests the hypolipidemic and hypocholesterolemic properties of *C. wightii* extract. The effect of guggul was very promising in experimental animal systems.

Immediately after trials guggul and its purified extract was established effective hypolipidemic agent in patients with ischemic heart disease, hyper-cholesterolemia, obesity and hyperlipidemia. In different trials with patients, a reduction of serum cholesterol (24 to 59%) and triglycerides (22 to 30%) was recorded. Hypolipidemic and antioxidant effects of guggulipid, a drug prepared from guggul, were demonstrated in patients with hypercholesterolemia.

Guggulipid used as adjunct to dietary therapy decreased the total cholesterol levels by 11.7%, low density lipoprotein (LDL) cholesterol by 12.5%, triglycerides by 12% and the total cholesterol/ high density lipoprotein (HDL) cholesterol ratio by 11%. 'Guggulipid', a purified ketonic fraction, is presently used in India and Europe for hyperlipidemia and hypercholesterolemia. In addition to its lipid lowering activity, guggul may also promote cardiovascular health through its ability to act as an antioxidant and to inhibit platelet aggregation.

The guggulsterones inhibited the oxidative modifications of lipid and protein components of LDL induced by copper (Cu) *in vitro* in a concentration dependent manner. Furthermore, guggulsterones also inhibited the formation of hydroxyl (OH^-) free radicals created in a non-enzymatic system in a concentration dependent manner.

Myocardial necrosis is associated with increased levels of lipid peroxides, xanthine oxidase activity and a lowering of superoxide dismutase, which may lead to increased formation of free radicals with subsequent cardiac cell damage. Guggulsterones, in a manner similar to two other cardioprotective drugs (propranolol and nifedipine), reversed this elevation of lipid peroxides and xanthine oxidase and the decrease in superoxide dismutase activity.

The extract of *C. wightii* along with that of *Terminalia arjuna, Inula racemosa* showed protection against isoproterenol induced myocardial necrosis in rats or that with *Allium sativum and A. cepa* showed protection against increased cholesterol and blood serum triglycerides, thereby, confering protection against atherosclerosis and myocardial infraction.

Mester et al. reported total inhibition of platelet aggregation *in vitro* induced by adenosine diphosphate, serotonin and adrenaline by isolated E- and Z-guggulsterones. Guggulsterone-Z and guggulsterone-E are responsible for lipid lowering properties in human blood and at least four mechanisms have been proposed to explain their activity.

First,guggulsterones might interfere with the formation of lipoproteins by inhibiting the biosynthesis of cholesterol in the liver. Second, guggulsterones have been shown to enhance the uptake of LDL by the liver through stimulation of the LDL receptor binding activity in the membranes of hepatic cells. Third, guggulsterones increase the fecal excretion of bile acids and cholesterol resulting in a low rate of absorption of fat and cholesterol in the intestine.

Finally, guggulsterones directly stimulate the thyroid gland. Guggul induced triiodothyronine

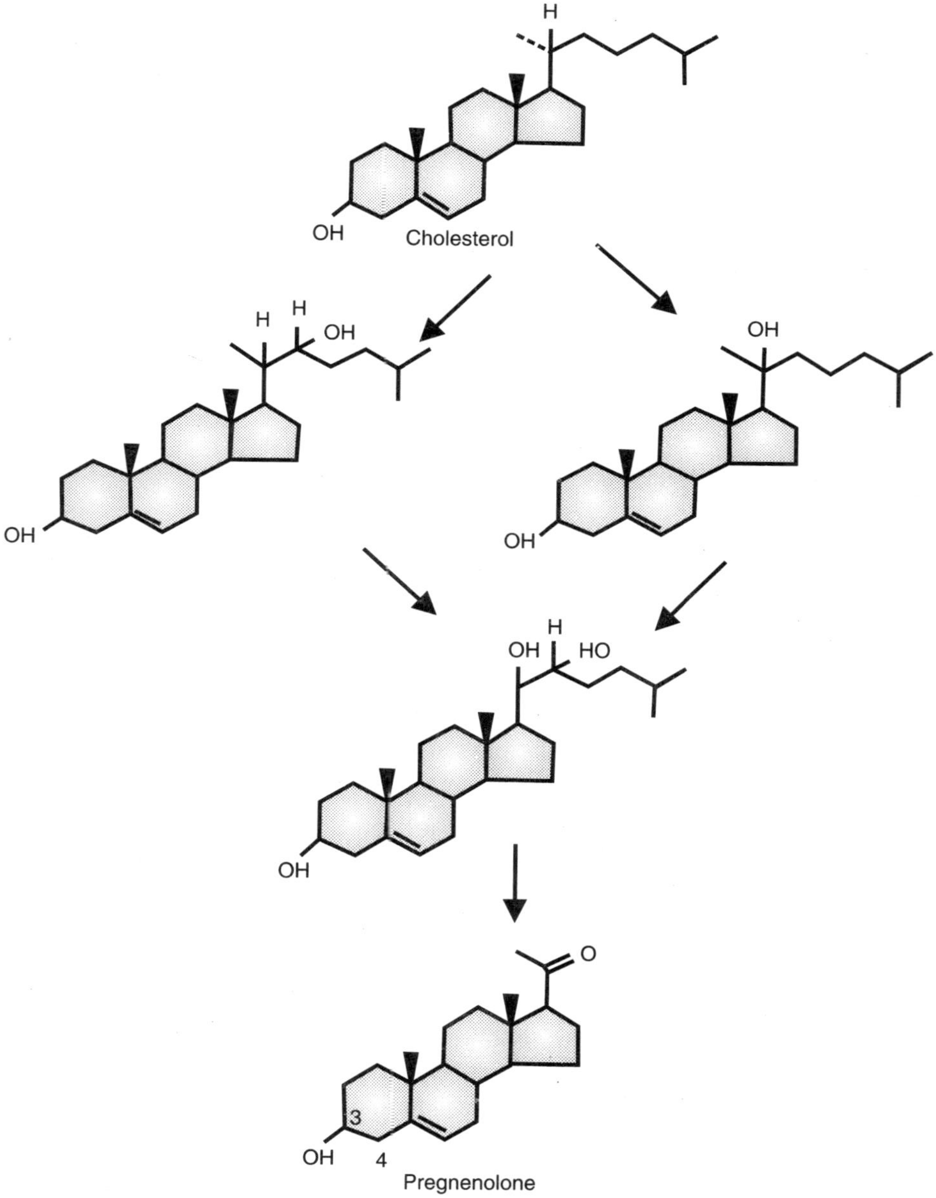

Figure 5.3: Metabolism of cholesterol to pregnenolone.

production with possible involvement of lipid peroxidation, demonstrated thyroid stimulatory effect of guggul administration in the experimental mice.

Because serum lipids, including cholesterol, are reduced in response to increased levels of circulating thyroid hormones, the effect of guggulsterones on the thyroid gland might explain the hypolipidemic activity and weight loss property of guggul. The anti-inflammatory effect of guggul from *C. wightii* on osteoarthritis has also been established. Anti-inflammatory effect is common in other plants of the family Burseraceae, *viz., Boswellia dalzielli, B. carteri, B. serrata and C. incisa.* The extract of *C. molmol*, another species from middle east, possesses anti-thrombosis activity, anti-ulcer and cyto-protective property, anti-inflammatory effect and cytotoxic and anti carcinogenic effect.

On the basis of non-mutagenic, antioxidative and cytotoxic potential of *C. molmol* extract, its use in cancer therapy was recommended. Similarly, sesquiterpenes responsible for hypoglycemic activity were isolated from *C. myrrha*. Dev reported that the steroid profile of C. *wightii* parallels the catabolism of cholesterol to C_{21} steroids.

Thus, there is sufficient evidence that both in mammalian tissues and in plants, the catabolism of cholesterol to pregnane derivatives proceeds by either of the two major pathways as shown in Figure elsewhere in this chapter. It is because of this property and increased demand for the natural product for hyperlipidemia, the plant has attained great importance in recent years.

GUM-RESIN PRODUCTION

In *C. wightii* the balsam (oleogum-resin) is present in 'balsam canals' in the phloem of larger veins of the leaf and in the soft base of the stem. The development and widening of gum-resin canal in young stem occurs schizogenously. The lumen of canal is surrounded by an epithelial layer of parenchyma containing dense cytoplasm and shows the presence of gum and resin droplets.

The walls of epithelial cells facing the lumen are thin and of fibrillar mesh. The resinous material is synthesized within the epithelial cells and is presumably transported into the canal lumen through the relatively porous wall.

Among various plant growth regulators applied on stem with lanolin paste, only kinetin increased the lumen size, while auxin and morphactin had adverse effect causing increase in number of epithelial cells. Gum is tapped in the winter season. Plants over 5 years old with a basal diameter more than 7.5 cm are suitable. Circular incisions of 1.5 cm deep are made on the main branches and stem at a uniform distance of 30 cm apart and at an angle of 60° with the stem.

The yellow, fragrant latex oozes out through the incisions and slowly solidifies into vermicular or stalactitic pieces which are collected manually. Subsequent collections of gum-resin are made at an intervals of 10–15 days. About 200–500 g dry guggul is usually obtained from a plant in one season.

Application of ethephon on the cuts enhances guggul production 22 times over that obtained in control. This technique developed by Bhatt et al. is inexpensive, safe and requires no skills, and hence can be used by the tribals very easily.

They established that guggul production is maximum with the onset of summer (a stress induced secondary product formation) as supported by observations with bright field and fluorescence microscopy. But in the long term, excessive production through ethephon application exhausts the plant and resultantly, kills the plant.

VEGETATIVE PROPAGATION

Fruit set and yield of fruits per plant are very low in natural conditions. Poor seed set, poor seed viability and harsh arid conditions are responsible for complete failure of plant establishment in nature from seed. Plants bear fruits in April to May and August to October.

About 27% fruits contained single embryo and 7% fruits contained 2 embryos while 66% fruits were without embryo. Therefore, attempts were made to propagate the material by conventional methods of stem cuttings and attempts are being made to propagate the plants through non-conventional biotechnological methods.

Conventional Methods

Rooting of stem cuttings has its own drawbacks in the arid environment like termites attack, desiccation and heat adversely affecting rooting. Rooting response of stem cuttings was shown to be improved by application of plant growth regulators, by selecting cuttings of suitable length and diameter and treating them with potassium salts.

However, such methods are not suitable for large-scale multiplication as stock material with sufficient biomass is not available as well as % response of the cutting is variable and affected by seasons.

Biotechnological Methods

Biotechnological research on *C. wightii* has been supported by central funding agencies since 1979 but nothing concrete has come out of these programmes which shows the difficult nature of the material. In nature, the plant is a very slow growing woody shrub.

Explants obtained from the mature plants (stem, leaf or petiole) produce fast growing, white and amorphous callus on MS medium containing kinetin and 2,4-dichlorophenoxy acetic acid (2,4-D). Resin exudation from the explants makes the process of sterilization difficult.

It is equally difficult to find tender stem explants for *in vitro* growth. Due to these reasons detailed investigations using explant as source material are hampered.

Clonal Propagation

Clonal propagation as a biotechnological approach is commonly applied for vegetative propagation of selected materials. Barve and Mehta described a method for clonal propagation of C. *wightii* using stem explants grown on Murashige and Skoog medium containing benzyladenine (BA, 4.0 mgl^{-1}), kinetin (4.0 mgl^{-1}), glutamine 100 mgl^{-1}, thiamine HCl 10 mgl^{-1} and activated charcoal 0.3%.

Shoots obtained from explants were incubated to elongate on medium containing lower concentration of BA (0.40 mgl^{-1}) and kinetin (0.4 mgl^{-1}). These elongated shoots were rooted by treating them with IAA and IBA for 24 h in dark and then transferred onto low salt basal medium with activated charcoal.

Six-week-old plants (5–6 cm in height) from half strength White's modified medium were used in transplantation. At hardening stage 60% of the transferred plants survived. Micropropagated *C. wightii* plants once established in soil showed vigorous and uniform growth with no morphological

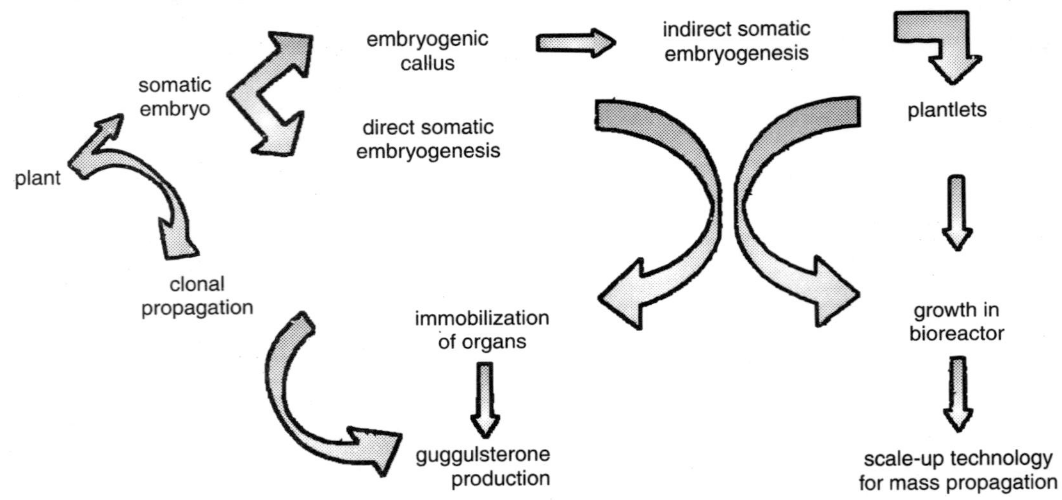

Figure 5.4: Schematic presentation of approaches for guggulsterones production and mass propagation of plants.

abnormalities. The lack of selected high yielding plants and limited number of plants produced by this method are limiting factors for use of this technique for large scale multiplication.

Somatic Embryogenesis

Somatic embryogenesis in callus cultures of *C. wightii* has been achieved. Somatic embryo formation was first observed in immature zygotic explants or intact ovules transferred on B5 medium. Though the frequency of explants producing embryonic culture was low, immature zygotic embryos were the only suitable explants to produce embryonic callus after reciprocal transfers between B5 medium containing 0.1 mgl^{-1} 2,4,5-trichlorophenoxy acetic acid and 0.1 mgl^{-1} kinetin and that devoid of it. All other media failed to produce embryonic callus.

Further, somatic embryogenesis in callus obtained from immature zygotic embryos was possible because of selection of embryonic cells. Embryonic cells were small, densely filled with cytoplasm and isodiametric as compared to non-embryonic cells, which were large, elongated and vacuolated.

Maximum growth of embryonic callus was recorded on MS-2 medium supplemented with 0.25 mgl^{-1} BA and 0.1 mgl^{-1} IBA. MS-2 salts supported higher growth of callus as compared to tissues grown on B5 medium containing same concentrations of plant growth regulators.

Embryonic callus transferred on MS-2 medium containing various combinations of IAA and BA produced globular, torpedo and a few early cytoledonary stage embryos. Maximum number of somatic embryos was observed on the medium containing 0.1 mgl^{-1} IAA and 0.25 mgl^{-1} BA. Torpedo and cotyledonary stage embryos obtained from experiments for development of somatic embryos of previous experiments were used.

Activated charcoal, ABA, and agar-agar were incorporated in MS-2 medium to generate stress and enhance maturation in somatic embryos. Maximum number of cotyledonary stage embryos were produced on the medium containing 0.5 gl^{-1} activated charcoal and 10 gl^{-1} sucrose. In this experiment embryos were placed on filter paper-bridge using liquid medium.

Cotyledonary stage somatic embryos kept on various maturation media were transferred onto MS-2HF medium. Somatic embryos grown on media containing plant growth regulators, irrespective of their concentration and combination, produced callus. A high percentage of embryos remained ungerminated, while about 10–25% produced secondary somatic embryos.

Therefore, proper selection of mature embryos was required for high percentage of germination. Somatic embryos showed precocious germination and callusing except those grown on MS-2HF medium, which could be maintained for several months.

By using static medium or liquid medium with filter paper bridge, about 25% torpedo staged embryos matured into cotyledonary stage embryos and out of these about 25% were converted into plantlets. Thickening of hypocotyl was observed in such plantlets without elongation of internodes. MS-2 medium containing 20 μgl^{-1} gibberellic acid was most effective for shoot elongation in such plantlets.

About 200 plantlets were successfully established in garden soil to verify the survivability of regenerants. Survivability was 95% for the plantlets. The embryo formation from zygotic embryo and ovule explants may be a case of induced polyembryony as already reported in this plant.

This achievement opened new avenues of research on *C. wightii* like cell culture and embryogenesis in bioreactor, formation of resin canals *in vitro* and production of guggulsterones in these organized cultures.

Guggulsterone Production

Unavailability of sufficient guggul from natural sources and destruction of plants from most of the localities, initiated the search for alternative methods of guggulsterone production. Cell culture is an excellent alternative to produce secondary metabolites. Cell suspension cultures of *C. wightii* were derived from leaf callus in MS medium containing 0.15 mgl^{-1} each of 2,4-D and kinetin. Cells were immobilized in calcium alginate beads.

The immobilized cells in stationary phase suspension cultures were less viable but they were active in the synthesis of guggulsterols. Guggulsterol production was 1.97% in stem explants, 0.22% in callus culture (2-month-old) and 0.32% in cell suspension culture (25-day-old).

Guggulsterone is produced in resin canals and hence unorganized cultures proved unproductive for this purpose. However, use of organized cultures, *in vitro* produced embryos and hypocotyls may prove better sources and are being evaluated.

PROSPECTS AND RESEARCH NEED

Commiphora wightii has become an endangered plant species because of high demand for its gum-resin. This plant has become a torchbearer of efficacy of plants used in the Indian system of medicine for treatment of complex human syndrome. Therefore, many more plants are being reinvestigated for their properties mentioned in the ancient system of medicine.

The complexity of molecules in mixture warrants their production by natural sources using biotechnological methods. Development of somatic cell cultures in static and liquid medium opened new avenues of research on this material because large quantities of aseptic material in organized form can be obtained. It is known that the organized material produces several-folds higher amount of active principle as compared to unorganized cultures.

This can effectively be used for immobilization of embryos, germinated seedlings, seedling parts, and so on. All these are available in aseptic condition in a material which is otherwise difficult to sterilize in large quantities due to presence of resin and hence sticky nature of the explants. Large amount of aseptic material is required for bioreactor culture and failure of the system due to contamination adds a lot to the cost factor of running the system.

Besides, the advantages of mass propagation and development of artificial seeds are evident from the results and need not to be emphasized again. The other new approach is the development of hairy root culture system using *Agrobacterium rhizogenes* for the production of active principle again on the lines as described above for the organized cultures.

Chapter 6

AGRICULTURAL BIOTECHNOLOGY

Most economists today are still trained in neoclassical economics. Its basic *comparative-static assumptions* of perfect competition, knowledge as a pure public good, and price-setting as market failure were very popular in the twentieth century because they enabled elegant formalizations of general and partial *equilibrium* models from the household economy to the world economy.

This *neoclassical* approach is based on the *assumption* that all goods and technologies that could possibly exist, do already exist. This philosophy of plentitude proves to be particularly inadequate in a knowledge economy where the exponential growth of knowledge leads to an *exponential* growth of the probability that new goods and technologies come into being and generate new markets.

This process is not just the primary source of wealth and prosperity but also generates a social welfare surplus that cannot be captured by the *innovating* company itself. Paul Romer, the father of new growth theory, was able to highlight the social welfare impact of new goods within the neoclassical economic model.

In that article "new goods" primarily refer to capital goods that are used as an input for the *improvement* of the production of already *existing commercial* goods or the *creation* of entirely new goods (e.g., products derived from *agricultural* biotechnology are new goods that improve the quantity and quality of agricultural production).

Explaining the Knowledge Economy

New growth theory emerged in the 1990s in response to the inadequate *assumptions* of *neoclassical* theory. In his paper, "Endogenous technological change" Paul Romer (1990) showed that knowledge, unlike other *production* factors such as land, labor, and capital, is a non-rival good that can be used by many at the same time without loss in value. Thanks to the *revolution* in information technology, this knowledge can be reproduced at almost no additional costs.

Yet, the creation of new knowledge is *expensive* since it requires large fixed costs spent on research and development (R&D). These costs also include the hiring of scarce and expensive human capital, the most *soughtafter* resource in the knowledge economy.

It is therefore not *surprising* that those who create new knowledge want to make its use partially excludable through intellectual property rights (IPRs). This *temporary* monopoly right allows the owning company to extract a rent by putting the price of the new knowledge-intensive product above its marginal production costs. It thus enables the company to compensate for the high fixed costs that were spent on R&D, and, at the same time, provides incentives to invest again in *improvement* of the product.

Monopolistic Competition

It is therefore *monopolistic* competition and not perfect *competition* (as portrayed in the idealized neoclassical theory) that creates new goods and new markets. The company that introduces a new good has the temporary power of setting the price of this good in the market (unlike companies in perfect *competition*, which are all assumed to be price-takers).

Neoclassical welfare *economists* often denounce price-setting power as the extraction of an undeserved rent by a monopolist at the expense of the *consumers*, who suffer a deadweight loss due to the higher price they have to pay for the product.

Even though empirical *research* by *economists* themselves showed that these deadweight losses are quite small, policy makers tended to identify monopolistic competition as market failure that needs to be addressed by government intervention. This thinking is, however, based on two contested assumptions:

(*a*) knowledge is a *non-excludable* public good funded by governments and produced at public universities and national research institutes and

(*b*) monopolies exist *primarily* because they repress competition through high barriers to market entry *generated* through collusion and political lobbying.

These two assumptions are not wrong but they are not the whole truth. Governments indeed fund the production of knowledge and make sure that it is widely *accessible*. But, only the private sector can convert new knowledge into successful markets for new goods, *technologies* and services, and this conversion also requires a fair amount of investment in R&D.

Moreover, the decision to invest in a new technology is particularly risky and expensive due to high R&D costs and a high degree of *unpredictability* with respect to the estimated demand, the length of the approval process, social acceptance, and potential liability costs. Therefore, a company will only *invest* in a new technology if there is a prospect of making a *considerable* profit.

Such a profit is possible if a company can obtain a temporary monopoly right through the protection of its intellectual property (the temporary right to exclude others from copying and selling the same product). This makes the non-rival good partially excludable. Yet, such a monopoly is not just extracting a rent from consumers through a higher price but also creates a new good that produces a social welfare surplus, which cannot be captured by the company itself (e.g., more employment, more tax revenues, more knowledge in the public realm through patent disclosure, economic spillovers leading to generic products that are also affordable to poorer consumers/producers, etc).

Thus the more useful knowledge that is generated in society the higher the likelihood that new goods will be created. Therefore, this new knowledge *economy* is no longer about *substituting* one existing good for another existing good but about the constant introduction of new goods.

Social Welfare Generated by New Goods

The social welfare that results from the introduction of a new good is not new but was already noted by a French engineer called Dupuit in the nineteenth century. He calculated the cost of building a bridge and the minimal toll the users of the bridge need to pay to reimburse the fixed costs for *building* the bridge. He was able to show that the *entrepreneur* who builds the bridge is constrained in his efforts to extract a *maximal* rent from the users, because if the toll is too high the user might simply not use the bridge (assuming that the users are acting in a competitive world with scarce resources themselves).

He therefore concluded that the entrepreneur can never capture all the benefits of building the new good "bridge". The same is true for a company that wants to develop a new *technology*. Instead of *extracting* an additional rent through a toll (as in the case of a physical good), it would do it through a royalty fee on the patented technology.

Adding a Dynamic Dimension to Welfare Economics

The creation of new goods that emerge from monopolistic *competition* can be illustrated by

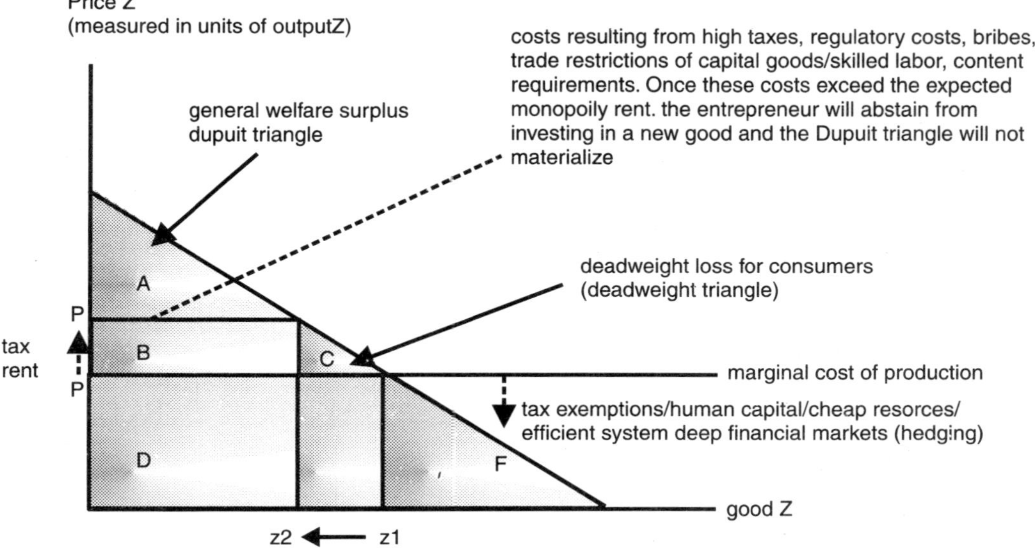

Figure 6.1: The welfare surplus generated by the introduction of a new good, according to Romer.

means of simplified version of a partial equilibrium model adopted from Romer. It represents a partial *equilibrium* model with the *x-axis* referring to the amount of production of good Z and the y-axis the price per unit charged by the company.

The marginal cost of production indicates any additional cost required to produce a next unit. This marginal cost curve is flat rather than increasing because it only represents the variable costs of production (below the line) that are assumed to remain constant with *increasing* production in view of the low and relatively stable *reproduction* costs of an innovation.

The price is higher than the marginal production cost because the company aims at reimbursing the high fixed costs spent on the development of good Z (not represented in the marginal cost of production) and making profits that allow for further *investment* in R&D.

Neoclassical *economists* interpret this graph as a typical case of a market that is dominated by a monopolist: there is only one producer of good Z, who has the power to determine the scale of production and set the product price in a way that *maximizes* the expected returns (the sloping demand curve illustrates how the price increases with *decreasing* output).

In order to illustrate the *monopolist* rent, the neoclassical economist would point at rectangle B, which represents the surplus the monopolist extracts through market power, and also triangle C, which represents the deadweight loss for consumers who have to pay a higher price for the good.

However, if this monopolist has obtained his *position* not through collusion but the investment in the development of a new good, an *additional* social welfare surplus, triangle A, comes into being.

This triangle has been ignored by economists prior to Romer. It represents the social welfare surplus that is generated through the new good. The monopolist cannot capture this triangle because if, as Dupuit already *recognized,* he raises the price above this level, he would lose the market (the price would be too high for potential buyers that are constrained by their own competitive markets).

Romer concludes that if *economists* became aware of this triangle A (which he also calls the Dupuit triangle), they would realize that the primary source of wealth and well-being in society is not based on perfect *competition* but on the introduction of new goods and technologies.

New technologies may create new *inequalities* and risks at the beginning, when only few have access to the relatively expensive technological innovation and accidents may happen due to lack of experience with the new technology. But in the long run many new *competitors* enter into the market, increase the total offer, constantly improve the safety of the technology, and lower the price.

Eventually, the product becomes cheaper and safer and thus generates a global mass market. At this stage, it could also become a tool of *empowerment* for the people who previously lacked access to the technology.

Role of Multinationals in Economic Growth

Currently, the main players in the global mass market are multinational *companies.* They are interested in selling their goods in all parts of the world. It is true that their product must be sold at a much lower price in *developing* countries with low *purchasing* power, but the fact that they have price-setting power allows them to exert price discrimination. They can charge higher prices in developed *countries* and lower prices in developing countries. Moreover, poor developing countries may be high-risk markets but they generally offer greater long-term growth potential.

There are certainly multinationals that tend to abuse the system by charging *prohibitive* prices in certain countries, aggressively push the strict *enforcement* of their patents and later on their extension, lobby for permissive regulation, and use the returns to enrich themselves rather than invest them in new R&D.

However, it can be assumed that those companies may eventually lose out in competition because they had spent their scarce financial resources on preserving short-term political and market power *instead* of investing them in R&D to ensures the long-term survival of the company in a competitive and *innovation-driven* market.

Responsibility of National Governments

Whether a multinational company decides not just to sell but also to develop and produce a new good in a particular country, depends to a great extent on national government policies. National governments can *discourage* such investment by imposing high profit taxes, trade restrictions on essential capital goods, prohibitive safety *regulation*, and inefficient and burdensome government *bureaucracies*.

Corruption and weak property rights can additionally increase the costs for the company until the point is reached where a company decides not to invest anymore in a developing country, despite cheap labor and abundant natural resources, because the increasing costs *exceed* the additional revenues expected through the higher price they intend to charge.

So the good will simply not be produced, which means that the respective country loses the social welfare benefits of the Dupuit triangle. Instead of just *taxing* and regulating companies and pushing up the bar of their marginal costs of production, governments can also serve as facilitators and encourage investment through tax exemptions for R&D, political stability, a *valuable* stock of domestic human capital, dependable public *infrastructure*, and a relatively open and well-developed financial market (that improves the possibility of hedging risks). In this context, governments in developed countries have many means available to facilitate private sector *investment*.

This allows them on the other hand to increase the cost of regulation of a company (e.g., by pushing up *environmental* and food safety standards). Poor countries, however, are in a different position: because the market is tiny and the state *budget* too small to improve investment conditions, additional regulations quickly erase the profits that foreign companies can expect.

This again leads to the loss of the Dupuit triangles, which are much larger than the deadweight loss triangles for *consumers* (triangle C). Paul Romer therefore concludes that "badly designed policy interventions do not come from their effects on the static allocation of resources between the activities in an economy that already exists.

Rather, they come from the stifling effect that the distortions have on the adoption of new technologies, the provision of new types of services, the exploitation of new productive *activities*, and on imports of new types of capital goods and produced inputs".

HOW TO USE KNOWLEDGE AND TECHNOLOGY FOR DEVELOPMENT?

As illustrated above, the primary contribution of companies to general social welfare may not occur through general taxes (as widely assumed) but through the generation of new goods and services. Yet, the problem is how to get the private sector to produce valuable goods for markets that are too small to be worth investing in? Often these markets are not served because of *generally* low *purchasing* power or small market size.

This is a serious problem because the Dupuit welfare triangles for such goods would be huge. For example, orphan crops with higher productivity and enriched nutritional quality (e.g., protein-enriched cassava) or orphan drugs against communicable diseases (e.g., *Malaria vaccine*) have the potential to improve public health in developing countries *enormously*.

Old welfare economics would suggest that the insufficient production of such goods is a typical example of market failure. New growth theory, in turn, would argue that it is state failure because governments fail to design appropriate *incentives* for the private sector to *collaborate* with universities in efforts to create new goods or improve existing goods that are of high value in the fight against poverty and for improvement of the natural environment.

Since neoclassical welfare economics shaped agricultural and development policies in the twentieth century and continues to do so, it is important to first understand the achievements, but also the unintended side effects, of this comparative-static *approach* during the Cold War.

COLD WAR ECONOMICS AND ITS IMPACT ON AGRICULTURAL POLICY AND RESEARCH

The rise of neoclassical welfare economics began after World War II. At that time, Western policy decision-makers were not just concerned about economic growth (which was strong anyway due to the reconstruction efforts) but also about the ideological *mindset* of its citizens. It was *assumed* that economic development may result in increased social inequality and that this might cause public disillusion with capitalism and a longing for *communism*.

Welfare economics was regarded as the ideal instrument to identify the aggregated preferences of a society (in terms of expected social outcomes) and to design policies that addressed such preferences (e.g., minimal social inequality). These *aggregated preferences* were portrayed in the form of a normative social welfare function. *Subsequently*, positive analysis was applied to maximize allocative efficiency of the public policy measures designed to achieve the resulting social objectives.

The celebration of the elegant formal language of general and partial *equilibrium* models in welfare economics and the belief that the public sector must assume the role of a rational social planner that addresses the problem of market failure influenced many areas of public policy. Yet, the great welfare economists at that time Paul Samuleson, Kenneth Arrow, and Robert Solow were too quick in *identifying* market failure when it came to the management of public goods.

Samuleson used the example of the lighthouse to illustrate the nature of a public good. He argued that since none of the shippers would be willing to build such a lighthouse (in view of the others who would just free-ride) it must be a public good that is based on non-rivalry and non-excludability and therefore should be provided by the public sector. Ronald Coase showed that this argument does not correspond to historical facts: lighthouses that were financed by user fees paid by the *shippers* existed in *Europe* in the eighteenth century, as he pointed out in his 1979 publication. Kenneth Arrow and other welfare economists portrayed the state as a rational social planner that looks at aggregate social preferences and, accordingly, allocates the scarce public resources in a pareto-optimal way (making at least *someone* better off without making anyone *worse* off).

Buchanan and Tullock discovered the flaws in such assumptions by highlighting the fact that the *democratic* decision-making process is not a rational process on an aggregated level as Arrow assumes. Political actors pursue their own selfinterest and are not driven by the desire to maximize social welfare. Moreover, there is no such thing as a rational social planner unless it refers to an allknowing dictator. Finally, Robert Solow managed to combine conventional growth theory with *neoclassical* welfare economics. He argued that knowledge must be funded by the public sector because it is a public good (*assuming* that it is non-rival and non-excludable).

As a result, technological change was treated as an exogenous factor that can be perfectly integrated into the neoclassical model of perfect competition, where companies are portrayed as passive price-takers. Romer finally *challenged* Solow's growth model with his paper on endogenous technological change.

He showed that it is not perfect competition but monopolistic competition that generates new goods and services. Moreover, his model was able show that companies also have to invest in R&D (something the Solow model did not account for. A more detailed discussion of the *theoretical* aspects of new growth theory and the knowledge economy has been published by Aerni in the ATDF Journal).

COLD WAR ECONOMIC COMMUNITY

In view of all these developments, one would therefore assume that economics has evolved by accepting this shift in paradigm. However, textbook economics continues to be welfare economics and technology is still largely treated as exogenous. Those economists who deal with issues such as technological innovation and monopolistic competition are generally advised to leave the economics departments and to join business schools. This ongoing reluctance to embrace the new *paradigm* may also be related to the entrenched interests within the academic community of economists.

It highlights once again the contradiction that is so obvious in social science. On the one hand, most social scientists claim to respect Popper's concept of *falsification*, but then who is really eager to falsify the theory that he or she helped create? Unlike in the natural sciences, where an increasing gap between theory and experimental outcome leads *eventually* to an adjustment of the theory, there is no such pressure in the social sciences.

Most theories that history has proved wrong or inadequate are still taught at universities and continue to expand in the form of *insider* research communities. These research communities have their own peer review process, which relies on reviewers from the inner circle of *professors* that are in charge of maintaining the dogmas of the theory and have an interest in keeping the theory alive.

The Mindset of Agricultural Economists and How it Influenced Agricultural Policy and the Green Revolution

The first wave of globalization in the second half of the nineteenth century led to a rapid decrease of transportation costs and erased many of the geographical barriers that previously protected local agriculture from foreign competition. This *threatened* the livelihoods of many farmers in the early stage of industrialization and governments started to be concerned about the capability of their respective countries to ensure the national food supply in times of war.

Political and economic (later also social and *environmental*) instabilities were identified as negative external effects that need to be addressed by the public sector. At the outset, state intervention was mainly justified in the name of managing the public good called "national food security". Agricultural *economists* were hired as social planners to ensure the *effective* management of this public good.

The planning models they used at a later stage (e.g., linear programming) to calculate how certain *normatively* set policy objectives can be achieved most effectively, were largely developed by scientists in the former Soviet Union. Even though these policy planners quickly realized that a democracy is not about the joint effort to design a *rational* policy but about the self-interested search for access to scarce public resources, it did not hinder them from sticking to the *principles* of welfare economics and ignoring the role of political economy.

Agricultural Policy During the Cold War

Agricultural economics embraced the theoretical concept of the so-called agricultural treadmill developed by Cochrane. In this concept, farmers produce a *homogenous* (presuming that there are no differences in quality) and inferior (because of its low income elasticity: the higher the income the lower the share of the *household* budget for food) commodity in the form of food.

They are portrayed as passive price-takers in a market of perfect competition. The role of technology is reduced to its potential to increase agricultural productivity (while its potential to improve food quality is not addressed). Since farmers are standing in perfect competition they are assumed to produce at the level where their marginal costs just equal their marginal revenues. According to the concept of the *agricultural* treadmill, it is possible that certain farmers adopt a new technology that allows them to lower their production costs and produce more efficiently.

This gives them a temporal advantage and thus a windfall profit. Yet, this advantage is quickly erased because all the *competitors* will have to follow suit if they want to stay in business. The agricultural treadmill is portrayed as the main reason for the surplus in food production and the decrease in relative food prices.

Unsurprisingly, agricultural economists concluded that the treadmill, and with it technological innovation, largely benefits food consumers at the expense of food producers. They argued that the agricultural *treadmill* produces a sort of market failure since farmers get poorer even though they produce more, due to the inferior prices. *Moreover,* the resulting intensification of agriculture will destroy the environment and family farming and *negatively* affect the quality of food.

Yet, in retrospect, even agricultural economists would admit that it was probably not the agricultural treadmill, but the market-distorting instruments of agricultural policies that provided the biggest incentives to adopt *monoculture* practices and degrade the environment and food quality. One only needs to go and watch the movie *We feed the world,* produced by Erwin *Wagenhofer* in 2005 (the most successful Austrian documentary movie ever) to get a picture of the unappealing endless number of greenhouses in southern Spain that focus almost *exclusively* on intensive tomato production.

Erwin Wagenhofer, who is an urban dweller with little knowledge of agricultural policy, blames the corporate world for all this misery. Yet, in fact intensive tomato production in Spain is a result of EU subsidies. The same goes for olive tree monoculture in Spain and Greece, overfishing in the *Atlantic Ocean, excessive* growing of low-quality wine in France, and many other *subsidized* products.

All these practices are not just harming the environment but they also discourage innovation and tend to make food quality worse - for why should these producers care about innovation or pleasing consumer taste if the money comes from Brussels anyway?

Agricultural Policy After the Cold War

In the 1990s, agricultural economists admitted that certain policies produced "suboptimal" results despite the rational social planning. They recommended a switch from production-tied subsidies to income-support subsidies. The new *objective* was the *maintenance* of a strong, healthy, and *environmentally* sustainable agricultural sector.

As a consequence, things like agrobiodiversity, food safety, decentralized *settlement,* and custodianship to cultural landscapes were declared to be the new public goods that are provided by farmers - after the old public good of *maintaining* food security became somewhat obsolete in view of the production surplus and the end of the Cold War. It provided the best justification to keep agricultural economists employed as social planners and continued to use all the old planning models that are focused on creating optimal allocative *distribution* in areas where the market presumably fails to do so.

But did the market really fail or are these bureaucrats increasingly managing state failure? There is increasing evidence that the new agricultural policies and the new *justifications* for government intervention in agriculture did not bring the *expected* improvements: Direct payments were designed to mitigate the structural change that was expected to result from slightly more open agricultural markets, as demanded by the WTO Agreement on Agriculture (AoA).

Yet, direct payments proved to be an obstacle to structural change because they increased the value of land and discouraged many farmers from becoming more innovative and competitive. At the same time, the new normative goals of agricultural policy to promote environmental, social, and economic *sustainability* through compliance schemes (e.g., agro-environmental measures/ labeling schemes in return for more direct payments and *premium* prices) once again turned out to be in the best tradition of *governmentsponsored* agricultural economic research, namely suboptimal. Environmental improvements were relatively meagre and largely achieved through more efficient input technologies.

Moreover, a large evaluation of agro-environment measures in Europe showed that such measures hardly ever contribute to an increase in *valuable* biodiversity. There also seems to be a correlation between the amount of direct payments a rural region receives and its economic decline. This is not surprising in view of the fact that a high dependence on direct payments is not an attractive way of life for the young people so they look for *opportunities* elsewhere to participate in the newknowledge economy.

Apart from this, the private sector is reluctant to invest in heavily subsidized regions because of the receiver mentality of the people and high production costs (pushed up indirectly through direct payments). In spite of the timid opening of agricultural markets, agricultural trade has hardly *increased* over the past two decades. One major *reason* for that is the WTO AoA itself.

It is primarily focused on a gradual improvement of market access rather than the reduction of domestic support measures. But, ultimately, it is domestic support measures that result in wrong market signals, *overproduction*, and subsequent market access restrictions. The fatal consequence was that the amount of domestic support did not decrease but was just moved from so-called "*actionable*" subsidies (placed in the amber box of the AoA to "*non-actionable*" subsidies (placed in the blue and green box of the AoA).

In WTO terminology, subsidies are identified by "boxes" which are given the colours of traffic lights: green (permitted), amber (slow down - i.e., to be *reduced*), red (*forbidden*). Export subsidies are prohibited and therefore fall into the red box. This is, however, not strictly applied in agriculture where export subsidizes are still tolerated. The amber box contains all *trade-distorting* domestic support measures.

Any support that would normally be in the amber box, is placed in the blue box if the support also requires farmers to limit production. Green box subsidies are meant to be non-trade-distorting. It was assumed that non-actionable subsidies would not be tradedistorting but it turns out that they are.

At any rate, this shift kept social *planners* employed and did not force anyone to look at theory. But are these policies sustainable and do they really benefit farmers? In consideration of what we now know, the answer is unlikely to be yes.

A parallel development with a similar *ambiguous* outcome happened in the international arena where the primary concern was to develop new agricultural technologies to enable developing countries to become self-sufficient in food production.

Theoretical Thinking Behind the Green Revolution

The same agricultural economists who argued in the 1940s that the Western states needed to maintain a healthy agricultural sector through *subsidies* and border protection argued that developing countries must be assisted in the development of new *varieties* and modern irrigation systems in order to boost food production and avoid hunger and starvation. It was assumed that the private sector would have no

interest in investing in technologies that would serve poor farmers in developing countries. Therefore, public investment in international agricultural R&D was declared to be a public good that must be managed by the public sector (following the Solow model). The *resulting* global public sector initiative is today widely known as the Green *Revolution*.

It was to a large extent a US-driven effort to improve food security in the non-aligned developing world as part of a global containment strategy against communism. The United States Agency for International Development (USAID) and the Rockefeller Foundation were the main financial contributors to the *establishment* of the first Centers of the Consultative Group of International Agricultural Research (CGIAR) in developing countries.

These CGIAR centers enabled Western scientists to work in wellequipped research centers in developing countries and design high-yielding varieties of major food crops such as maize, wheat, and rice. The new varieties were subsequently distributed in rural areas through government agencies. The private sector was hardly involved, even though it later benefited from the scientific knowledge generated through this international undertaking.

The research at these centers (CGIARs) contributed to significant increases in agricultural *productivity* and technology transfer to local universities and national research institutes in developing countries.

International Agricultural Research During the Cold War

There is no doubt that the Green Revolution greatly contributed to global food security through the excellent international agricultural research that was conducted at CGIAR centers during the Cold War. However, the interaction between Western scientists, who developed high yielding varieties, and local farmers in developing countries who adopted these varieties through the national seed *distribution* programs, was rather poor.

This led to some long-term problems such as inadequate use of pesticides, *insufficient* operation and maintenance of irrigation systems, little seed choice for farmers, and *monoculture* practices. In addition, farmers in marginal regions did not benefit to the same extent from these new hybrid varieties that were mainly designed for favourable agricultural conditions with access to fertile soil, irrigation, markets, and essential inputs. Left-wing development activists point at these *unintended* side effects of the Green Revolution and denounce them as the destructive forces of science and business, and they conclude that *environmentally* destructive monoculture *practices* must be part of the capitalist logic.

Yet, as highlighted in the case of the documentary of Erwin Wagenhofer, these undesirable side-effects are a result of too little rather than too much private sector involvement. For example, public sector researchers based at CGIARs did not have to bother much about the real and complex set of problems that *farmer* face in the field or the *particular* consumer taste of different cultures.

They could just focus on plant variety traits that would increase yields and then select the elite varieties and hand them over to national agencies for *distribution*. As a result, the private sector may have had little interest in investing in the development and commercialization of new *varieties* in developing countries and in competing with the public sector, which would distribute the seeds almost for free. Thus, the private sector largely stayed out of the Green Revolution.

This explains, for example, why the greatest bottleneck in many developing countries is probably the absence of a local seed industry. It also explains why many Filipino consumers prefer to buy rice from Thailand, which is the greatest exporter of high-quality Indica Rice, but actually never *embraced* the largescale adoption of high-yielding rice varieties. They say it simply tastes better than the rice varieties that were bred by the International Rice Research Institute (IRRI) and widely adopted by *Filipino* farmers.

International Agricultural Research After the Cold War

After the end of the Cold War, foreign aid was cut in almost all state budgets of developed countries and public sector funding for international agricultural *research* decreased significantly. Right-wing politicians argued that there was no need for further investment in CGIAR research because the Green Revolution has already largely achieved its purpose of *eliminating hunger*.

This argument is quite cynical considering the fact that there are still over 800 million people worldwide suffering from hunger and malnutrition. Left-wing politicians, in turn, were using the familiar but flawed argument that there is *enough* food around but that it is badly distributed. This led them to the *conclusion* that there is no further need for investment in technology but instead just a need to bring the food to the poor.

Agricultural ministries in developed countries that still do not know how to get rid of production surpluses would most certainly welcome the idea. Yet, the fatal consequences of such *dumping* policies are widely known: local farmers in developing countries that cannot compete with donated food are forced to abandon farming because of lack of revenues.

Thus, such policies are likely to worsen food self-sufficiency and increase dependence on Western food aid. Even though the "*distribution* problem" argument is still widely used by teachers in high-schools, it is rejected even by *left-leaning* development activists who now embrace the paradigm that farmers in developing countries need to be assisted in growing their own food in a *sustainable* way.

Yet, the problem with Western non-governmental organizations (NGOs) that pursue this approach in developing countries is that they generally dismiss the role that business and new technologies have in agricultural development, arguing that it would *introduce* a capitalist logic that is not compatible with local traditions. They believe that farmers should rely on their *traditional* lowinput and low-tech practices.

They may assist them in finding slightly better techniques of soil fertility and integrated pest management but, in general, farmers are encouraged to use the agricultural practices they would use anyway. Subsequently, these *Western* NGOs help them to export the harvested agricultural products to developed countries where they are sold under different kinds of *environmental* and social *labeling schemes*.

Such a strategy resembles the top–down approach of the Green Revolution. Both strategies assume that there is a sort of market failure because business does not care about the poor. This would produce negative externalities such as increasing social inequality, hunger, and malnutrition that must be addressed by responsible Westerners. The only difference is that one approach looks at modern technology as the solution whereas the other approach sees it as the main problem.

However, the ideological mindset of anti-technology NGOs is likely to harm poor farmers in developing countries more than the previous overemphasis on public sector R&D. Farmers need to become actively involved in the process of technological change and they need to learn how to take *advantage* of the emerging knowledge economy. This will eventually lead to more *self-confidence* and *entrepreneurship* and result in increases in agricultural productivity and nutritional quality of the traditional food crops.

This is especially true for Africa, which did not benefit from the first Green Revolution. It is often argued that local entrepreneurship in developing countries is hampered by the absence of land titles in the *informal* sector. Yet it is still unclear whether the *assignment* of land titles effectively results in more entrepreneurial activity independently of the other local circumstances.

More important is the creation of an entrepreneurial infrastructure that lowers local market transaction costs and opens access to new markets. Local *entrepreneurs* can further be encouraged to become low-tech innovators through the establishment of petty patent systems. In 2001, the UNDP Human Development Report was titled *Making new technologies work for human development.*

It attempted to counteract the misconception of the supposedly negative role of technology and the private sector in sustainable development and was promptly attacked by sustainable development activists. This is a pity because this report merely reminded policy-decision makers that there is Principle 12 in the UNCED Rio Declaration, which emphasizes the important role of new technologies in sustainable development.

It seems that neither agricultural economists that helped shape the Green Revolution, nor Western NGO leaders that advocate participatory approaches in agricultural development can see any benefit in getting the private sector more involved in agricultural development and encourage local entrepreneurship. This may be related to the fact that they tend to use theoretical concepts that might have looked reasonable in the Cold War economy, but they are rather outdated in the new knowledge economy.

THE NEW KNOWLEDGE ECONOMY AND THE NEW RULES OF THE GAME

The two major driving forces of the new knowledge economy are the revolutions in information technology and biotechnology that took off in the 1970s and 1980s. Both revolutions started initially at universities and were *strongly* supported by the public sector.

However, when the first prototypes of commercial interest emerged, the university-based inventors decided to seek intellectual property protection for their *inventions* in order to set up their own businesses in the form of spin-off firms.

Some of them eventually established highly successful multinational companies themselves, others were able to partner with *multinationals* in the commercialization of the technology, others focused on licensing out their patented technology to whoever was interested in using it, and others simply lost out to entrepreneurial young outsiders that quickly grasped the *economic* potential of certain clumsy prototypes and improved them to a level where they could become commercial *successes.* Both, the IT industry and the biotechnology industry have matured over the past decade.

As a consequence, the costs of IT and biotechnology products and tools have decreased *significantly* and are now reaching a far wider customer base. Unlike in the old economy where most developing countries merely played the role of suppliers of primary *commodities* and lacked the critical base of domestic human capital to make use of modern technology to develop their homegrown *technologies,* the new knowledge economy allows them to participate in the global economy to a far greater extent.

Effects of Information and Communication Technologies

Thanks to all the new *communication* and information technologies, new knowledge spreads more quickly and widely, international research networks become much more *extensive* and effective, outsourcing business activities from simple accounting to R&D has become an integral part of the strategies of *multinational* companies, and global venture capital is increasingly invested in talented techno-entrepreneurs in developing countries.

The resulting rise of many developing countries in science, culture, business, and political power makes the jargon of the North–South dialogue of many Western development activists look increasingly old-fashioned. South–South business *investments* and research *collaborations* are growing five times as fast as their North–South equivalents. Moreover, many big companies in the South are starting to even gobble up companies in Europe and the USA.

Effects of the New Tools of Biotechnology

There is a widespread prediction that the biotechnology revolution, powered also by the information technology revolution, will eventually transform a rather dirty agrochemical and *petrochemical* industry into a cleaner biology industry. The potential economic, social, and environmental welfare benefits of this *transformation* are enormous, and this time it is likely that developing countries with a critical domestic knowledge base will be at the forefront of new product developments.

If mankind is serious about protecting the natural environment and ensuring access to food, the growing demand for food over the next 50 years should not be met by further *colonizing* of pristine ecosystems but rather by raising productivity on existing farmland. Agricultural biotechnology is not just ideally positioned to meet this challenge but also likely to produce new food products that are safer, more *nutritious*, and tastier.

The potential environmental and health risks of biotechnology must be taken seriously, but after 10 years of experience and innumerable public risk assessment studies there is no *indication* that existing GM crops pose any risks that go beyond the risks known from conventional crops. Moreover, the ethical concerns raised about the current techniques of genetic engineering could quickly be overturned by the emergence of completely new transformation techniques and the *advancements* in genomic research.

But, one ethical concern will certainly not go away and that is the crucial aspect of social equity. The private-sector-driven biotechnology revolution may result in enormous social inequalities because the least developed countries that have simply no means, no critical knowledge base, and no attractive markets to participate in this emerging sector may once again be left out. As a result, the new products would merely improve the needs of affluent societies that can promise a high return on investment while the basic needs of the poor remain *unaddressed.*

Management of Public Goods in the New Knowledge Economy

As mentioned earlier, an exponential increase in knowledge leads to an exponential increase of the probability that new products and services will come into being. These new goods and services generate innumerable new Dupuit welfare triangles - but only for those societies that do not discourage their introduction and those that have sufficient purchasing power and market size to attract them.

Therefore, there is a high likelihood that the knowledge economy will increase global inequality unless national governments and international organizations design policies that ensure that the new technologies will also benefit and eventually empower people in the least developed countries. However, it would be a mistake to address the challenge by simply embracing a second Green Revolution because, as *explained* above, the *underlying* principles of welfare economics are no longer applicable to the rules of the new knowledge economy. The belief that public goods should be provided exclusively by the public sector ignores the fact that the private sector *increasingly* contributes to the production of public goods (e.g., clean technologies, more *efficient* use of natural *resources*).

The public sector should therefore not assume tasks that the private sector might provide in a more efficient way and in better quality (more focused on consumer/client needs) but should learn how to better play the role of a facilitator of private sector activities that generate large Dupuit welfare *triangles*. As shown in Sect. 1, the generation of these *social* welfare triangles requires high fixed costs that are spent on large investment in R&D, physical infrastructure, and product development.

Companies are often unwilling to invest such high fixed costs in the development of a new good unless the resulting market is expected to be profitable. This also explains why the first prototypes of new technologies were almost always designed in *university* rather than in corporate laboratories. *Throughout* the history of technology we can always observe the same pattern: there is the curiosity-driven researcher funded by the public sector who has no immediate interest in business.

But, there is also the bold entrepreneur who uses the knowledge generated by the curiosity-driven researcher to design a new prototype that is patentable. The prototype is subsequently adjusted to market needs, and finally commercialized on a large scale. Both *characters*, the inventive researcher and the entrepreneur, are needed to *create* social welfare triangles. Sometimes the curiosity-driven researcher and the profit-oriented entrepreneur can be one and the same person.

But often the inventor is not necessarily a good entrepreneur and the good entrepreneur is not *necessarily* good at inventing. 7, without the *entrepreneur* who is primarily focused on creating a new market and making large profits as a temporary monopolist, the fruits of science would never become real. This is one of the most important insights, and it was ignored during the Green Revolution when research, product development, and distribution of new agricultural technologies was entirely managed by the public sector.

The private sector is simply better at *converting* knowledge into useful goods and services. The public sector should mobilize this strength of the private sector for the management of public goods rather than try to duplicate it. Today it is *important* that the public sector first identifies a list of biotechnology products that could potentially generate large social and environmental welfare benefits (e.g., biofuel generated from cellulose, drugs for diseases that are extremely rare, vaccines that protect people from sudden outbreaks of *communicable* diseases).

It should then offer university research teams funding to develop a first prototype of such a new product. At the same time, it could offer a generous award to the research team that first develops a dependable prototype that is sufficiently attractive to be licensed out to the private sector.

Yet, it should not be the university but the government that does the licensing *negotiations* because it would otherwise just increase the costs of patent lawyers at universities and distract *researchers* from what they do best, namely research and development. This would not prevent researchers from becoming *entrepreneurs* themselves. In this case the technology patent would not be licensed out but used as the first asset of a new spin-off firm.

The researcher would then entirely assume the role of an entrepreneur, who must ensure that the product can be sold on the market at a profitable price. Often the private sector is *discouraged* from using a new prototype because inexperienced researchers overestimate the value of their invention and underestimate the fixed costs and risks that companies face when *commercializing* a new technology with uncertain market potential.

The government that initiated the research initiative to achieve certain social and environmental objectives may have a real interest in encouraging the private sector to use the prototype and convert it into new goods and new markets. The government might therefore be willing to forego the licensing fee entirely in return for certain *reservations* when it comes to the *commercialization* of the product.

It could ensure that there is ongoing research collaboration between the firm and the university and that the new products are affordable to the poor in developing countries (which may still be profitable for the company if it is free to charge a higher price in developed countries and able to prevent parallel imports). If the prototype is still not attractive to the private sector because the market is too small to make a profit, the government can design additional incentives such as fast track regulatory approval and tax credits for product research.

Once the private sector is willing to embrace the product because it expects to make a profit thanks to the *additional* incentives, it will be much more *efficient* and more end-user responsive than the public sector could possibly be.

Global Governance to Produce Global Public Goods

Often governments in developing countries may not have the means to offer sufficient incentives on their own to induce companies or research institutes to come up with products that would produce high social welfare triangles for their country. For example, improved orphan crops could save thousands of lives and *significantly* improve the health of the poor, but neither the local private sector nor the national governments in developing countries have the means and the know-how to successfully invest in such *improvements*.

At the same time, multinational companies might have the know-how but do not have an incentive to invest (due the small market size for the good and the low purchasing power). International donors could address these *constraints* by creating incentives for the private sector to produce such goods by offering a generous prize for the first company or research organization that was able to produce such a good or contract an advance purchase that would boost expected demand.

Some people would denounce this as creeping privatization but the fact is that the new technologies that were derived from the information and biotechnology *revolutions* make it *increasingly* cost-effective to include the private sector in the *management* of public goods. Generally, these technologies permit smaller *producers* and more scope for competition.

Future Role of CGIARs in Promoting Agricultural Biotechnology

From 1996 to 2004, biotechnology crops have reduced the volume of pesticide spraying globally by 6%, equivalent to a decrease of 172500 t. The technology has also significantly reduced the release of greenhouse gas *emissions* from agriculture; a reduction equivalent to removing 5 million cars from the roads. The increase in farm income that resulted from the *adoption* of GM crops is equivalent to adding 3–4% to the value of global production of the four main *biotechnology* crops.

Moreover the adoption of Bt cotton in many different developing countries turned out to have significant economic, health, and *environmental* benefits for small- and large-scale farmers alike. All these facts just refer to GM crops and do not take into account the large economic and environmental gains that have been *achieved* by using all the other tools of modern biotechnology such as tissue culture, marker-assisted breeding, gene silencing, and genome mapping. But why then do politicians often argue that agricultural biotechnology does not offer any benefits to society and the environment?

This may be largely based on a generally hostile public opinion and vested interests that prefer the status quo in agriculture. Yet, it also seems that agricultural economists are not really able to provide convincing arguments about why *agricultural* biotechnology will also benefit the poor and the environment. Agricultural economists, who still stick to the principles of *neoclassical* welfare economics, assume that the private sector has, once again, no interest in addressing the needs of the poor and that therefore the public sector needs to step in and initiate a new public-sector-driven Green Revolution.

The skepticism about private-sector involvement may be related to the general distrust of *monopolistic* competition that drives the process of technological innovation. In agriculture, it adds to the already existing *scepticism* related to the agricultural treadmill hypothesis, which treats technology as exogenous and implies that benefits from introducing technology in agriculture would not go to farmers but *primarily* to the seed and *agrochemical* industry and to the food consumers.

This clearly contradicts the numbers of Brookes and Barfoot, who calculated an increase in global farm income through the adoption of GM crops of a cumulative total of $27 billion for the period 1996-2004, derived from a combination of enhanced productivity and efficiency gains. Obviously *agricultural* biotechnology must be more than just an agricultural treadmill.

Moreover, it is wrong to reduce farmers to passive price-takers who struggle to survive in perfect competition. Farmers were always innovators and interested in collaborating with *researchers*; but national agricultural policies can either encourage or discourage innovative farmer activities.

Farmers as Innovators

The land grant college system in the USA was set up in the nineteenth century with the purpose of promoting applied science and *stimulating* economic activities in the rural areas. The state universities, which were established all over the country, had the explicit mandate to cooperate with local farmers and support their efforts in finding solutions to specific crop problems, but also to develop agricultural innovations with a *commercial* potential.

This collaboration produced technological innovations, new agricultural products, and new companies in agribusiness. Apart from stimulating economic growth it also contributed to the social empowerment of rural areas in the USA. A similar development happened in Switzerland at the end of the nineteenth century.

The first agricultural law was passed in 1893 with specific emphasis on *improvement* agricultural research and development and, in 1898, national agricultural research institutes were set up to meet this challenge. This successful *partnership* between the university researcher and the farmer has largely been abandoned in Europe and the USA but New Zealand started to rediscover this success story after it decided to liberalize agriculture in the 1980s.

The Royal Institutes of New Zealand were semi-privatized and its agricultural research activities only get funding if committed to making the farming sector more competitive and sustainable through innovation. This implies a close collaboration with business and the farming *community*. Even though genetically engineered crops are not yet approved for commercialization, agricultural biotechnology is at the center of this new agricultural policy in New Zealand.

New Zealand's biotechnology industry *generated* an estimated revenue of NZ$811 million in 2005, with over NZ$250 million in exports. The industry has helped ensure the continued international *competitiveness* and efficiency of New Zealand's food and beverage sector.

This focus on technological innovation did not just create a more sustainable and competitive agricultural sector (compared to the previous subsidy-based agricultural system) but also boosted the farmers self-confidence. Farmers do not feel victims of a new knowledge-based economy but an *integral* part of it. The agricultural treadmill would have predicted a different outcome.

Crop Research Networks

Some would argue that New Zealand is an exception. It has invested a lot in knowledge and human capital, is well-governed, has excellent infrastructure and highly developed input and financial markets. Poor developing countries where none of this applies would face a much bigger *challenge* to make technology compatible with *sustainable* development, especially when it comes to the improvement of orphan crops that are largely grown by subsistence farmers.

These farmers would first of all not benefit from private-sector *innovations* because there is no incentive to invest and, if they did invest, farmers would lack the knowledge to use new technology in a sustainable way. The *arguments* may sound reasonable but they underestimate the power of creative solutions.

The *Cassava* Biotechnology Network (CBN), which started in 1988 as a global initiative to mobilize the development and application of biotechnology tools for the improvement of cassava agriculture, is an excellent example to illustrate how creative thinking can employ agricultural biotechnology for the benefit and empowerment of local subsistence farmers. Cassava is a typical orphan crop that is produced mostly by *smallholders* on marginal and *submarginal* lands in the humid and *subhumid* tropics.

It is a starchy root crop that grows in a wide range of environments and is very tolerant to drought and acidic soils. Cassava has the advantage of flexible harvesting (the root tuber can be preserved in the soil for up to one year) and this makes it the crop of last resort for many poor farmers that prefer to harvest cassava whenever there is a shortage of food or animal feed. CBN is based at the Centro *International* de Agricultura Tropical (CIAT) in Colombia and consists of a loose network of stakeholders that represent cassava research, cassava farming, *cassava* business, and international donors with an interest in cassava agriculture.

The triennial CBN meetings serve as the major platform of information *exchange*. The aims are to share knowledge on cassava, identify new challenges in research, improve farmer adoption and marketing of cassava, and set up new research *collaborations* that are focused not primarily on research but on the development of new and useful products for cassava farmers. CIAT, which belongs to CGIAR research centers that were set up during the Cold War, is the major driving force behind this network.

It responded to the financial crisis of the CGIAR system in the 1990s by basically reinventing itself as an engine of *innovation* in the area of orphan crops. CIAT realized that the old supply-driven system of international agricultural research was not really addressing the needs of the end-users and treated them in a rather *paternalist* way.

It was a purely public sector initiative, which may be able to sponsor excellent research but does not know how to make useful products and disseminate them efficiently. One goal of CIAT was therefore to get more involvement of the private sector and farmers to find out more about the effective demand for *certain* innovations in cassava agriculture.

Once they have identified the areas of research that would meet the biggest demand, they look for the best partners worldwide to collaborate on joint research projects. Thanks to the advances in modern information and communication technologies, international research *collaborations* have become much cheaper and also more effective.

This also explains why crop research networks such as CBN are probably best placed to facilitate an efficient exchange of knowledge and experience in the area of cassava research, production, and marketing worldwide. The interesting thing is that it is not just the development community but also the private sector and advanced research institutes in developed countries that have an interest in participating in such networks and in learning more about the advances in cassava science, the *opportunities* to create new markets out of cassava innovations, and ways to stimulate consumer demand for this neglected crop.

Unsurprisingly, the fundraising activities for certain joint projects are not limited to the mainstream official donor community but also include the private sector and governments in developing countries. In fact, two thirds of the members of the CBN are from developing countries. This indicates that South–South technology transfer initiatives may eventually become as important as North–South technology transfer.

Agricultural Biotechnology as a Tool of Empowerment

CBN is primarily designed to improve cassava as a food crop. Cassava is efficient in carbohydrate production but has a very low protein content. This causes a major problem of malnutrition if there is a high dependence on cassava-based daily meals.

Agricultural biotechnology proved able to enhance the protein content of cassava and was also used to analyze the biochemical pathways of b-carotene-rich cassava cultivated by indigenous tribes in the Amazon. Another problem of food security is the *creeping* genetic erosion of cassava, which results in ever decreasing yields. Cassava is a vegetatively propagated crop and the planting material must therefore be exchanged in the form of cassava *stakes* rather than seed. Stakes are often highly contaminated with viruses and affected by genetic erosion. These problems largely account for the very low yields that cassava subsistence farmers reach in the field. Tissue culture techniques, some of the earliest tools of agricultural biotechnology, proved to be an excellent solution to address this problem. They allow the cheap and effective *reproduction* of clean cassava stakes. Moreover, tissue culture technology has been constantly improved over the past *decades* and the price of a tissue culture laboratory has dropped *significantly* over the past years.

CIAT's Biotechnology Research Unit (BRU) has developed lowcost cassava in-vitro rapid multiplication techniques in collaboration with a Colombian farmer organization called FIDAR (Fundcacidn para la Investigacidn y Desarollo Agricola). This comprises small tissue culture laboratories, cold chambers, and greenhouses, built mostly with local material.

The use of local material made the end product six times cheaper than the official market version. *Subsequently*, men and women were trained to learn how to use their traditional knowledge about the best local cassava varieties and reproduce them in a tissue culture laboratory. The project proved to be very successful and induced *especially* women to set up local businesses and specialize in the local selling of high quality cassava stakes.

In interviews conducted in 2003 with these women it was striking how this project also boosted their self-confidence. Suddenly, high technology ceased to be a magic practice that could only be *handled* by Western scientists but became a practical tool in daily life. It proves how the value of indigenous knowledge can be enhanced through the application of the new tools of agricultural biotechnology; and it shows that agricultural biotechnology can be a tool of empowerment. CBN is, however, not just promoting cassava as a food crop but also as a cash crop.

This is because cassava as a food crop hardly generates new markets as long as it only serves the immediate needs of those who produce it. Moreover, any production surplus that exceeds the demand of the family of the *subsistence* farmer and his neighbors is likely to turn into waste because of the absence of markets.The absence of markets is largely related to infrastructure problems and the lack of *entrepreneurs* who are interested in developing a market for cassava.

Agricultural biotechnology has a potential to stimulate the development of new cassava markets by designing cassava varieties that are more productive and taste better. Moreover, the new techniques of agricultural biotechnology also have the potential to accelerate the bulking process of cassava, prevent its quick postharvest deterioration, and shorten the time of effective *fermentation* (*detoxification*).

This would definitely help to overcome some of the constraints on cassava trade and attract more local entrepreneurs. Most of the ongoing CBN research projects designed to make cassava a more attractive cash crop and increase consumer demand through better marketing are in collaboration with partners in the private sector. CBN is not interested in giving its product *innovation* away for free but wants the farmer to pay a price he can afford so that he learns to appreciate the value of innovation.

The farmer's willingness to buy also signals that there is an actual demand for the product. Moreover, it changes the farmer's attitude: he is now taken seriously as a businessman and is not just a mere receiver of charity products. All in all, the CBN projects prove that CGIARs can very well play a new role as brokers of new public–private partnership projects for the benefit of those who are most *vulnerable* in the knowledge economy.

CBN could be still more successful if the Europeans could finally overcome their mental barrier towards agricultural biotechnology and support this success story.

FINAL REMARKS

Current economic analysis of the value of agricultural biotechnology is still weighing the risks and benefits of existing GM crops for farmers and consumers. Even though, as shown in this article, these economic analyses largely confirm the economic and *environmental* benefits of existing GM crops, they largely ignore the fact that agricultural biotechnology advances at an unprecedented speed, continuously improving the economic and environmental performance of existing agro-biotechnology products and substituting earlier agrochemical products with new and cleaner biotechnology-based products.

The value of this constant introduction of new goods into the economy is ignored by comparative–static welfare economics because the associated general and partial equilibrium models assume that all goods that could possibly exist do already exist. The fatal consequence is that the social welfare produced through the introduction of new goods is simply ignored. Welfare economics and agricultural economics in *particular* are therefore too much focused on the potential risks of new technologies and underestimate the benefits for society at large.

The *assumption* of a comparative–static economy is especially flawed in the new knowledge economy where an exponential growth of knowledge leads to an exponential increase of the probability that new goods come into being. This largely explains why national agricultural policies that still rely on the principles of classical welfare economics have largely *discouraged* farmers from becoming more competitive and innovative (e.g., in *Europe*) while new policies that have embraced the ideas of new growth theory are encouraging farmers to become innovative and *competitive* (e.g., in New Zealand).

New growth theory policies also proved to be more effective in improving the environment and social welfare. If new technologies are not only to serve the attractive markets in affluent societies but also to contribute to a better life in poor developing countries, it is time to learn from past mistakes and also to design new development policies that take into account bottom–up solutions.

As shown in the case of the Cassava Biotechnology Network (CBN), agricultural biotechnology proved to be perfectly *compatible* with such bottom–up solutions that involve the local farmers as well as the local private sector. Moreover, modern information and communication technologies offer new forms of decentralized international collaboration that enable a stronger involvement of local participants and a more *effective* international network of research and product development.

In this context, CBN *represents* another example of a new approach that is no longer based on the old principles of welfare economics but has embraced new growth theory and thus enabled the marginal farmers in developing countries to *participate* more effectively in the new knowledge economy.

Chapter 7

RECOMBINANT DNA TECHNOLOGY

The genus *Malus* includes a variety of species all closely related and easily crossable. Agriculturally relevant is the species *Malus x domestica*, the edible apple. Peculiarities that are relevant to its use are self-incompatibility, leading to high heterozygosity, and the difference between apple as a crop and *Malus* in natural conditions. Commercial orchards are uniform as they generally consist of few varieties (*cultivars*) with each individual tree being the result of a rootstock and a grafted scion, both obtained by vegetative *propagation*.

These original selections, made possible centuries earlier, were intensively replicated, and are *currently* present worldwide. In contrast, in natural conditions each tree is a result of a meiotic event and assembles characteristics from the mother tree and the pollen donor tree, so that in nature no two trees will ever be *genetically* equal. Vegetative propagation by grafting has allowed the selection of high-quality apples through the centuries, by propagating a single tree developed from a seed which coincidently had more *appealing* fruit and growth *characteristics* than other locally known varieties.

Some of the popular cultivars, such as Golden Delicious, Granny Smith, or the more local Gravensteiner and Boskoop, are chance seedlings. Few cultivars are the result of an oriented breeding program. Although agronomically this genetic uniformity and the *maintenance* and distribution of a particular *genotype* is desirable, genetic uniformity is deadly for the resistance toward any natural enemy.

In fact the most successful commercial apple cultivars have lost the efficacy of their resistance genes toward the most frequent fungal pathogens, first of all *Venturia inaequalis*, causal agent of scab, and secondly *Podosphaera leucotricha*, causal agent of apple mildew. Still efficacious resistance genes can be found in wild *Malus* species, but commercial apple cultivars need constant protection through *fungicides*, wherever the climatic conditions are favourable for either of these diseases.

On average 10 to 15 fungicide applications are necessary during a season to produce scab-free fruits, with costs from US $500 to 1500 per hectare per year. Common commercial cultivars are also susceptible to the sporadic, yet highly damaging, fire blight infections caused by the bacterium *Erwinia amylovora*. Fire blight is in some parts of the world a constant threat.

Control measures against fire *blight* are relatively inefficient; only the antibiotic streptomycin is reliable, but is *banned* in several countries. Not only the high costs of controlling diseases, but also the increasing concern from consumers and environmentalists over *pesticide* residues in organic (*copper residues*) and conventional production have fostered the breeding of disease resistant *cultivars*.

Therefore, most current apple breeding programs are oriented toward reducing the need for pesticides, without losing the high plant and fruit quality. Even though the advantage of a resistant cultivar is evident, resistant cultivars do not yet dominate the market. The reason is simple: after each cross the seeds produced will lead to individuals differing from the mother, father, and each other; each seed will be genetically unique, and if selected will be a new cultivar with considerably different properties from those of the mother and father trees.

This is a relevant limitation in apple breeding; each time we would like to introduce a new trait by breeding, a completely new genotype with new characteristics will be generated. Backcrossing with one of the ancestors is not possible (or only to a very limited extent) in apple breeding, and therefore the original cultivar with the incorporation of the trait of interest, mostly resistance to a pathogen, cannot be re-created. To eliminate unwanted genome segments a very time-consuming *pseudo*-backcrossing with different domestic cultivars is necessary.

Even if modern DNA analysis methods, such as genetic maps, identification of DNA markers linked to traits of interest, and marker-assisted breeding, can help and accelerate the process, a new cultivar with fruits *possessing* a new taste, storage and conservation capacity, and a new tree form and growth patterns will be created. Last but not least this new cultivar also needs acceptance by the consumers, who often stick to a few *favourite* cultivars.

Moreover, if not a single gene (single trait) but several have to be introduced, such as resistance to scab and mildew, to fire blight, or several genes for a single trait (e.g., several resistance genes against a single pathogen to achieve durable resistance), it becomes an almost impossible endeavor. Under these premises, the introduction of specific gene(s) into a particular cultivar which already has all the necessary qualities, except the trait in question, is attractive. Recombinant DNA technology (gene technology) promises to do exactly this.

DEVELOPING THE TECHNOLOGY

Apple was an early target for the emerging recombinant DNA technology. *Agrobacterium tumefaciens*, a disarmed Ti binary vector, and leaf fragments or callus cultures used in the original experiments were and still are the materials of choice. The first transgenic apple plants contained a cassette with the genes for nopaline synthase and neomycin *phosphotransferase* (nptII, conferring resistance to the antibiotic *kanamycin*). The selection of the transformed cells, regeneration, and rooting were made in the presence of the antibiotic. The incorporation of the genes into the genome was confirmed by Southern blot analysis. The nopaline synthase gene continued to be expressed in *greenhouse*-grown plants several months after removal from the in vitro growth conditions.

The transgenic apple plants showed a normal phenotype, except for a somewhat reduced capacity to root, which is an important trait if the transformant is a rootstock but irrelevant if the target is a scion cultivar. Other *laboratories* followed or paralleled the work of James et al., all showing that apple can be genetically transformed and improved, stepwise, the *transformation* efficiency. Several papers summarize various aspects of the methodology.

Technical problems such as stability of the expression were analyzed, often also using foreign genes such as the gene stilbene synthase. As the various cultivars have different requirements for almost all steps, including *regeneration* and *rooting,* several reports deal with special *protocols* for a particular cultivar.

Not only *A. tumefaciens* was used for transformation; some attempts were made with *A. rhizogenes,* mainly driven by the interest in the *rol* genes. Most studies start from wounded leaf sections, following the customary technique ameliorated by *crushing* rather than by cutting. Apical internodal explants from etiolated Royal Gala apple shoots had a high efficiency of producing transgenic shoots.

For transient expression studies protoplasts have been used. For gene activation, in many cases researchers relied on the wellcharacterized constitutively *expressing* 35S promoter from *cauliflower mosaic* virus (CaMV 35S). Some attempts have been made to find and use other promoters.

Although not mentioned specifically in *conjunction* with a target gene, promoters linked with targeted expression patterns have been identified, for example the 940 *extA* promoter which is active in young tissue, two promoters, RBCS3C and SRS1, suitable for the expression of transgenes in green photosynthetic tissues of apple described by Gittins et al., and others with particular expression patterns which are under development and characterization. A major technical problem for genetically modified (GM) apple is the use of the *nptll* selection gene which is still used exclusively.

Recently, some groups started to develop alternatives. As described in the section on herbicide resistance, the herbicide resistance genes are not true alternatives for several reasons, including patent *ownership* restrictions. For obvious reasons the selection system has to be positive and, as stated in the Introduction, in apple removing it through backcrossing is not feasible.

A selection system acceptable to the general public or an alternative post-selection elimination process should be developed. Currently, the most promising alternative to *antibiotic* resistance and probably a more *acceptable* system is the use of the gene *phosphomannose* isomerase (PMI).

PMI-transformed cells are able to use mannose as C source, which the untransformed apple cells cannot. First successes were reported even if the efficiency seems not as good as with the *nptll;* however, by adapting various parameters, similar and even higher *efficiency* can be obtained (Iris Szankowski, personal *communication*). Clearly a "clean vector technology for marker-free transgenic" apples is the ultimate aim. Plant Research International in The Netherlands recently proposed such a technology, and the developers were able to obtain transgenic apple plants free of selection marker genes.

 INSECT RESISTANCE

After establishing the technology, the time came to produce transformants with a potential value. The target had to be a trait which can be sufficiently enhanced through a single or relatively few genes and gives a potentially commercially interesting advantage to the transgenic apple plants. Essential targets were resistance to diseases or pests; however, the main constraint was the availability of functional genes. Resistance to the main insect pests is partial in apple and thought to be regulated by *quantitative*, additive genes and a few single, major genes.

Only the rosy leaf curling apple aphid gene *sd1* is well characterized and mapped. However, neither *sd-1* nor any other resistance gene has been cloned. Therefore, insect pest resistance must rely on genes from other species. A gene encoding a cowpea trypsin inhibitor (*CpTI*) and the gene *crylA(c)* from *Bacillus thuringiensis* encoding an intracrystalline protein were used to transform apple and tested against *Cydia pomonella* (codling moth) attack. *CrylA(c)* transgenic plants have been field grown since *1992*. In the first obtained transformants only very low levels of *CrylA(c)* were expressed; however, using a synthetic version of the gene together with the promoter CaMV *35S*, the problem was *seemingly* overcome.

Currently, the further development of codling moth resistant GM apple is carried out by private nurseries and no published data on the pest resistance of the transgenic apple plants are available. Similar efforts were reported by Cheng et al. who introgressed the Bt endotoxin; however, the intended resistance to *Adoxophyes orana* (tortrix moth) was not induced. Genes expressing the biotin binding proteins avidin and strepavidin were also used to infer resistance to transformants of the cultivar Royal Gala to *Epiphyas postvittana* (the light brown apple moth).

In these experiments between 80 and 90% of the feeding larvae died in a period of 3 weeks, compared to 14% on the control plants. The effect on larval weight in feeding *experiments* made in vitro on tissue-cultured apple shoots was inversely proportional to the *expression* of the two proteins with a threshold for activity.

 FUNGAL DISEASE RESISTANCE

Contrary to insect resistance, fungal disease resistance can be governed by single major genes and by quantitative additive genes. Since the *1980s* great efforts have been made in Europe, the USA, and New Zealand to map the major resistance genes against scab caused by *V inaequalis* and mildew caused by *P. leucotricha*. These two diseases absorb the *majority*, if not the entirety, of the *fungicide* treatments necessary to produce high-quality apples.

However, in the early 1990s, once the apple *transformation methodology* was well established, no apple own resistance gene had been cloned. So the choice of the genes was directed toward foreign genes, with potentially toxic or inhibitory effects on *V inaequalis*. Mildew, although a *dominant* problem in countries like England, was never a target.

Suitable genes from other species were those encoding chitinases and glucanases from the fungus *Trichoderma*, a well-known biocontrol agent of fungal diseases, and those encoding lytic enzymes from *Lepidoptera* and antimicrobial peptides from phages (AMP). In 1998, several studies were presented by the *research* group at Cornell University.

Wong et al. reasoned that, since chitin is a major component of fungal cell walls and higher plants do not produce sufficient chitinases, the constitutive expression of chitinolytic enzymes like endochitinase and chitobiosidase from the biological control agent *Trichoderma harzianum*, which have been shown to have antifungal activity, may increase host resistance to scab.

Two out of three endochitinase (*ech42*) gene transgenic lines of Royal Gala tested showed detectable levels of the enzymes, and were more resistant than the untransformed Royal Gala when micrografted shoots were spray inoculated with scab. Unfortunately no detailed values are reported. Similarly, Bolar et al. reported the transformation of the cultivar McIntosh with the genes *ech42* and *Nag 70*, an exochitinase gene encoding N-acetyl β-D-glucosaminidase, from the biocontrol agent *T. harzianum*.

The expression of *ech42* varied considerably between lines, few reaching high levels. Plant growth was inversely correlated with its expression level. The number of scab lesions and extent of scab sporulation on micropropagated plants in greenhouse tests was relevantly reduced at high levels of *ech42* expression. *Nag 70* gave a lower level of scab resistance, but with no vigor reduction.

The authors hypothesized that pyramiding both genes, each with a relatively low level of expression, may lead to high resistance whilst retaining good vigor. Transgenic lines with both genes (low *ech42* and high *Nag 70* expression) had a high level of scab resistance without significant growth reduction. The authors reported that selected lines were evaluated in the field for resistance, growth, and fruit quality; however, no data are currently published. Aldwinckle et al. reported transfer of native (attacin E) and synthetic (SB-37) genes from the saturniid moth (*Hyalophora cecropia*), *lysozyme* genes from hen egg white, and T4 bacteriophage, all enhancing scab resistance to a variable degree.

A similar program with the same genes was started in 1997 at the Institute for Fruit Breeding in Dresden Pillnitz, Germany, and resulted in several transgenic lines transformed with the lysozyme from the bacteriophage T4 and/or attacin E. ; however, this program concentrated on subsequent effects on fire blight infections. Concurrently, the research group from Leuven (Belgium) led by Keulemans attempted a different approach. As plants produce defense-related proteins, such as PR proteins and *antimicrobial* peptides, constitutive expression of these may enhance *resistance.*

They introgressed, separately, the genes coding for Rs AFP2, an antifungal peptide from radish, and Ace AMP1, an antimicrobial peptide from onion, under the control of the CaMV 35S promoter into the cultivar Jonagold. Expression of Rs AFP2 reached high levels, up to 0.8% of the total protein fraction. Ace AMP1 expression was much lower: up to 0.15%. The proteins retained their antifungal activity after *appropriate* extraction against the test organism *Fusarium culmorum*, a wheat pathogen.

Transformants were micrografted and than chip budded in order to reach a sufficient number of replications for the resistance tests. Artificial inoculation with *V. inaequalis* conidia showed that some lines presented a reduced susceptibility expressed as a lower number of lesions with reduced and retarded *sporulation*. Unfortunately this line of research was discontinued and no accessible publication reporting data on the disease resistance is available. The fourth group involved in producing GM apples is the French group at the INRA Angers led by *Chevreau.*

The Angers station is well known for its apple breeding program, so transgenic apples are considered an additional improvement to the traditional breeding program. One line of research is similar to the one mentioned above, i.e., introgression of endochitinase *ech42* and exochitinase *Nag 70* genes into Galaxy, a scab-susceptible cultivar, and Ariane, a cultivar carrying the *Vf* resistance gene. The theory is that in new commercial cultivars the *Vf* resistance gene requires supplementation by additional resistance factors in order to maintain resistance durability.

The transformed lines obtained were tested with the common genotypes of *V inaequalis,* against which *Vf is* still efficacious, and with race 6 which can overcome *Vf.* Half of the lines showed very limited growth in the greenhouse. The others were tested for scab resistance and two Galaxy lines (with a wild-type growth) were found to be less *susceptible.* Both lines had an increased endochitinase activity and equal, or slightly higher, *exochitinase* activity *compared* to the wild type.

The Ariane lines, including the greenhouse acclimatized plants, had a reduced growth rate. One line, which only had a shoot height reduction of 35%, showed a scab reduction of 25%, whereas another line with 50% shoot reduction achieved an 80% reduction of scab infection. High *exochitinase* activity was in all cases correlated to plant *vigor* reduction.

The *endochitinase* activity in the plants leads to changes such as increase in *peroxidase* and glucanase activity and higher *lignification.* As the authors discuss in detail, these changes have a relevant impact on growth, and therefore it is difficult to evaluate any direct effect on the fungus. *Antimicrobial* peptides from plants may not create the side effect caused by fungal chitinases. *Puroindolines [PinA* and *PinB]* are antifungal *cysteine*-rich proteins present in wheat seeds.

Both genes are characterized and have already been used to transform rice. Transformant lines expressing these genes showed an increased resistance to major fungal *pathogens* of rice. A similar experimental setup to that for the chitinase experiments was used to test the potential of *PinB* in apple. The introduction by genetic engineering of the gene into two cultivars had no effect on growth, *regardless* of the *expression* level, which reached up to 0.24% of the total soluble *proteins.*

The results of the scab susceptibility test are rather curious. Strain 104, race 1, which represents the common *V. inaequalis* population on the commercial cultivars, is not affected by the *PinB* at any expression level; strain EU D 42, race 6, is inhibited progressively with the increasing amount of *PinB.* So the two strains must exhibit differential tolerance to *PinB.*

If the gene is to be used to render *Vf* resistance more durable, it would be of interest to check other strains to know if this effect is correlated to the race or, more likely, to another independent factor. A different approach to obtaining scab resistant plants was attempted by the joint team of the Department of Fruit Tree and Woody Plant Sciences of the University of Bologna, Italy, and the Plant Pathology group at ETH Zurich, Switzerland. Starting from an EU project which offered an excellent linkage map of apple and molecular markers mapped in the region of the scab resistance *Vf* , the positional cloning of *Vf* was initiated.

A contig of BAC clones spanning the region between the two *Vf* molecular markers M18 and

AM19, one on each side of *Vf* , was constructed and allowed the identification of four genes, named *HcrVf1* to *HcrVf4*, coding for receptor-like proteins with a high homology to the *Cladosporium fulvum* (*Cf*) resistance gene family of tomato. The genes have an extracellular leucine-rich repeat domain and a transmembrane domain. The gene *HcrVf2* under the control of the CaMV 35S promoter was introduced into the susceptible cultivar Gala using the *nptII* gene for selection.

First in vitro tests evaluating the progression of the scab infection (penetration and stroma formation by the fungus), and later greenhouse scab inoculations of lines containing a single functional copy of *HcrVf2*, unambiguously demonstrated that the four lines carrying *HcrVf2* were at least as resistant to scab as the conventionally bred *Vf* cultivar Florina.

As the resistance *Vf is* known to be overcome by race 6 and 7, it was of interest to see if the introgressed gene really conferred the same type of resistance, i.e., recognition of the avirulent *V. inaequalis* genotypes and induction of the defense cascade, or the plants were resistant through some *artifact*. Plants transformed with both *HcrVf2* and *nptII*, those transformed with *nptII* only, and wild-type Gala and Florina (*Vf*) were challenged with a field inoculum (mixture of genotypes) known to have the ability to cause scab on Gala but not on *Florina* and with an *inoculum* derived from *Malus floribunda* 821, the original donor of *Vf* .

The *HcrVf2 line* was, as expected, resistant to the field inoculum similarly to Florina, whereas all the other plants showed typical scab lesions with abundant sporulation. Inoculation with race *7* from *M. floribunda 821* resulted in sporulating lesions on all plants; however, the inoculum was less *aggressive* and the sporulation was less abundant than the field *inoculum*.

Moreover, the *HcrVf2* line still retained some resistance, being slightly more resistant than the line transformed with only *nptII* and the original Gala. This experiment demonstrated that resistance in *HcrVf2* transformed lines reflects that of a *Vf* carrying plant.

A further step has been made by identifying the promoter sequence of the *HcrVf1, 2,* and 4 and demonstrating their functionality. Current work is focused on testing *HcrVf1, 2,* and 4 under the control of their own promoter and *nptII*, and on the testing of an alternative selection system (unpublished results).

SELF-INCOMPATIBILITY

A further area of interest is the fertility of *Malus;* each genotype, i.e., cultivar is self-infertile. Fruit setting depends, therefore, on insects which have to bring foreign pollen to the flower. For that reason orchard planning must include plants of another cultivar or wild *Malus* at regular intervals (pollinators). Under adverse climatic conditions pollination may be hampered and can lead to *irregular* fruit production.

Self-incompatibility (SI) is controlled by a single multiallelic locus (S). When one of the S alleles matches one from the other partner, pollen tube growth is blocked. Several of the S alleles have been cloned and sequenced. Silencing this gene should allow self-pollination.

The research group in Leuven transformed the cultivar *Elstar* with one of its own S alleles (S3) with sense, antisense, and incomplete antisense sequences. Some lines with the sense and the incomplete antisense sequences showed good self-pollination. Three years of field trials showed that the production of fruits from two *transgenic* lines produced by selfing was *comparable* to that reached with cross-pollination. The inactivation of the SI *mechanisms* was also achieved by suppressing the S RNA expression in Elstar, rendering both alleles S3 and S5 dysfunctional. The two lines had no negative *feature* and were identical to the wild-type Elstar.

A self-fertile cultivar can improve and facilitate apple production and breeding, so this is a trait of great relevance. However, the use of the selection gene *nptII* and the CaMV 35S promoter may make the two lines unacceptable for practical use.

HERBICIDE RESISTANCE

Herbicide resistance may have two applications: it could be used as an alternative selection system to the overwhelmingly used *nptII* and as a practical application in nurseries during rootstock production. In fruit production there is little advantage, as herbicides damage trees only in exceptional situations and when normal safety precautions are not observed.

The gene for resistance to the herbicide bialaphos (bilanafos) (bar, encoding phosphinothricin acetyltransferase) was introduced into *M. prunifolia* var. *ringo*. The herbicide resistance also served as the selectable marker. Transgenic plants were fully resistant to bialaphos. The transformation and selection of the phosphinotricine (PPT, Basta) resistant apple clonal rootstock No. 545 offered some interesting information.

PPT compared to *nptII* was inefficient as a selectable marker, giving a high frequency of escapes and a transformation frequency of 1.3% compared to 3.5-14% obtained with *nptII* and the same rootstock. However, once an *appropriate* line was selected, the plants were resistant to the commercially applied dosages of Basta.

FIRE BLIGHT RESISTANCE

Probably the first and most important target of transgenic apple was fire blight resistance, which was pioneered by the Cornell University group led by Aldwinckle. As for the experiments performed to infer resistance against fungi and insects, several genes with different characteristics were used to *transform* apple cultivars and rootstocks with the aim of increasing their resistance to fire blight.

Attacin A and E genes were used to transform the rootstock cultivar Malling 26. In both in vitro and greenhouse trials the plants from the transformed line were more resistant. The GM plants supported a tenfold higher *inoculum* dosage before reaching the 50% *lethal* factor.

GM lines of the cultivar *Galaxy*, with the genes for attacin E and T4 lysozyme, have increased resistance to fire blight. A set of German cultivars (Elstar, Pinova, Pilot, Pingo, Pirol, and Remo) have been similarly transformed with T4 lysozyme gene.

Unfortunately, we found no further report on the outcome of the mentioned fire blight resistance tests. Norelli et al. overviewed the ongoing work and additionally reported transformation of Gala, Royal Gala, and Galaxy with the cecropin analogs SB 37 and Shiva 1, and a hen egg white lysozyme (HEWL). However, the best results were obtained with the attacin E.

One particular line displayed only 5% shoot blight compared with approximately 60% of the untransformed Royal Gala control plants in 1998 field trials. A short overview of the field trials with the various *lytic* protein transgenic lines was presented by Aldwinckle. Instead of using the constitutive promoter CaMV 35S, Liu et al. genetically engineered into Royal Gala a modified cecropin SB37 gene (MB39), fused to a secretory coding sequence from barley alpha amylase and placed under the control of a wound-inducible osmotin promoter from tobacco. Seven *diploid* transgenic lines were produced (*transformation* efficiency of 1.7%) of which three lines, all having multiple insertions, had a higher *proportion* of shoots with low infection scores than Royal Gala.

A similar approach was attempted by the German team of the Max Planck Institute in Lenburg and the Institute for Fruit Breeding in Dresden. The extracellular polysaccharide (EPS) amylovoran produced by *E. amylovora* is an essential pathogenicity and survival factor, building a capsule around the bacterium.

E. amylovora bacteriophages produce an EPS depolymerase to degrade this capsule to finally reach the bacterium. The EPS depolymerase gene was inserted into an apple rootstock JTE *H* by Agrobacterium-mediated transformation and nptII as selection gene. In the first pathogenicity tests, four out of five transgenic plants were completely resistant to fire blight.

A further extension of the work used the cultivar Pinova as the host plant. In vitro, 61 out of 83 transgenic lines were significantly more resistant to fire blight. Some transgenic lines were transferred to greenhouse conditions and tested for increased resistance. The authors do not report any data, but state that the correlation between the in vitro data and the greenhouse data is a mere $r = 0.5$.

The same gene has been used in pear and only two out of 15 lines showed consistent increase of fire blight resistance in vitro and in greenhouse tests. The two lines were also those with the highest EPS depolymerase expression. The pathogen-induced plant resistance approach starts from the theory that the pathogen secretes substances which are recognized by the host and may initiate the defense cascade.

It has been shown that by applying these substances prior to inoculation with a pathogen, an incompatible reaction can be induced. The *E. amylovora* effector protein, harpin, when sprayed on flowers protects against subsequent *E. amylovora* infection, probably by inducing the systemic acquired resistance responses. However, high dosages lead to cell death.

The gene encoding for harpin N (*hrpN*) under the control of the promoter Pgst1, which is induced in *E. amylovora* challenged leaves and under the weak constitutive promoter *nos* with and without a signal peptide (SP) which should direct the harpin N to the intercellular space, were introduced into the rootstock M.26.

Transgenic plants expressing *hrpN* under the control of the *nos* promoter have been obtained as well as a few lines under the control of Pgst1. In an overview Aldwinckle reports that some lines showed an increase of resistance in the field. Also, the cultivar Galaxy has been transformed with the constructs *Pgst1-SP-hrpN.*

Field trials for fire blight resistance are currently under way and it will be highly interesting to know the outcome. In all the above reported approaches to create a fire blight resistant plant, the resistance induction was entrusted to *a non-Malus* gene. Besides the harpin N, *E. amylovora* also produces the pathogenicity effector protein *dspE,* which interacts directly with four leucine-rich repeat (LRR) receptor-like serine/threonine kinases of apple (DIPM).

If this interaction does not take place *E. amylovora* will not be able to infect that host, i.e., in the absence of the host DIPM receptor proteins the specific plant is a nonhost. Constructs with complete or partial hairpin *DIPM* gene sequences controlled by the CaMV 35S promoter were made. Several GM lines were recovered and in most the corresponding *DIPM* genes were silenced. However, no line with all four *DIPM* genes silenced was recovered.

Preliminary fire blight resistance tests in growth chambers indicate that some lines are more resistant to artificial infection of the shoot tips. As the authors point out, "resistance due to silencing of a native apple gene(s) is likely to be more acceptable to regulators, growers, and consumers than the addition of any foreign genes." Overexpression of an apple own gene, linked to defense, was proposed by Aldwinckle.

The *NPR1* protein in *Arabidopsis* is linked to the defense reaction and when overexpressed enhances resistance to *Pseudomonas syringae* and *Peronospora parasitica*. Aldwinckle's group incorporated the apple ortholog *MpNR1* with the promoter Ppin2 (*E. amylovora* induced) into Galaxy and M.26. Two transgenic lines of Galaxy were clearly more resistant to fire blight, with 36% compared to 93% (controls) of the shoots blighted in a growth chamber test with potted plants.

FRUIT RIPENING

When discussing the pros and cons of replacing popular cultivars with high susceptibility to diseases with new cultivars possessing improved disease resistance, one of the principal arguments against any change is the lack of adequate storage and shelf life. A number of genes have been identified which play a determinate role in the ripening and softening of fruit; however, ethylene remains a key factor.

Ethylene is synthesized in plants from S-adenosyl methionine with the aid of two enzymes: 1-aminocyclopropane1-carboxylic acid synthase (ACS), which converts S-adenosyl methionine to 1-aminocyclopropane carboxylic acid (ACC), and 1-aminocyclopropane carboxylic acid oxidase (ACO), which oxidizes ACC to ethylene. Hrazdina et al. transformed Royal Gala with an antisense sequence of one of the ACS genes.

Transgenic lines were planted in the field and evaluated. GM plants developed normally and the fruits were, in most lines, similar to the nontransgenic controls. There was no significant decrease in ethylene production per fruit, nor in the other relevant parameters (soluble solids, pH, and penetration force) when the average value of the various parameters of all plants of a line was considered.

However, the authors write that "on the percent basis ethylene production in apples from the individual trees from different transgenic lines was downregulated after 69 days of storage by 11 to 97% with the majority of fruits falling in the 60–80% range." Apparently the time course of ethylene presence was changed so that, at the last evaluation date, the fruit from the transgenic lines produced only half the amount of ethylene compared to the control fruits.

The authors infer that the data indicate that the structural integrity of the fruit may be maintained simply under refrigerated storage and controlled humidity. Similar results were achieved by Galli et al. with fruit from lines of transgenic Royal Gala with downregulated ethylene synthesis. The fruits remained firm and had an increased resistance to shriveling, splitting, and spoilage following extended storage at room temperature. In a very careful and accurate study the research group of the Department of Pomology, University of California, Davis, evaluated the impact of the suppression of ethylene on the fruit quality using transgenic apple trees with ACC synthase or ACC oxidase enzyme activity suppression.

Sucrose and fructose levels were lowered, malic acid degradation was reduced, and the volatile aroma ester and alcohol fractions were similarly reduced. In a collaborative work with Horticulture Research International, East Malling, England, some incongruence still remains, as the results suggest that sugar and acid composition are not under the direct control of ethylene and alcohol volatiles seem not to be influenced.

However, for practical application, the firmness of the fruits from transgenic lines with suppressed ethylene biosynthesis remained almost constant after storage (shelf life), whereas in the control Greensleeves fruits it decreased dramatically. These studies can be viewed from an application point of view but they also contribute highly to our understanding of the ripening processes in apple.

Others do not have such an evident potential of application, but merit being mentioned as they contribute to basic knowledge. Cheng et al. demonstrated the plasticity of the apple photosynthesis system by using antisense inhibition of sorbitol synthesis in GM apple; Kanamuaru et al. were able to determine that S6PDH is a key enzyme regulating partitioning between sorbitol and sucrose in apple leaves. Atkinson et al. overexpressed polygalacturonase and obtained a range of new phenotypes, altering leaf morphology, plant water relation, stomata structure and function, as well as leaf attachment.

Underexpression of polyphenol oxidase (PPO) (catechol oxidase), the enzyme responsible for enzymatic browning of apples, by use of an antisense PPO gene clearly led to reduced calli browning and shoots had a similarly lower tendency of browning through the PPO activity.

ALLERGENS

Malus has four allergenic proteins, Mald 1 to 4. They cross-react with the birch pollen Bet v 1 specific IgE antibodies, so that an apple allergy is common in patients with a birch pollen allergy. Apple cultivars have different allergenic potential, Golden Delicious being a cultivar with a high potential. A hairpin construct from Mald 1 was introgressed by *Agrobacterium*mediated transformation with the selection gene nptII in the cultivar Elstar.

Six plantlets displayed a significant reduction of Mald 1 production (at least tenfold) and induced significantly less reaction in patients than the control plantlets. Specific RNAi silencing can therefore solve a diet problem for birch pollen allergenic patients.

ROOTING ABILITY

Vegetative propagated plants depend on a high rooting capacity or have to be grafted on rooted "rootstocks". Clearly such a rootstock cultivar has to have, besides its specific growth characteristics (mostly vigor leading to dwarfing or allowing a high tree), a good rooting ability. Apple *rootstocks* are propagated by stool layering, seldom by rooting of cuttings as some are recalcitrant to root from cuttings even with the use of auxin.

Root-inducing genes have been characterized in *Agrobacterium rhizgenes* (*rolA*, *rolB*, and *rolC*), and contribute to causing "hairy root" disease in the host.

The bacterium introduces parts of its DNA containing the *rol* genes into the host plant, which reacts by producing additional roots and often assuming a dwarfing stance. Almost all transgenic works with the *rol* genes have their origin at the Department of *Horticulture* at the Swedish University of Agricultural Science in Alnarp.

Transforming the apple rootstock M.26 with *rolA* controlled by its own promoter resulted in four lines with variable growth reduction and wrinkled leaves. Incorporation of the *rolB* gene, also under the control of its own promoter, into the rootstock M.26 increased auxin sensitivity and rooting ability. To think of commercialization of a rootstock with increased ability to root, it is necessary to demonstrate that no negative effect will be transmitted to the graft scion.

Zhu and Welander used the cultivar Gravenstein as a test scion. Its growth characteristics were not influenced under unlimited nutrient conditions. The specific root length of the M.26 rootstock was significantly reduced, from which it can be speculated that under limited nutrient supply, as in orchard conditions, it may induce tree dwarfing. M9 is another very popular rootstock with excellent dwarfing characteristics; however, it roots badly from cuttings. GM M9 with the *rolB* gene roots extremely well.

In vitro rooting ability went from almost nil to almost 100% with 3.5 to 9.5 roots per cutting. Also, the rootstock Jork 9 does not root readily from cuttings. Incorporation of *rolB* dramatically increased the rooting of shoots in the absence of externally added auxin. The control untransformed shoots were able to reach almost the same rooting ability with the addition of indole-3-butyric acid (IBA).

The rootstock Jork 9 and the various transgenic lines have been used to study plant response to the rooting process, natural or auxin induced. Welander et al. report that "the permission for field trial on the transformed rootstocks has been obtained from the Board of Agriculture in Sweden."

Plants of different transformed clones and the untransformed controls of M.26 and M9/29 have been produced and grown in the field for 2 years. Five apple cultivars commonly used in Sweden and Europe have been budded onto the rootstocks to evaluate the influence of transgenes on the growth and development of the grafted cultivars.

The *rolC* gene has been introduced into a Japanese rootstock (*M. prunifolia* var. *ringo* Asami Mo 84 A), which roots well but has not the desired dwarfing ability.

The aim was to determine whether *rolC* can reduce growth characteristics without altering the rooting characteristics. Some transformants had shortened internodes, some reduced height, some both and some were normal; rooting was intensive in some transformants. The authors state that a few lines may be suitable candidates for dwarfing rootstocks.

Radchuk and Korkhovoy transformed the scion cultivar Florina with *rolB* with the theory that multiplying and growing it on its own roots would reduce costs and accelerate production of plant material. They obtained various lines with enhanced rooting ability.

From the data for 2 years of greenhouse experiments they report no change to the above-ground growing characteristics compared to those of the original untransformed Florina.

ACCEPTANCE AND RISK ASSESSMENT

Apple is vegetatively propagated and seeds only have a role in breeding. Transformation of the apple into a weed through a selective advantage gained from a foreign gene is not plausible under any imagination. Environmental risk is restricted to the gene products and is not inherent to apple. Apple is a fresh product often consumed raw, and therefore consumers are particularly attentive to any manipulation.

Currently it is improbable that foreign genes will be acceptable under European law where transgenics are highly regulated and have to be declared as such. So the arguments listed in 1996 by Koller et al., leading to the opinion that transgenic apple carrying foreign DNA will not be commercialized in the near future, are still valid. Almost all work cited in this review, excluding the few specially mentioned exceptions, relied on the selection gene *nptII,* on non-apple gene promoters (CaMV 35S amongst others), and on *Agrobacterium tumefaciens-mediated* transformation. Often, for experimental purposes, genes not influencing the target trait were used, such as the gene producing β-glucuronidase (GUS).

To our knowledge no environmental risk studies specific to apple have been published. The only argument studied was the possibility that particular insects may become resistant to the Bt toxin. The development of Bt resistance was analyzed using simulation experiments in the agrosystems apple and clover, both hosts for various leafrollers.

Probably the researchers are still concerned with producing acceptable GM apple cultivars with a commercial interest and possibly having environmental benefits, such as reduction of pesticide use. However, an in-depth discussion should be conducted on any gene used.

For example, genes expressing the biotin binding proteins avidin and strepavidin, as indicated above, and used to infer insect resistance in fruits would also bind biotin for the consumer and may lead to reduced vitamin B absorption similar to a surplus of egg consumption. What is acceptable, however, is debated and some argue that only applederived genes will have the possibility of encountering the favour of the producer and consumer.

In this context the research by Kassardjian et al. on purchase behaviour is interesting, mainly as a guideline that includes innovative ideas on how to elicit consumers' willingness to purchase GM apples. The goal of their research was to try a new methodology (thought-listing technique and questionnaires) to elicit consumer willingness to pay for GM food in New Zealand.

However, it is not that the researchers are inactive; in popular journals, interviews, and web sites arguments for and against the use of GM apples are presented and debated.

CONCLUSION

In the next few years many "apple own" resistance genes will be sequenced and transferred into some test cultivars, probably under the control of their own promoters. RNA interference technology will be able to block some unwanted traits. Pathogen-derived genes inducing host resistance will be available. Apple own promoters expressing genes only where and when desired will no longer be a mirage but will appear slowly on our horizon.

With the "clean vector technology" allowing the removal of the selection marker, it will be possible to produce cisgenic apples, i.e., apple plants which will be modified exclusively with *Malus* genes and controlled by *Malus* promoters. Many of the discussions about risks, phantom or truly demonstrated, will be obsolete.

However, the insertion site will not correspond to the original site of the gene and this could lead to epigenetic effects. Major effects are readily discovered and such lines are discharged; subtle effects are more difficult to discover, so we have to strive to devise smart experiments allowing for selection of lines which harbor no surprises. The apple cropping system based on artificial vegetative multiplication of the particular genotype and its planting over large areas has rendered this crop susceptible.

The possibility offered by DNA recombinant technology can be used to replace nonfunctional resistance alleles by not-overcome alleles (*gene therapy*). The benefits of GM apple resistant to various diseases will be real, not only for the owner of a patent but also for the producer, environment, and consumer. The reduction of fungicide and, in some parts of the world, antibiotic use alone justifies all the efforts. It remains to be seen how long it will take until a broad acceptance by the public is achieved.

8 Chapter

TRANSCRIPTIONAL FACTORS

Exposure to microbial pathogens results in massive transcriptional reprogramming of the plant genome. *Historically*, genes and proteins induced in response to pathogen challenge were labeled as pathogenesis-related (PR). At least 14 classes of PR proteins are recognized, several of which have the potential for direct *antimicrobial* activity, including chitinases, glucanases, and cationic peptides.

In general, ectopic (over) expression of individual or pairs of *PR* genes in transgenic plants does not substantially augment disease resistance, and there has been no reported characterization of *PR* gene mutants. Accordingly, the specific contribution of *individual* PR proteins to disease resistance remains elusive. The *magnitude*, complexity, and dynamics of *pathogen* challenge on plant gene expression are currently being revealed by genome-wide *transcript* profiling studies.

In addition to classical *PR* genes, several hundred, if not thousands of genes encoding products implicated in almost every aspect of plant physiology have been shown to be affected. Such changes in gene expression likely represent a combination of plant defense and disease *susceptibility* responses. Although the precise function of most genes modulated by pathogen challenge remains unknown, it is clear that the timely, coordinated *transcriptional* control of large sets of genes is crucial for plant disease *resistance*.

Transcription of a gene is ultimately determined by the combination of the *cis*acting transcriptional regulatory elements that it possesses and the repertoire of active *trans-acting* transcription factors (TFs) present in the cell. Whereas a gene's complement of cis-acting elements is "*hardwired*" in the genome, the abundance and activity of numerous TFs are modulated by signal *transduction* events initiated following pathogen recognition by the plant cell. Such TFs will include key regulators of the plant's inducible defense responses against pathogens and are the subject of this chapter. I will review how these TFs were identified and are functionally analyzed and regulated following pathogen challenge. To place the *discussion* into proper context, a brief overview of different plant defense systems and the importance of *model* systems in the study of TFs *mediating* plant defense responses are first *provided.*

Three broad types of trans-acting TFs can be distinguished: general (or basal) TFs, *sequence-specific* TFs, and cofactors. With the exception of the large body of literature on the cofactor NONEXPRESSOR OF PATHOGENESIS-RELATED GENES 1 (NPR1), most studies have focused on the sequence-specific *DNAbinding* TFs.

Accordingly and unless otherwise stated, these will be referred to here simply as TFs. Whenever possible, I reference recent reviews or representative original publications in an effort to minimize the number of references.

INDUCIBLE PLANT DEFENSE SYSTEMS

Plant defense against pathogen attack involves recognition of *pathogen*-derived molecules called *elicitors.* One of the best characterized defense systems, *race-specific* resistance, is governed by genetic interactions between genes encoding a class of elicitors called avirulence factors and plant resistance (R) genes. *Avirulence* factors are *polymorphic* among *isolates* of a single pathogen species and trigger a very rapid and specific response in plants expressing the corresponding R gene product.

In these cases, *host-pathogen interaction* is said to be incompatible, the pathogen is avirulent, and the plant host is resistant. Absence of specific genetic recognition results in a compatible interaction, in which the *pathogen* is said to be virulent and the host susceptible, and *disease* ensues.

Under these circumstances, plants rely on a basal defense response triggered by the recognition of pathogen-associated *molecular* patterns (PAMPs) that are conserved among several *microbial* species. Basal defense responses are not as specific or *rapid* as those mediated by R-gene-avirulence gene recognition. They do not prevent disease but restrict pathogen spread. Accordingly, *mutations* in components of the basal defense system result in *hypersusceptibility* to virulent pathogens.

Regardless of whether pathogens are detected through avirulence *determinants* or PAMPS, the signaling events that are triggered rapidly converge into a limited number of interacting pathways, or networks, that rely on small molecules, including salicylic acid (SA), jasmonic acid (JA), and ethylene (ET), as *secondary* messengers. *SA-dependent* signaling is required for resistance to certain pathogens that derive energy from living host cells (*biotrophs*; see *Oliver* and *Ipcho* for *definitions*) and for *mediating* a type of broad-spectrum, inducible disease resistance known as systemic acquired resistance (SAR).

Pathogen activation of the *PR* genes *PR-1, PR-2,* and *PR-5* depends on SA, and these genes serve as marker genes for SA-dependent signaling events. JA and ET *signaling* are generally required for resistance to necrotrophic pathogens (i.e., derive energy from killed cells) and for an SA-independent form of induced systemic resistance called ISR.

Expression of *PDFJ.2,* which codes for a *cationic antimicrobial* peptide, is often used as a marker for JA signaling in *Arabidopsis,* while expressions of *PR-3* and *PR-4* are popular markers for ET-mediated signaling in tobacco.

The SA- and JA/ETdependent signaling *pathways* appear to interact in a complex fashion; the primary mode of interaction is mutual *antagonism.*

However, examples of positive interactions have also been reported. Interpathway communication has been speculated to help plants fine-tune and prioritize defense *responses* upon *encountering* multiple signals.

As is discussed later, it is common for TFs to be regulated by multiple signaling *molecules,* suggesting that they play an important role in integrating different signaling *pathways.*

Table 8.1: Transcription Factors Implicated in Mediating Plant Defense Responses

Family name	*Family size (Arabidopsis)*	*Cognatecis-element*	*Comments*
ERF	56	GCCGCC (CGG-box)	Part of AP2/ERF superfamily; contains single -60 a.a. DNA-binding domain consisting of three-stranded β-sheet with parallel running a-helix; binds DNA as monomers
WRKY	74	(T)GACC/T (W-box)	Contains one or two -60 a.a. DNA binding domain consisting of fourstr-anded β-sheet and novel zinc binding pocket; binds DNA as monomers
Whirly	3	GTCAAAAA/T(elicitor	Highly conserved β-sheet surface response element [ERE]) within family members (Whirly domain) involved in DNA binding; acts as tetramers; bind single stranded DNA
TGA	10	TGACGTa *(as-1* element)	Part of bZIP superfamily; extended α chelix contains basic DNA-binding domain and leucine-zipper dimeriza-tion motif; act as homo- and heterodi-mers
R2R3MYB	125	Various	Largest group of plant MYB superfa-mily; contains two -52 a.a. domains with helix-turn-helix structure

Model Systems

Early model systems for studying TFs mediating defense responses included tobacco, potato, and cell suspension cultures from parsley and soybean. The rapid and efficient genetic transformation systems in *tobacco* were ideal for functional testing of candidate TF genes. Protoplasts and cell cultures were also convenient systems for functional assays based on transient or stable expression of foreign genes.

Furthermore, they provided large amounts of relatively uniform starting material well suited for studying physiological and biochemical responses induced by the addition of defined chemical elicitors of defense responses. The availability of a well-defined chemical elicitor was a contributing factor in stimulating research on TFs in *potato tubers*. These early model systems also provided ample starting *material* for gene *isolation* and *biochemical* analysis of TFs.

Well-defined host-pathogen interactions were also critical for the establishment of model systems. The characterization, in the 1980s and early 1990s, of several host-pathogen systems involving *Arabidopsis thaliana* permitted application of this popular model species to the study of molecular plant pathology. The genetic and genomic resources available in *Arabidopsis*, including the full genome sequence, are particularly relevant to studies of TFs. Because TFs regulate gene expression on a genome-wide basis, *Arabidopsis* resources maximize the potential for identifying and characterizing target genes.

Many TFs are encoded by large multigene families. Accordingly, the availability of annotated fullgenome sequence information greatly *facilitates* identification of closely related family members with potentially redundant functionalities, while resources for reverse genetics allow identification and analysis of mutations in such genes. The large collection of *Arabidopsis* mutants affected in defense responses also represents a formidable analytical tool.

SEQUENCE-SPECIFIC TRANSCRIPTION FACTORS

Isolation

The major classes of TFs implicated in mediating plant defense gene expression are listed in Table elsewhere in this chapter. This section summarizes some of the approaches exploited to isolate them. Detailed information on structure and evolution of the TF families is provided in the following section. Each of the methods *described* here has specific *advantages* and *disadvantages*.

Those based on DNA-binding, protein-protein interactions, and mutant phenotypes do not require a priori knowledge of gene or protein sequence and do not rely on differential gene expression. With the exception of genetic *approaches*, none of the methods listed relies on the recovery of recognizable mutant phenotypes. However, none of the methods, by itself, provides as *convincing* evidence for implicating a TF in mediating defense responses as does the *genetic* approach.

Several of the methods can be applied to model species as well as crops. *Genomic* approaches and those based on sequence similarity are well suited for identifying from crop species putative *homologs* of TFs initially characterized in model systems.

DNA Binding

The ability to bind short stretches of DNA of defined sequence with high affinity is a characteristic feature of TFs and has been exploited extensively as a means of isolating these proteins and their corresponding genes.

The founding members of all the families of TFs listed in Table elsewhere in this chapter, with the exception of MYB proteins, were isolated based on their ability to bind to cis-acting elements required for gene expression in response to microbial pathogens or signaling molecules such as SA, JA, or ET. Southwestern hybridization, which involves screening expression libraries with radiolabeled oligonucleotide probes containing consensus cis-acting DNA sequences, was *exploited* to isolate the ethylene response factors (ERFs), WRKY proteins, and different classes of basic *leucine* zipper (bZIP) proteins, including TGA *factors* and G/HBF-1, as well as *homeobox proteins*.

Biochemical purification of proteins capable of binding elicitor responsive or silencing elements found in the potato *PR-10* promoter was used to isolate a Whirly *protein* and the novel silencing element *binding* factor (SEBF), respectively.

Partial peptide sequencing of *purified* proteins was subsequently used to clone the corresponding genes by degenerate PCR. The yeast one-hybrid system has also been exploited to isolate TFs, including MYB and MYC type proteins, based on their ability to bind specific *cis-acting* elements.

Genetic Approaches

The isolation of mutants compromised in their response to pathogens or the growth regulators SA, JA, and ET represents the most productive strategy for identifying genes that regulate plant defense responses. However, only a small fraction of the genes recovered using this approach encodes TFs.

Examples include the *Arabidopsis LESION SIMULATING DISEASE]* (*LSD]*), which encodes a novel zinc finger protein; *BOTRYTIS SUSCEPTIBLE]* (*BOS]*), which encodes an R2R3MYB protein; and *JASMONATE INSENSITIVE]* (*JINI*), which encodes an MYC protein.

Many plant TFs belong to large multigene families. Consequently, it is very probable that functional redundancy between related members of a family limits the applicability of conventional genetic screens as a means of isolating TF genes. This is well illustrated by a recent study of *Arabidopsis* TGA TFs, which revealed that knockout of three related genes (*TGA2, TGA5*, and *TGA6*) was required before differences in PR gene expression and disease resistance were apparent.

Alternatively, loss of TF function may be lethal to the plant in certain instances. Mutagenic strategies based on ectopic gene expression, such as activation tagging, may overcome the limitations of conventional, loss-of-function genetic screens for identifying TF genes mediating plant defense responses.

Protein-Protein Interaction

Protein-protein interaction screens, such as the yeast two-hybrid system, using components of signal transduction pathways mediating plant defense responses have also led to identification of TFs. In many cases, the TFs recovered (or closely related factors) had been previously identified by other means. Nevertheless, their physical interaction with proteins implicated in mediating defense responses provided valuable functional information.

Examples include TGA factors found to interact with NPR1 and the ERF proteins Pti4/5/6 recovered on the virtue of their ability to interact with the tomato R protein Pto.

Gene Expression

Methods used to identify genes based on their differential expression following pathogen infection or elicitation, including differential screening, suppressive subtractive hybridization, differential-display reverse transcription-PCR, and cDNA amplified fragment length polymorphism (cDNA-AFLP), have also led to identification of TF genes.

Furthermore, large-scale transcript profiling studies, including those making use of microarrays, serial analysis of gene expression (SAGE) or massively parallel signature sequencing (MPSS), are revealing that large numbers of TF genes are differentially expressed under these conditions. Data from many of these studies are available in public databases and tools facilitating the search of these databases are emerging.

In addition to the TFs listed in Table elsewhere in this chapter, results from large-scale transcript profiling are implicating many other classes of TFs in plant defense responses.

Sequence Similarity

The isolation of relatively few TF genes using the preceding methods enabled rapid cloning of numerous genes encoding related factors on the basis of sequence similarity. Methods exploited for these purposes include screening libraries at reduced stringency with DNA probes or with degenerate oligonucleotides, as well as PCR with degenerate oligonucleotides.

Genomic Approaches

Large-scale expressed sequence tag (EST) projects under way in most major crops are now identifying large numbers of TF genes expressed following infection with several pathogens of economic importance. Furthermore, whole *genome* sequencing efforts in *Arabidopsis*, rice, and, more recently, poplar, are revealing the entire repertoire of TF genes in these organisms.

Although the rice *genome* is close to three times the size of the *Arabidopsis* genome, both encode similar numbers of TFs (1300 to 1500).

Both species also contain large families of genes encoding classes of TFs implicated in mediating defense responses, such as ERF, WRKY, and R2R3 MYB proteins. The sizes of other gene families, such as those encoding Whirly and TGA factors, are considerably smaller.

It is noteworthy that, even in *Arabidopsis*, only a small fraction of the total number of TF genes has been functionally characterized. Some families of TFs, such as the R2R3MYB proteins, contain members with very diverse functions; only a fraction of genes within these families may be involved in mediating defense gene expression.

Conversely, some classes of TFs may be specialized for modulating gene expression in response to pathogen challenge.

This may be the case for the WRKY family, in which 49 of 72 *WRKY* genes tested were found to be regulated in response to pathogen infection or SA treatment.

Structure and Evolution of Transcription Factors

Transcription factors are modular, consisting of one or more separate DNA-binding and effector (i.e., transcriptional activation or repression) domains. DNA-binding domains are by far the most conserved portion of TFs and are commonly used as the basis for classifying these proteins. Outside the DNA binding domains, family members may have little or no sequence similarity.

The ERF Family

ERF proteins contain a novel DNA-binding domain of about 60 amino acids previously identified in the product of the floral homeotic *gene APETALLA 2 (AP2)*. Accordingly, AP2 and ERF TFs are commonly grouped into a single family (AP2/ERF family) that includes three other *subfamilies*: the *dehydration*-responsive element-binding (DREB) factors, related to ABI3/VP1(RAV), and others.

To date, only members of the ERF subfamily have been implicated in plant defense responses. Based on sequence similarity, the ERF family can be further divided into a number of subclasses. The structure of the AP2/ERF domain from the *Arabidopsis* ERF1 protein in complex with its targetDNA was solved by NMR. It consists of a three-stranded antiparallel β-sheet and one α-helix running almost parallel to the β-sheet. DNA contact is achieved through *arginine* and tryptophan residues in the β-sheet of the ERF1 monomer.

Whereas amino acid residues making contact with the DNA are highly conserved among ERF proteins, they are not present in AP2 proteins, reinforcing the view that AP2 and ERF proteins are distinct. The *three-dimensional* structure of the ERF domain is related to those of the Tn916 and 2 integrases and the human methyl-CpG binding domain MBD, even though these proteins do not appear to share any *amino acid* sequence similarity. The ERF/AP2 domain was *initially* thought to be plant specific.

However, genes capable of encoding this domain have recently been identified in the ciliate *Tetrahymena*, the cyanobacteria *Trichodesmium erythraeum*, and two bacteriophages. None of the nonplant AP2/ERF domain proteins are TFs. Instead, they are predicted HNH homing *endonucleases*, a class of proteins with catalytic and DNA-binding activities responsible for the lateral transfer of intervening sequences from genes into cognate alleles *lacking* them.

It has been proposed that AP2/ERF TFs originated through lateral transfer of an HNH-AP2/ERF homing endonuclease gene from bacteria or *bacteriophages* into plants.

The WRKY Family

WRKY proteins contain one or two conserved domains of approximately 60 amino acids harboring the conserved sequence WRKYGQK at its N-terminal end and a novel zinc finger-like motif. This family of TFs was originally classified into three groups; however, reclassification into five groups has recently been proposed based on comparison of a more extensive set of proteins. As revealed by NMR, the solution structure of the C-terminal WRKY domain of the *Arabidopsis* WRKY4 protein consists of a four-stranded antiparallel β-sheet with a novel *zinc-binding* pocket at one end.

Proper folding of the domain depends on the presence of zinc ions, which had previously been shown to be required for the DNA-binding activity of WRKY proteins *in vitro*. The hallmark WRKYGQK motif is localized within the N-terminal-most β-strand and has been proposed to be involved directly in DNA binding. Based on partial structural similarity, an evolutionary relationship was proposed to exist between WRKY domains and the drosophila GCM TFs. Similar to ERF proteins, WRKY proteins were originally thought to be plant specific, but have recently been identified in other eukaryotes. More specifically, WRKY like domains have been identified in the slime mold *Dictyostelium discoideum* and the unicellular protist *Giardia lamblia.*

They have also been reported in ferns, mosses, and green algae. No functional information is available on *WRKY* genes from nonplant and lower plant sources. Thus, it appears that the structural framework of DNA-binding domains from so-called plant-specific TFs were in fact established before *divergence* of the plant kingdom.

The Whirly Family

The potato Whirly factor StWhyl is the only sequence-specific TF from plant for which the crystal structure has been solved. The active TF consists of four protomers packed perpendicularly against each other.

Each protomer is made up of two antiparallel b-sheets and a helix-loop-helix motif. Interaction of protomers occurs through this motif, with the b-strands protruding outwards and resulting in the whirligig appearance that inspired the family name.

Sequence comparison between Whirly-like proteins from different plant species indicates that the region forming the β-sheet surface of StWhyl is the most highly conserved and has been named the Whirly domain.

his region is thought to be the major surface involved in binding DNA. Interestingly, the StWhyl tetramer binds to single-stranded DNA and StWhyl shares limited sequence similarity with single-stranded binding proteins from a number of sources.

However, genes encoding Whirly proteins have only been found in higher plants (angiosperms and gymnosperms) and the unicellular green algae *Chlamydomonas reinhardtii.*

The TGA Factor Family

TGA factors are a class of bZIP TFs originally isolated based on their ability to bind to the SA-, JA, and auxin-inducible *activating sequence-1* (*as-1*) element found in the cauliflower mosaic virus 35S promoter or the related ocs element in the octopine synthase promoter.

As such, this class of factors is occasionally referred to as ocs-element binding factors (OBFs). Although no proteins closely related to TGA factors are found outside the plant kingdom, the bZIP domain is found in TFs from all eukaryotic kingdoms. bZIP proteins typically function as homodimers and/or heterodimers. When bound to DNA, the bZIP domain of each monomer exists as a contiguous a-helix.

The Nterminal basic region consists of approximately 16 amino acids that bind in the major groove of double-stranded DNA. The C-terminal consists of heptad repeats of leucines or other bulky hydrophobic amino acids and is amphipathic.

This region *mediates* dimerization, forming a *parallel coiled coil* called the leucine zipper. TGA factors also contain an additional, novel domain important for dimerization. Three groups of TGA factors (I, II, III) can be distinguished based on sequence similarity.

The R2R3MYB Family

The MYB domain is a conserved region of about 52 amino acids that displays a helix-turn-helix structure capable of intercalating into the major groove of DNA. One to three copies of this domain (Rl, R2, R3) are typically present in MYB proteins. Compared to other eukaryotes, plant genomes encode a large number of MYB *proteins*; most contain two MYB repeats (R2R3).

Zinc Finger-Related Proteins

Zinc fingers are protein domains that use conserved cysteine and/or histidine residues to coordinate a zinc iron, yielding a compact *"finger"*-like structure. The specific arrangements of *cysteines* and histidines define different types of zinc finger domains, some of which have the potential to bind DNA and others that mediate protein-protein interactions. Plant TFs *implicated* in mediating defense responses appear to contain novel zinc finger domains.

The pepper CAZFP1 contains two novel C_2H_2-type zinc fingers, while LSD 1 defines a novel type of plant-specific C_2C_2 zinc finger. Neither LSD1 nor the related protein LSD One-Likel (LOU) has been shown to bind DNA, and it is possible that they function as *scaffolds* instead of TFs.

Others

The structures of TF domains conserved across eukaryotic kingdoms, such as the homeobox and MYC basic helix-loop-helix (bHLH), have been summarized elsewhere. Some homeobox TFs implicated in regulating defense response genes also contain a leucine zipper. Interestingly, the plant-specific SEBF shares sequence similarity with nuclear-encoded *chloroplast* RNA-binding proteins, suggesting that it may also be involved in RNA *processing*.

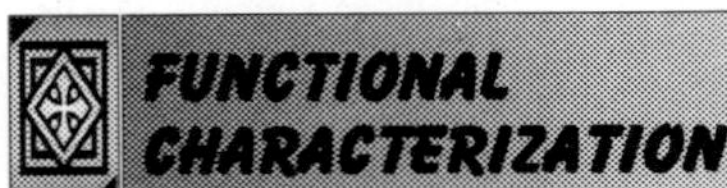

FUNCTIONAL CHARACTERIZATION

Following their isolation, additional analyses are typically required to further elucidate the role of TFs in mediating plant defense responses. This section briefly *summarizes* information obtained on characterizing their DNA-binding sequences, transactivation and *transrepression* properties, gene expression patterns, and subcellular *localization*.

Research aimed at identifying target genes, *studying* post-translational regulation, and assessing the role of TFs for disease resistance by genetic or transgenic approaches is discussed in subsequent sections.

DNA Binding Preferences

In general, the DNA-binding targets of TFs are initially determined *in vitro*, using approaches such as the electrophoretic mobility shift assay (EMSA) or PCR-based oligo selection (see Carey and Smale for a detailed account of methods used to study TFs). *Transcription* factors to be tested are typically produced *in vitro* or highly purified from plant tissues.

It is also possible to make use of cell extracts in combination with antibodies against specific TFs to "supershift" the DNA-binding activity. At a minimum, studies should demonstrate specificity of DNA binding. This is *typically* achieved using DNA probes with mutations at key residues or by addition of excess nonlabeled competitor DNA. All ERF, WRKY, Whirly, and TGA factor *proteins* tested to date have the ability, *in vitro*, to bind to their cognate *cis-elements* listed in Table elsewhere in this chapter. Of note, two ERF proteins, Tsil from tobacco and CaPF1 from pepper, were shown to have dual specificity in binding to the GCC-box and the *DRE-box*.

Binding of Tsil and CaPF1 to the DRE-box appears to be biologically relevant because overexpression of these TFs results in the *constitutive* expression of DRE-box-containing genes and enhanced tolerance to abiotic stress. Most ERFs have not been tested for *binding* to the DRE-box and, accordingly, it is not known how common this dual specificity may be.

However, Tsi 1 and CaF'P1 are not closely related family members, suggesting that other ERF proteins may share this property. As a group, MYB proteins appear to have broader DNA-binding *specificities*. The tobacco *Mybl protein* was shown to bind MBSII (GTTTGGT)- and MBSI (TAACTG)-related elements in the promoter of the *PR-1a* gene; NtMYB2 was found to bind to a wounding and elicitor-responsive L-box element (TCTCACCTACC) present in the promoters of genes involved in *phenylpropanoid biosynthesis*.

The above groups of TFs bind to double-stranded DNA. In contrast, the *Whirly proteins* StWhy1 and AtWhy1 (and probably all family members), and SEBF preferentially bind to single-stranded DNA. Unwinding of DNA to generate single strands may be a means of relieving the torsional stress associated with gene *transcription*. It has been proposed that the Whirly proteins and SEBF may stabilize unstable regions of melted DNA *in vivo*. Relatively few studies have attempted to determine the preferred binding sequences of TFs implicated in *mediating* plant defense responses.

When compared, the binding specificity of individual members of a TF family displays slight differences, which are likely of *regulatory significance* in determining target site selection and gene expression parameters *in vivo*.

Transeffector Properties

A key function of TFs is to recruit the transcriptional machinery to specific gene promoters. This is achieved through direct interactions with one or more general TFs or indirectly through cofactors (*coactivators*) that do not bind DNA. Several classes of coactivator have been described. Many comprise large multiprotein complexes that possess chromatin *remodeling* and/or modifying activities.

These activities facilitate access of TFs and the basal transcriptional machinery to specific gene promoters. Conversely, transrepression domains may interact with co-repressors possessing chromatin remodeling and/or modifying activities that impede access of the basal transcriptional machinery. Transcription factor interfaces responsible for recruiting the preceding classes of proteins are called effector (transactivation or transrepression) domains.

The specific amino acid sequences and structural features required to create effector domains are not as well defined as those responsible for DNA binding. Many transactivation domains are rich in acidic amino acids; others are rich in glutamine or isoleucine.

Repression domains have been characterized as being charged, rich in alanine or in alanine and proline. Proline-rich domains have been implicated in transcriptional activation and repression. It is also possible for individual TFs to display transactivation and repression properties.

The cell-specific concentration of a TF as well as the presence and concentration of interacting proteins are factors that may determine its ability to transactivate vs. transrepress. Sequence analysis of TFs implicated in plant defense responses reveals that most contain features characteristic of effector domains. Several of these proteins have been experimentally shown to possess transactivation or transrepression properties.

To facilitate analysis, tests are usually performed in yeast cells or transient plant-based assays, although stably transformed plant tissues have also been used. It is also common in these tests to fuse the TF to heterologous DNA-binding domains, such as the one from the yeast GAL4 protein, which have well characterized DNA-binding elements. Each of the ERF, WRKY, and TGA families contain members that transactivate and others that transrepress.

The potato StWhyl has been shown to transactivate. However, several Whirly factors lack obvious effector domains, a feature that was considered as a main source of divergence within the family. SEBF possesses transrepression activity. The intact CaZFP1 protein did not display transactivation potential in yeast.

Otherwise, the transactivation properties of R2R3MYB and zinc finger proteins implicated in mediating defense responses have not been assessed. A number of studies have attempted to localize protein regions important for the transactivation or transrepression activity of TFs implicated in plant defense responses. In some cases, sequences commonly associated with effector domains were found to be important.

In other cases, mutational analyses were not sufficiently detailed to resolve individual amino acids required or indicated a role for multiple parts of the proteins. Comparisons of effector domains between family members suggest distinct modes of transactivation within individual TF families. The transrepression properties of ERF proteins have been studied in detail.

In addition to repressing basal transcription in transient assays, NtERF3, AtERF3, and AtERF4 also repress transactivation of other TFs. Repression occurred in a dose-dependent fashion and was effective against ERF and non-ERF transcriptional activators. Repression also required DNA binding of the repressor protein, but did not rely on competition for DNA binding sites with the transcriptional activator.

The transrepression activity of ERF proteins was localized to a conserved amphiphilic motif L/FDLNL/F(X)p (the EAR motif) with the capacity to suppress transactivation when fused to heterologous DNA-binding domains. Several ERF proteins containing the EAR motif cluster into the same subfamily.

Subcellular Localization

As would be expected for proteins with a role in regulating gene expression, TFs implicated in mediating plant defense responses that have been analyzed to date localize to the nucleus. Functional nuclear localizing signals have been identified in WRKY and TGA factors. SEBF, which shares sequence similarity to chloroplast RNA-binding proteins, was found to be localized to the nucleus and chloroplasts.

The SEBF cDNA has the capacity to encode a putative transit peptide. Interestingly, the N-terminus of several Whirly proteins is also predicted to be chloroplast transit peptides. These results suggest that SEBF and Whirly factors could play a role in coordinating defense responses in both compartments.

Control of subcellular protein localization, in particular nuclear import and exclusion of TFs, can be of considerable regulatory importance. However, there have been no reports indicating that nuclear localization of TFs is regulated in response to pathogen challenge.

Gene Expression

Many, if not most, of the TF genes implicated in regulating gene expression during defense responses are differentially regulated following pathogen challenge. Collectively, and even within individual TF gene families, the range of expression patterns observed is complex, showing differences in directionality (upor downregulated), kinetics (immediate early, early, late expression), amplitude (strong, weak) and duration (transient, sustained).

For any given TF gene, these parameters may be influenced as a function of the infecting pathogen and the plant host's ability to resist the pathogen. In general, there does not appear to be any correlation between the directionality of gene expression and the transactivation properties of the encoded protein (i.e., genes encoding transactivators are not necessarily upregulated and those encoding transrepressors are not all downregulated). Regulation of TF gene expression may occur at the site of infection as well as in systemic, noninfected tissues.

Gene expression may also be affected by abiotic stresses and can show developmental regulation in the absence of biotic or abiotic stress. It is also common for expression of these TF genes to be modulated by one or more of the defense-related growth regulators, with simultaneous exposure to *combinations* of growth regulators having *synergistic* effects on gene expression. For example, many *ERF* genes are regulated by ET, with ET and JA having *synergistic* effects.

However, SA has also been shown to induce the expression of some *ERF* genes and SA/ET synergism has been reported. *Transcription* of certain *ERF* genes as well as the CAZFP1 gene encoding a novel zinc finger is induced by SA, JA, and ET.

Many *WRKY* genes are also regulated by SA or JA. *Two MYC* genes implicated in JA signaling were found to be induced by this growth regulator, while the tobacco *Myb1* gene associated with tobacco mosaic virus (TMV) infection is induced by SA. Some genes encoding TFs implicated in mediating defense responses do not appear to be modulated in response to pathogen challenge or defense-related growth regulators. To further elucidate the signaling pathways mediating TF gene expression, several groups have exploited mutants compromised in defense-related signaling.

Of note, the expression of several *WRKY* genes depends on NPR1, and an *MYB* gene (*AtMYB30*) associated with the hypersensitive response is constitutively expressed in several *lsd* mutants that spontaneously form diseaselike lesions. Little is known about the mechanism regulating expression of TF genes implicated in defense responses. Several immediate-early *WRKY* genes contain W-boxes in their promoters, and it was recently shown using chromatin immunoprecipitation that the parsley *WRKY3* and *WRKY1* genes are *in vivo* targets of WRKY1.

This suggests possible autoregulatory control of *WRKY* gene expression. Transcriptional regulation of the *Arabidopsis ERF1* gene has been studied in some detail. Expression of *ERF1* requires the ETHYLENE INSENSITIVE 3 (EIN3) TF, which specifically binds a primary ethylene response element (PERE) distinct from the GCC-box, in the *ERF1* promoter *in vitro*.

In the absence of ET, EIN3 is continuously degraded through the action of two closely related

F-box proteins, EBF1 and EBF2 (EIN3-binding *F* box protein 1 and 2). Exposure to ET or mutation of *EBF1* and *EBF2* leads to increased levels of EIN3 and increased levels of *ERF1*.

It is not known how ET modifies EIN3 or EBF proteins to prevent EIN3 degradation, but it has been speculated that ubiquitination and degradation of EIN3 may be triggered by its phosphorylation status. In support of this hypothesis, a rvitogen-activated protein kinase (MAPK) signaling cascade proposed to operate upstream of EIN3 has recently been identified.

Constitutive activation of the MAPK kinase involved (SIMKK) leads to constitutive expression of *ERF1*. The consequences on disease resistance of elevated *ERF1* expression in the *EBFI/EBF2* mutant and SIMKK plants have yet to be determined. Treatment of tobacco calli with inhibitors of protein kinases and phosphatases substantially inhibited the levels of *ERF2-4* transcripts, suggesting that protein phosphorylation is also important for expression of these genes.

Curiously, expression of these genes as well as several *Arabidopsis ERFs* was induced by the protein synthesis inhibitor cycloheximide (CHX). It was proposed that CHX may be preventing the *de novo* synthesis of a labile transcriptional repressor or mRNA degrading enzymes. It will be interesting to determine whether the *putative labile* repressor is associated with EIN3 *degradation*.

FUNCTIONAL ANALYSIS OF MUTANT AND TRANSGENIC PLANTS WITH ALTERED LEVELS OF TRANSCRIPTION FACTORS GENES

Definitions and Limitations

Convincing proof of a TF's participation in mediating defense responses usually requires the characterization of plants containing altered levels of it. The gold standard for such analyses is the study of loss-of-function mutations. The more traditional, forward genetic approach involves screening populations of mutagenized plants for a given phenotype (e.g., altered PR gene expression or disease resistance).

The alternative, reverse genetics, involves identifying mutations in genes of known sequence and subsequently determining the resulting phenotypes. Such strategies have become increasingly popular with onset of the genomic era. Reverse genetics offer the researcher more control in targeting genes of interest and facilitate creation of plants harboring mutations in multiple (related) genes.

However, there are no guarantees that alleles recovered by reverse genetics will yield visible phenotypes. In fact, data from *Arabidopsis* and nonplant model systems suggest that few do. It is also important to recognize that unrelated, secondsite mutations may be generated by mutagenic treatments. Several transgenic strategies have also been exploited to reduce levels of TF gene expression. These include antisense and RNA interference (RNAi) technology.

Compared to antisense, RNAi is usually more effective at reducing levels of target gene transcripts. The creation of dominant-negative TFs has been a popular means of interfering with TGA factor function. Basically, transgenes are designed that will produce proteins lacking DNA binding activity but retain the ability to dimerize with biologically active, endogenous factors. Expression of the dominant-negative TF sequesters the endogenous dimerization partners, preventing them from binding DNA and regulating gene expression.

One advantage of the dominant-negative approach is that it may overcome issues of functional redundancy in cases where related TFs have the *potential* to *dimerize*. Transgenic ectopic (over) expression has represented the most popular means of functionally testing plant TFs. It is not *affected* by functional *redundancy* and can be applied to crop species in which tools for reverse *genetics* are not available.

Even when loss-of-function phenotypes are available, ectopic expression has been shown to yield valuable new *information*. A twist to the ectopic expression approach involves use of TF fused to strong, *constitutive transactivation* domains, such as the one from viral particle 16 (VP16). This strategy may be particularly *informative* in cases in which the *transactivation* potential of a TF is weak or tightly *regulated*.

It is noteworthy that *overexpression* of TFs can trigger artifacts, so results need to be *interpreted* with caution. One potential problem is that of squelching. As discussed in other Section of this chapter, effector domains represent interfaces for protein–*protein interactions*.

Accordingly, increasing the quantity of a TF through transgenic ectopic expression can lead to sequestering of interacting proteins, such as *cofactors*, which may be limiting and required for the function of unrelated TFs. Expression of TFs at unphysiologically high levels may also result in binding of low-affinity DNA elements not normally bound by the TFs and the *subsequent nonspecific activation* or repression of gene sets.

As a result of nonspecific binding or squelching, cellular homeostasis may be disturbed, yielding pleiotropic effects. Dominant-negative approaches described earlier are also prone to yielding pleiotropic effects. Transient gene expression methods have also emerged as important tools for *functionally* testing TFs.

Of note, virus-induced gene silencing (VIGS) offers the potential to test large numbers of candidate genes rapidly, without the need to generate stable *transgenic* lines.

ERFs

Individual *ERF* genes have been overexpressed in a variety of transgenic plants including *Arabidopsis, tobacco, tomato,* and *hot pepper*. In most instances, these modifications resulted in enhanced levels of disease resistance to a selected range of pathogens, including *Botrytis cinerea, Fusarium oxysporum*, and *Plectosphaerella cucumerina*, which elicit the JA/ET-defense pathways in the infected host.

Enhanced resistance was not compromised by mutants blocking early steps of ET signaling, indicating that the downstream events in this pathway were constitutively activated. Testing of pathogens in which resistance is not mediated primarily by the JA/ET signaling pathways yielded variable results.

For example, overexpression of tomato *Pti4* in *Arabidopsis* enhanced tolerance to the biotrophic bacterial pathogen *Pseudomonas syringae*, but overexpression of *Arabidopsis ERF1* increased susceptibility to this pathogen. Such results may reflect examples of negative or positive cross-talk, respectively, between the JA/ET and SA signaling pathways mediated by transgenic expression of the ERF protein. It is also notable that overexpression of tomato *Pti5* in *Arabidopsis* failed to increase resistance to *P. syringae* or *Erysiphe orontii*, but this TF was effective against the former pathogen when overexpressed in tomato.

Table 8.2: Effect of ERF Gene Overexpression on Disease Resistance and PR Gene Expression

Gene	*Host*	*Disease resistance*	*Marker gene expression*
Tomato *Pti4*	*Arabidopsis*	*Erysiphe orontii* *Pseudomonas syringae pv. tomato* DC3000 (tolerance)	*PR-1, PR-2, PR-3, PR-4, PDF1.2,Thi2.1*
Tomato *Pti5*	*Arabidopsis*	Noenhanced resistance to *E. orontii* or *Pst* DC3000	*PR-1, PR-2, PR-3, PR-4, PDFI.2* (weaker than 35S:Pti4)
Tomato *Pti6*	*Arabidopsis*	Noenhanced resistance to *E. orontii* or *Pst* DC3000	*PR-1, PR-2, PR-3, PR-4, PDFI.2, Thi2.1* (weaker than 35S:Pti4)
Tomato *Pti5*	Tomato	*P. syringae pv. tomato*	Enhanced expression of *GluB* and *Catalase* after pathogen challenge SA-regulated *PRlal* and *PRlbl* not expressed
VP16:Pti5	Tomato	*P. syringae pv. tomato* (resistance similar to *Pti5* alone)	Higher levels of *GluB* and *Catalase* than observed with *Pti5* alone
Arabidopsis ERFI		*Arabidopsis* Plectosphaerella cucumerina Enhanced susceptibility to *Pst* DC3000	*Botrytis cinerea* and *Basic chitinase* and PDFI.2
Arabidopsis ERFI		*Arabidopsis* *conglutinans* and *F oxysporum f.* sp. *lycopersici*	*Fusarium oxysporum sp.* *PDFI.2*
Tobacco *Tsil*	Tobacco	*P. syringae pv. tabaci*	*PR-2, PR-3, PR-4,*

(Table Contd.)

Gene	Host	Disease resistance	Marker gene expression
			Osmotin, SAR8.2, PR-I
Tobacco *Tsil*	Hot pepper	Pepper mild mottled virus, Cucumber mosaic virus, *Phytophthora capsici,* and *Xanthomonas* camperstris pv. vesicatoria	*PR-1, PR-2, PR-4, PR-5, PR-10, PinII, LTPI, SAR8.2*
Pepper *CaPFI*	*Arabidopsis*	*P syringae pv. tomato*	*GST & PDFI.2* DC3000d
Tobacco *ERF5*	Tobacco	Tobacco mosaic virus No enhanced resistance to *syringae pv. tabaci, P. syringae pv. pisi*	No enhanced expression of *PRIa, PRIb, PR3 P.*
Tomato *TSRFI*	Tomato	*Ralstonia solanacearum syringae pv. tomato*	*PR2, PR3* and P
Gene	Host	Disease resistance	Marker gene expressionb
Tomato *TSRF1*	Tobacco	*R. solanacearum*	*PR1, PR2 & PR3*
Tomato OPBP1	Tobacco	*Phytophthora parasitica* var. *nicotianae* and *P. syringae* pv. *tabaci*	*PR-5d, PR-1a*
Pepper *CaERFLP1*	Tobacco	*P. syringae pv. tabaci*	*β-glucanase, osmotin,* HMG-CoA reductase, cysteine protease

These apparently conflicting results may be attributed to distinct functionalities between proteins (in the case of ERF1 and Pti4) or of the proteins in different host plants (*Arabidopsis* vs. tomato), as well as to variations in experimental designs between studies. Enhanced resistance to pathogens conferred by overexpression of *ERF* genes was usually correlated with enhanced levels of marker *PR* gene expression following *pathogen* challenge or with their *constitutive* expression.

Expression of a tomato Pti5:VP16 fusion protein was found to elicit higher constitutive levels of marker *PR* genes than the native Pti5, but did not substantially increase disease resistance compared to native Pti5. Lack of enhanced disease resistance in *Arabidopsis* plants overexpressing *Pti5* or *Pti6* was correlated with weak to no constitutive expression of marker genes. Most marker genes analyzed in the preceding studies contained GCC-boxes in their *promoters* and were known to be responsive to ET or JA. In some cases, only a subset of GCC-box-containing marker genes tested was activated in diseaseresistant transgenic plants. SA-inducible genes were also tested and found to be activated in some cases, but not in others. As discussed later, transcript profiling of transgenic plants overexpressing *ERF* genes revealed that these TFs potentially regulate large sets of genes. Overexpression of *Pti4* and *ERF1* in *Arabidopsis* was correlated with developmental *abnormalities*.

The most common phenotype observed is reminiscent of the so-called ethylene triple response, referring to the inhibition of hypocotyl and root elongation and an exaggerated curvature of the apical hook observed in wild-type seedlings exposed to ET.

Seedlings of *Arabidopsis* plants overexpressing *Pti4* or *ERF1* display the inhibition of hypocotyl elongation in the absence of ET; those overexpressing *ERF1* also display inhibition of seedling root elongation. In no case was the curvature of the apical hook found to be affected. Interestingly, *HOOKLESS1* (*HLS1*), a gene regulating the *apical hook* curvature, contains a GCC-box in its promoter.

Although ERF proteins can bind *in vitro* to a promoter containing a multimer of the *HLS1* GCC-box, *HLS1* expression is not activated in *Pti4* or *ERF1* overexpressing plants. Plants overexpressing *Pti4* and *ERF1* also display phenotypes associated with ET exposure at the adult stage, including smaller size, greener leaves, and, in the case of *ERF1* overexpressors, inhibition of cell *enlargement*, wilting, and death before *bolting*.

Together, these results indicate that subsets of ET responses are constitutively activated by overexpression of these *ERF* genes. Overexpression of the other *ERF* genes tested to date has not been reported to induce developmental abnormalities. The phenotypic consequences associated with loss of *ERF* gene function have yet to be determined. Analysis of the SIGnAL T-DNA database revealed the presence of insertions in or near many members of the *Arabidopsis* AP2/ERF family.

However, there have been no published reports of their *characterization*. Similarly, antisense expression of the tomato *Pti5* was reported to have no effect on *race-specific* resistance, although no data were shown.

WKRY Factors

Similar to the ERF family, there are few reports describing the phenotypic consequences of mutations in *WRKY* genes. Mutations in more than *40 Arabidopsis WRKY* genes have been identified by reverse genetics, but most do not appear to display visible mutant phenotypes. The consequences of altering levels of WRKY factors on disease resistance have been reported for only five *Arabidopsis* genes.

VIGS has also been applied to study the involvement of WRKY genes from solanaceous plants in R-gene mediated resistance pathways. Overexpression of *WRKY70* enhanced resistance to virulent strains of the bacterial pathogens *Erwinia carotovora* and *P. syringae*, but antisense-mediated reduction of *WKRY70* resulted in enhanced susceptibility to these pathogens.

Overexpression was correlated with increased expression of SA-inducible PR genes and decreased expression of JA/ET-regulated defense-related genes. Conversely, antisense suppression of *WRKY70* resulted in constitutive expression of JA/ET-regulated marker genes. Together, these results indicate that WRKY70 acts as a positive regulator of disease resistance and it was proposed to represent a convergence point for the SA and JA signaling pathways.

Overexpression of *WRKY18* also resulted in enhanced resistance to *P. syringae* and increased expression of SA-inducible PR genes.

However, these phenotypes were only observed in older plants (e.g., 5 weeks old). Interestingly, mutations at the *NPR1* locus had little effect on PR gene expression in the *WKRY70* overexpressors; however, they abolished potentiation of *PR* genes and enhanced disease resistance observed in older plants overexpressing *WKRY18*. Despite increasing levels of *PR* genes, overexpression of *WRKY6* had no appreciable effect on resistance to virulent or avirulent strains of *P. syringae*. Transposon-induced mutation of *WRKY6* also induced changes in gene expression consistent with a possible role in defense responses, but was not associated with any visible mutant phenotype.

Thus, although results implicate WKRY6 and WRKY18 in mediating the expression of defense genes, they also indicate that neither gain nor loss (in the case of WRKY6) of these TFs by itself is sufficient to affect disease resistance. In all cases reported, stable transgenic overexpression of *WRKY* genes resulted in developmental abnormalities, including stunted growth, altered leaf morphology, and changes in flowering time.

Overexpression of *WRKY6* was also associated with development of necrotic areas on leaves and loss of apical dominance. Reduction of *WRKY70* levels resulted in larger plants and early flowering (overexpression of *WRKY70* delays flowering), but reduction or loss of *WRKY6* did not result in any developmental abnormalities.

Expression of *WRKY6* and *WRKY18* transgenes also decreased levels of corresponding endogenous genes, suggesting that these TFs are involved in regulating expression of their genes and, possibly, other members of the family. Transcript profiling of plants with altered levels of *WRKY6* and *WRKY70* also supports the notion that these TFs can activate and suppress target gene expression. Transient overexpression of *AtWRKY29* in *Arabidopsis* leaves reduced disease symptoms caused by *P. syringae* and *B. cinerea*.

However, the effects of altering levels of this *WRKY* gene have not been tested in stable transgenic or mutant plants. VIGS has also implicated the tobacco *WRKY1*, *WRK2*, and *WRKY3* genes as being required for full N gene-mediated resistance to TMV.

Whirly Factors

AtWhy1 is the only Whirly factor functionally characterized to date. Two missense mutant alleles in this gene were recovered by targeting induced local lesions in genomes (TILLING). Based on the crystal structure of StWhy1, one mutation mapped to the single-stranded DNA-binding domain, and the other was located in the region implicated in tetramerization.

Mutant plants exhibited reduced AtWhy1 DNA-binding activity and *PR-1* expression following SA treatment. Both mutant backgrounds were hypersusceptible to a virulent strain of the biotrophic oomycete *Peronospora parasitica*. The atwhy1.2 mutant also showed intermediate susceptibility to an avirulent strain of *P. parasitica* and was unable to mount an effective SAR response following treatment with SA.

Thus, AtWhy1 constitutes a positive regulator of SA-dependent disease resistance. However, SA-induced Whirly DNAbinding activity does not require functional NPR1, suggesting that Whirly activation occurs through a distinct, NPR1-independent pathway.

TGA Factors

The most compelling evidence implicating TGA factors as mediators of defense responses comes from the analysis of the triple *tga2tga5tga6* loss-of-function *Arabidopsis* mutant. This mutant is compromised in SAR against virulent strains of *P. syringae* and *P. parasitica.* Interestingly, basal resistance to these pathogens was not compromised.

The mutant also failed to express *PR-1* in response to SA, but displayed higher basal levels of *PR-1* in the absence of SA elicitation. Although no developmental abnormalities were reported, seedlings of the triple mutant were hypersensitive to the toxic effects of SA, a phenotype also observed in *npr1* mutants. Attempts to study the role for TGA factors in mediating disease and *PR* gene expression using dominant-negative versions of TGA factors have yielded conflicting results.

Expression of a dominant-negative *Arabidopsis TGA2* gene in *Arabidopsis* compromised basal resistance against *P. syringae pv. maculicola;* expression of a similar dominant-negative version of the same *Arabidopsis* gene in tobacco enhanced SAR against *P. syringae pv. tabaci.* In these studies and others, the transgenic plants were also monitored for expression of marker genes containing *as-1* elements. Two types of genes were considered, based on the timing of their expression following SA treatment: early genes and late genes. *PR-1* is considered a late gene, and was the only maker gene evaluated in one study.

Loss of basal resistance observed by transgenic expression of dominantnegative *TGA2* was correlated with reduced levels of *PR-1* in *Arabidopsis,* while enhanced SAR in tobacco was associated with increased levels of *PR-1a* but reduced levels of early gene transcripts.

In contrast, a dominant-negative tobacco *TGA2.2* resulted in reduced levels of early and late genes, while (over)expression of wild-type *TGA2.2* increased levels of early genes but had no affect on *PR-1a.* Finally, plants expressing wild-type or dominant-negative versions of the closely related tobacco *TGA2.1* gene displayed enhanced and reduced levels of early genes, respectively, but showed no changes in *PR-1a* levels.

Based on the observation that dominant-negative *TGA2* has opposite effects on early and late gene expression, it was proposed that TGA factors have both positive and negative roles in mediating plant defense responses. Additional evidence in support of a dual role for TGA factors comes from RNAi analysis, which revealed a negative role for TGA4 and a positive role for TGA5 in regulating a reporter gene under the control of a multimerized cis-element related to *as-1.*

The consequences of (over)expressing wild-type or dominant-negative *TGA2.1* and *TGA2.2,* or the *TGA4* and *TGA5* RNAi transgenes on disease resistance have yet to be determined. In the only study that appears to have observed altered disease resistance as a consequence of (over)expressing TGA factor genes, Kim and Delaney found that transgenic *Arabidopsis* plants overexpressing *TGA5* displayed enhanced resistance to a virulent strain of *P. parasitica,* but reduced levels of *PR* genes.

In contrast, neither sense nor antisense overexpression of *TGA2* affected resistance to this oomycete. Overall, most analyses to date have focused on group II TGA factors, which include TGA2, TGA5, and TG6. The only study demonstrating a role for group I TGA factors in mediating disease resistance comes from VIGS of tomato TGA1 homologs, which compromised Pto-mediated resistance to *P. syringae pv. tomato* harboring *avrPto.*

R2R3 MYB Proteins

The R2R3MYB gene *BOS1* was recovered in a genetic screen aimed at identifying *Arabidopsis* genes required for resistance to *B. cinerea*. Loss of BOS1 function resulted in enhanced susceptibility to necrotrophic fungi, including *B. cinerea* and *Alternaria brassicicola*. The *bos1* mutant displayed more severe disease symptoms in response to infection by the biotrophic pathogens *P. parasitica* and *P. syringae;* however, no detectable increase in growth of these pathogens was observed.

The mutant accumulated more reactive oxygen species (ROS) in response to *B. cinerea* infection and was also more sensitive to a number of abiotic stresses. Expression of marker genes for the SA and JA signaling pathways were not altered in the *bos1* mutant; however, *BOS1* transcripts were found to accumulate following infection with *B. cinerea*.

Increased *BOS1* expression was blocked in the coi1 mutant that mediates JA signaling, implicating BOS1 as a mediator of this defense pathway. The *Arabidopsis MYB30* gene was initially identified based on its differential expression during a type of programmed cell death called the hypersensitive response (HR) associated with incompatible interactions.

Constitutive overexpression of *MYB30* in *Arabidopsis* and tobacco resulted in elevated expression of HR marker genes, accelerated HR against avirulent pathogens, and development of HR-like lesions upon infection with virulent pathogens. Importantly, it increased resistance to virulent and avirulent biotrophic pathogens.

Conversely, antisense suppression of *MYB30* suppressed HR marker gene expression, delayed the HR against avirulent pathogens, and decreased resistance against virulent and avirulent pathogens. Constitutive expression of *MYB30* did not result in spontaneous lesion formation, indicating that factors other than MYB30 are required to initiate the HR. The tobacco *MYB1* gene was also shown to be required for N-mediated resistance to TMV using VIGS technology.

Zinc Finger-Related Proteins

The *Arabidopsis LSD1* gene was isolated in a screen aimed at identifying mutants that misregulate cell death responses. Loss-of-function *lsd1* mutants initially display a normal HR in response to infection with virulent pathogens, but cannot limit the extent of the cell death, leading to a phenotype known as runaway cell death. Runaway cell death in the *lsd1* mutant is also observed following treatment with SA and SA analogs and depends on production of superoxide.

It requires NPR1 as well as the positive regulators of disease resistance ENHANCED DISEASE SUSCEPTIBILITY 1 (EDS1) and PHYTOALEXIN DEFICIENT 4 (PAD4), both of which are putative lipases. *lsd1* mutants are more resistant virulent strains of *P. parasitica*, but neither SA nor NPR1 is required for this phenotype.

These results indicate that LSD1 is a negative regulator of basal defense responses, somehow interpreting ROS-dependent signals triggered by the HR. LOL1 was identified based on its similarity to LSD1 and the two proteins appear to have opposite roles in mediating cell death and disease resistance. Reduction of *LOL1* expression in the *lsd1* mutant suppresses runaway cell death, while conditional high-level expression of LOL1 is sufficient to trigger cell death in the absence of pathogen infection or SA treatment.

Furthermore, stable moderate overexpression of *LOL1* enhanced resistance to virulent strains of *P. parasitica*, while reduction of *LOL1* levels enhanced susceptibility to this pathogen. Overexpression of the pepper *CaZFP1* gene in transgenic *Arabidopsis* resulted in enhanced resistance to a virulent strain of *P. syringae* and enhanced tolerance of drought.

Overexpressing lines also displayed developmental abnormalities. Expression of SA-regulated PR genes or the JA-regulated *PDF1.2* gene was not altered by *CaZFP1* overexpression, suggesting that the observed enhanced resistance was mediated through the activation of other pathways.

MYC Proteins

The *Arabidopsis jin1* mutants were identified in an ET-insensitive genetic background as being insensitive to JA and represent alleles of *AtMYC2*. Mutant plants display enhanced resistance to necrotrophic pathogens, including *B. cinerea* and *P. cucumerina*. All *jin1* mutant alleles are semidominant and retain the capacity to produce the N-terminal end of the protein. It has been speculated that these could interfere with interacting proteins.

POST-TRANSLATIONAL REGULATION

In addition to regulation at the transcriptional level, there is overwhelming evidence for the post-translational regulation of TFs in response to pathogen challenge. Most studies have focused on reversible phosphorylation/dephosphorylation, which is a prevalent means of modulating the activity of eukaryotic TFs.

Reversible oxidoreduction of key cysteine residues is emerging as an important mechanism for regulating mammalian and microbial TFs, and recent studies have implicated this type of control in the regulation of plant TFs mediating defense responses. Physical interactions between TFs and other proteins are also likely to be of regulatory significance.

Post-Translational Modifications

In one of the few studies aimed at monitoring changes in the nuclear pools of TFs following defense response elicitation, Turck et al. visualized changes in the pattern of WRKY proteins by two-dimensional gel electrophoresis. Several proteins migrated as adjacent pearl strings, suggesting different post-translationally modified forms of the same protein. The number of proteins and the complexity of modifications rapidly increased following treatment with an elicitor.

However, the nature of the putative post-translational modifications has not been determined. As detailed later, the property of TFs most frequently shown to be influenced by post-translational modification is their ability to bind DNA. In a few cases, protein–protein interactions were shown to be affected. Few studies have directly linked post-translational modification to changes in transactivation potential.

However, several have shown that transactivation is regulated in response to treatment with elicitors or growth regulators implicated in mediating defense responses; this suggests possible post-translational control. In many cases, the significance of posttranslational modification for disease resistance is unknown.

Phosphorylation

The ability of TFs present in plant extracts to bind to several cis-acting elements implicated in mediating defense gene expression is dramatically altered by treatment with phosphatase or protein kinase inhibitors. In addition, constitutive activation of a MAPK signaling cascade in tobacco resulted in substantially more binding activity to the W-box.

Consistent with this result, two WRKY factors (WKRY22 and WRKY29) were identified as downstream components of an MAPK signaling cascade conferring basal resistance to bacterial and fungal pathogens in *Arabidopsis.*

Transcription factors and components of MAPK signaling cascades were also identified as being required for N-mediated resistance to TMV in tobacco and Pto-mediated resistance to *P. syringae* (*avrPto*) in tomato. However, in none of these studies was it demonstrated that the TFs were directly phosphorylated by MAPKs.

One study suggested that the MAPK cascade may lead to phosphorylation and inactivation of a specific inhibitor of WRKY proteins; results from another were consistent with regulation occurring at the point of *WRKY* transcription. The rice ERF OsEREBP1 is the only plant TF implicated in mediating defense gene expression that has been shown to be phosphorylated by a MAPK.

Activity of this MAPK, BWMK1, is rapidly induced by a pathogen-derived elicitor, ET, SA, and JA. Overexpression of BWMK1 in transgenic tobacco results in constitutive expression of PR genes and enhanced resistance to *P. syringae* and the oomycete pathogen *Phytophthora parasitica.* BWMK1 phosphorylates OsEREBP1 *in vitro,* resulting in enhanced binding to the GCC-box.

Furthermore, transient coexpression of *BWMK1* and *OsEREBP1* in *Arabidopsis* protoplasts induced expression of a reporter gene under the control of the GCC-box to higher levels than either gene alone. Therefore, it is likely that OsEREBP1 is regulated *in vivo* by a pathogen-activated MAPK signaling cascade. The tomato ERF Pti4 is also regulated by phosphorylation.

Pti4 physically interacts with the *Pto* R-gene product, which encodes a serine-threonine protein kinase Pto kinase activity is required for interaction with Pti4 as well as for race-specific resistance against *P. syringae* pv. *tomato* harboring *avrPto.*

Phosphorylation of Pti4 by Pto enhances its DNA binding to the GCC-box element *in vitro.* This effect is highly specific because two protein kinases related to Pto (Fen and Pti1) do not phosphorylate Pti4, and Fen does not stimulate Pti4 DNA-binding activity. Pti4 contains ten threonine residues: two each in the putative DNA-binding and transactivation domains and one near the nuclear localization signal. Pto phosphorylates at least four of these residues (but no serines).

At this time it is not known which residues are phosphorylated *in vitro* or *in vivo.* It is also not known whether Pti4 may be regulated by other protein kinases during compatible plant–pathogen interactions. As discussed in other section of this chapter, overexpression of Pti4 confers enhanced basal resistance to pathogens independently of the Pto–avrPto interaction.

Phosphorylation has also been implicated in the regulation of bZIP proteins. SARP (salicylic acid response protein) is a cellular factor immunologically related to TGA factors implicated in mediating the rapid induction of *as-1* binding activity following treatment with SA.

In the absence of SA, it was proposed that SARP is sequestered by an inhibitory protein called SAI (SA-inhibitor) and that release of SARP from SAI is triggered by SA-induced phosphorylation of either or both proteins. Inhibitors of casein kinase II (CK2) suppress *as-1* binding, suggesting that CK2 may be responsible for activation of SARP. More recently, the *Arabidopsis* TGA2 was shown to be phosphorylated *in vivo* by a CK2-like activity induced by SA. Phosphorylation of TGA2 appeared to suppress its ability to bind to the *as-1* element. This is in contrast to previously published studies showing that phosphorylation by CK2-like kinases enhanced *as*1 binding *in vitro*. Importantly, no difference in *PR-1* expression was observed in transgenic plants overexpressing TGA2 compared to those expressing site-directed mutant with putative CK2 phosphorylation sites removed.

Thus, the regulatory significance of TGA2 phosphorylation by CK2-like kinases remains elusive. G/HBF-1 is a non-TGA bZIP TF that binds to an SA and elicitor-responsive element in the promoter of the *chalcone synthase15* gene from soybean. Elicitor treatment does not alter levels of G/HBF-1 transcript or protein; however, it rapidly induces activation of a protein kinase capable of phosphorylating G/HBF-1.

Furthermore, phosphorylation of G/HBF-1 *in vitro* increases its DNA-binding activity. The identity of the G/HBF-1 kinase is not known. DNA-binding activity of Whirly proteins is also regulated by phosphorylation. In particular, a kinase related to mammalian protein kinase C has been implicated. Even though Whirly factors are constitutively present in the nucleus, their associated DNA-binding activity is not detectable until after pathogen challenge or treatment with SA or an elicitor.

Chromatographic purification of StWhy1 from uninduced nuclei activates its DNA-binding activity, possibly by removing an unidentified inhibitor protein. The possible role of phosphorylation in mediating the interaction of Whirly proteins with inhibitor proteins is unknown.

Oxidoreduction

Binding of nuclear factors to the *as-1* element is regulated by redox conditions. A recent study has started to unravel the possible mechanism involved. The interaction between *Arabidopsis* TGA1 and NPR1 in plant cells is positively influenced by treatment with SA. Two conserved cysteines located in the Cterminal region of TGA1 (C260 and C266) play a key role in regulating this interaction.

Although wild-type TGA1 and NPR1 do not interact in the yeast two-hybrid system, mutation of these residues permits interaction with NPR1 in yeast and in *Arabidopsis* cells regardless of SA induction. Using a novel labeling strategy that distinguishes between protein sulfhydryls and disulfides, it was demonstrated that the redox status of cysteines in TGA1 and/or the closely related TGA4 shifted considerably following SA treatment to become predominantly reduced.

Thus, strong interaction with NPR1 is correlated with the reduced state of TGA1 and/or TGA4 cysteines. Changing redox conditions altered the mobility of *in vitro* produced TGA1 in nonreducing gel electrophoresis, consistent with the formation of an intramolecular disulfide bridge under oxidizing conditions; this could impede interaction with NPR1.

Changing redox conditions has been shown to influence the DNA-binding activity of several TFs; however, the ability of TGA1 to bind the *as-1* element *in vitro* was unaltered by vast molar excess of redox-regulating compounds. Instead, redox regulation of TGA1 DNA binding required the redoxregulated recruitment of NPR1, which was proposed to act as a cofactor in stimulating TGA1 DNA-binding activity.

Protein Turnover/Proteolysis

TGA factors were shown to have different stabilities during tobacco development; TGA1 and

TGA3, but not TGA2, were rapidly degraded in mature leaves. Degradation of TGA3 appeared to be mediated via the proteosome. However, the regulatory significance, if any, of TGA factor degradation in response to SA or pathogen challenge was not assessed. The role of targeted protein degradation in the regulation of *ERF1* transcription was discussed in other section of this chaptrer. Some preliminary evidence also suggests regulation of defense-related TFs by proteolysis.

The tobacco ERF3 was shown to interact with a ubiquitin-conjugating enzyme (NtUBC2) in the yeast two-hybrid system. However, NtUBC2 did not affect the transrepression capacity of ERF3 in transient assays. The *Arabidopsis* R2R3MYB protein BOS1 contains a consensus sumolyation motif, suggesting that it may be regulated by SUMO, a ubiquitin-like modifier associated with protein stabilization.

Finally, mutations of the *Arabidopsis* Fbox protein CORONATINE INSENSITIVE1 (COI1) block JA-mediated defense responses, resulting in increased susceptibility to several necrotrophic pathogens. Although the targets of COI1 are unknown, it has been proposed that it may mediate removal of TFs tagged by JA-dependent phosphorylation.

Protein–Protein Interactions

Transcription of eukaryotic genes is achieved by synergistic interactions between combinations of TFs. This combinatorial control relies on precise juxtapositioning of specific TFs through DNA-protein and protein–protein interactions using functional groupings of cis-regulatory elements called enhancers (or silencers).

Enhancers are responsible for a subset of the total gene expression pattern. Accordingly, a typical gene may contain several enhancer elements. The stable nucleoprotein complex containing enhancer DNA with associated bound TFs and interacting proteins is sometimes referred to as an enhanceosome. Little is known about the organization of cis-acting modules required to create pathogenresponsive enhancer elements in plants.

However, individual cis-acting elements implicated in defense gene regulation frequently cluster, and synergistic interactions between closely spaced elements have been reported (e.g., W-boxes), suggesting cooperation between TFs. Several examples of protein–protein interactions have already been discussed in this chapter, including Pti4–Pto and TGA factors–NPR1.

TGA3 has also been shown to interact with calmodulin, suggesting a possible role for calcium signaling in regulating TGA factors. BZI-1, a bZIP protein not part of the TGA factor family, interacts with a novel protein (ANK1) containing ankyrin repeats. As revealed by analysis of dominant-negative transgenes, BZI-1 is required for resistance to TMV. ANK1 does not appear to act as a cofactor of BZI-1 and was speculated to function in the cytosol rather than the nucleus. Although the combinatorial model of gene expression stresses the importance of interactions between TFs, few have been reported between factors implicated in mediating plant defense responses.

TGA4 was shown to interact with the ERF protein AtERF and with members of the Dof family of TFs, which encode proteins with a single zinc finger. The Dof proteins (OBP1-3) are capable of binding to a cis-element in the CaMV35S promoter and enhance the DNA-binding activity of TGA4 *in vitro.* The genes encoding OBP1-3 are inducible by SA. Overexpression of OBP3 results in numerous developmental phenotypes, but its effects on defense gene expression and disease resistance were not assessed. PRHA, a homeobox protein capable of binding to an elicitor responsive element of *a PR-10* promoter, also interacted with two putative cofactors. The limited number of reported protein interactions may be attributed in part to limitations of the yeast two-hybrid system for studying TFs. First, the natural autonomous transactivation or transrepression properties of many TFs complicate their analysis in systems that rely on reconstitution of an active TF to detect reporter gene expression.

Second, yeast cells may not be competent to carry out post-translational modifications required for interactions. Finally, the yeast two-hybrid system monitors binary interactions between proteins. As indicated at the beginning of this section, TFs are likely to function as part of larger protein complexes requiring multiple protein–protein and protein–DNA interactions.

Approaches aimed at recovery and characterization of protein complexes *in vivo*, such as tandem affinity purification (TAP) strategies, coupled with mass spectroscopy, promise to be useful tools in identifying proteins interacting with plant TFs.

TRANSCRIPTIONAL TARGETS AND NETWORKS

Even for the best characterized plant TFs, the identity of very few target genes is known. Nevertheless, such information is very important in assigning function to individual TFs. Large-scale transcript profiling studies have identified motifs related to W-box, GCC-box, and ERE as enriched in genomic DNA upstream of genes found to be differentially expressed following exposure to pathogens, elicitors, or defense-related growth regulators, the W-box being the most prevalent.

These genes can be viewed as putative targets for WRKY, ERF, and Whirly factors, respectively. There have been no reports of enrichment for MYB binding sites, and *as-1* like elements appear to be under-represented in promoters of these genes. Several novel motifs are also enriched in promoters of these genes; however, the identity of the TFs binding to these sites, if any, is unknown.

As discussed earlier in this chapter, TFs bind to short stretches of DNA (5 to 10 bp) that can accommodate limited sequence variability. Sequences outside the consensus cis-element may also influence binding—in particular, the presence and spacing of neighboring TF recognition sites. Furthermore, only a fraction of TFs within a family is likely to bind to any given promoter containing cis-elements that conform to the consensus binding sequence for that family.

For example, a number of genes containing GCC-boxes in their promoters are not differentially regulated in plants overexpressing ERF factors. Accordingly, it is not possible to identify target genes accurately based uniquely on DNA sequence information. Genome-wide transcript profiling of plants containing altered levels of a TF offers a means of identifying putative targets for that specific factor.

However, due to the existence of transcriptional cascades, these approaches cannot distinguish direct and indirect targets. Nevertheless, they have confirmed that many genes differentially regulated in plants with altered levels of *WRKY6* contain multiple W-boxes in their promoters; those ectopically expressing *Pti4* are enriched for GCC-boxes. One study also reported an enrichment of MYBbinding sites in *Pti4* overexpressors.

Many of the genes found to be differentially regulated in plants expressing altered levels of *WRKY* or *ERF* TFs genes encode proteins with functions potentially relevant to disease resistance.

These include classical PR proteins, proteins implicated in oxidative stress responses (e.g., P450s, glutathione-S-transferase), calcium signaling, and cell wall modification.

Numerous genes identified in these plants, including many without known functions, are also differentially regulated in wildtype plants following treatment with pathogen, elicitors, or defense-related growth regulators, suggesting a role in mediating disease resistance. Comparison of overall gene expression profiles between plants with altered *WRKY70* levels and signaling mutants suggests that this TF controls expression of a substantial number of genes regulated by the JA- and SA-dependent pathways (~60 to 40%, respectively); a positive correlation was observed between WRKY70and JA-dependent expression, but the inverse was true for SA-dependent expression.

Similarly, more than one third of the genes induced by treatment with ET and JA were constitutively expressed in *ERF1* overexpressors. The overlap increased to 80% if only genes classified as "defense related" were considered. Another class of target genes frequently recognized in plants with altered levels of *WRKY* and *ERF* genes includes those that encode TFs. Thus, transcriptional repression or, alternatively, activation and subsequent translation, of such genes could lead to modulation of additional sets of target genes, some of which may again encode TFs.

The result is a transcriptional cascade in which TFs at the top of the hierarchy have the potential to regulate entire developmental or metabolic programs. Such proteins are commonly referred to as "master switches" or "controllers." Well-known examples include homeotic genes in drosophila and plants; mutation of these genes leads to the dramatic development of organs at inappropriate positions.

Closer to the bottom of the hierarchy would be TFs controlling more focused aspects of specific programs. Research in well-characterized systems, such as yeast, has revealed the existence of a number of regulatory loops, or networks, in which the activity of one or more TFs influences that of another. Results obtained with WRKY70 and ERF1 suggest that these TFs could be master controllers of defense-related pathways involving JA/SA and JA/ET signaling, respectively.

Neither TF controls the expression of all genes known to be regulated by any given signaling molecule; instead, they appear to control branches of these pathways implicated in defense responses. ERF1 has been proposed to integrate JA and ET defense pathways and WRKY70 to mediate cross-talk between SA and JA pathways.

Based on expression of marker genes, the *Arabidopsis* MYC2 was recently proposed to discriminate between different JA-mediated defense responses. It will be interesting to see if this claim is substantiated by genomewide transcript profiling. Chromatin immunoprecipitation (ChIP) permits identification of direct targets for TFs in real time and space.

Using this approach, StWhy1 was shown to bind to the potato *PR-10* promoter in response to wounding and elicitor treatment, while the recruitment of *Arabidopsis* TGA2 and TGA3 to the *PR-1* promoter was shown to depend on SA and NPR1. ChIP analysis of parsley WRKY1 revealed it was transiently recruited to regions containing W-boxes in the promoter of its own gene, as well as those of the immediate-early genes *WRKY3* and *PR1-1* following elicitor treatment.

Occupancy by WRKY1 was associated with reduced expression of *WRKY1* but enhanced expression of *PR1-1*. When not occupied by WRKY1, these promoters are otherwise constitutively occupied by different WRKY factors. It was proposed that activation of *WRKY1* and *PR1-1* involves elicitor-dependent post-translational modification of WRKY factors already occupying W-boxes in these promoters; the newly produced WRKY1 would be subsequently recruited to autoregulate expression of its own gene and activate late gene expression.

Post-translational modification of these WRKY factors may rely on MAPK signaling cascades as described in other section of this chapter. ChIP analysis of a subset of *Arabidopsis* genes differentially expressed in response to Pti4 overexpression revealed that about 60% (11 of 18) were direct targets of this ERF.

Three promoters that were not immunoprecipitated with the Pti4 antibody contained GCC-boxes, further emphasizing the notion that the presence of cognate cis-elements in a gene promoter does not necessarily make it a target *in vivo*.

Interestingly, seven of the promoters immunoprecipitated lacked GCCboxes within 1 kb upstream of the coding region. It was suggested that Pti4 may be capable of binding to a novel cis-element other than the GCC-box or be recruited to promoters indirectly through other DNA-binding proteins. New strategies such as ChIP chip and sequence tag of genomic enrichment (STAGE) now permit identification of direct targets of a TF on a genomewide basis.

Although neither method has yet to be applied for the characterization of plant TFs implicated in mediating defense responses, a whole-genome *Arabidopsis* tiling array is available that would be suitable for ChIP chip.

NPR1

Genetic Analysis of NPR1

Isolation of npr1 Mutants and Cloning of NPR1 Gene

The *npr1-1* mutation was found to block SA-inducible *PR* gene expression following treatment with the SA analog 2,6-dichloroisonicotinic acid (INA). Additional *npr1* alleles were subsequently recovered in different genetic screens aimed at identifying genes required for expression of *PR* genes in response to SA (*salicylic acid* insensitive1, renamed npr1-5), basal resistance against virulent *P. syringae* (*enhanced disease susceptibility* 5 and 53, renamed npr1-2 and npr1-3, respectively), and INA-induced SAR against *P. parasitica* (*non-inducible* immunity1; nim1).

The nim1 mutants have not been redesignated as npr1 alleles and continue to be referred to by their original names. All npr1/nim mutants, with the exception of nim1-5, are recessive; however, the dominant phenotype observed in nim1-5 plants is likely attributed to a second site mutation.

Because the npr1 phenotype cannot be rescued by exogenous SA, it was proposed that NPR1 functions downstream of this metabolite in the signaling pathway. Mutations in NPR1 compromise basal resistance against biotrophic pathogens such as *P. syringae*, *P. parasitica*, and *E. cichoracearum*. They cannot mount effective SAR against *P. syringae* or *P. parasitica* or ISR against *P. syringae*.

Also, different npr1 mutants have been found to be more susceptible to some incompatible races of *P. parasitica*, *P. syringae*, or *E. cichoracearum*. Loss of NPR1 function does not appear to affect age-related resistance against *P. syringae* or basal resistance against necrotrophic pathogens, including *A. brassicicola* and *B. cinerea*.

However, unlike the wild-type, npr1 mutants do not show enhanced resistance to *B. cinerea* following treatment with SA. Although npr1 mutants fail to express well-accepted marker genes for the SAsignaling pathway (*PR-1, PR-2, PR-5*) in response to treatment with SA or SA analogs, transcripts for these genes continue to be expressed in response to pathogen challenge (albeit at reduced levels) or in combination with mutations at other loci.

These observations indicate that one or more SA-dependent, NPR1-independent defense pathways exist in *Arabidopsis*. Despite being unresponsive to exogenous SA, npr1 mutants accumulate high titers of this metabolite following pathogen challenge and npr1 seedlings grown in the presence of SA bleach and die after developing cotyledons.

It has been proposed that NPR1 may be involved in feedback regulation of SA accumulation. Cloning of the NPR1 gene revealed that it encodes a protein with two identifiable protein–protein interaction motifs: a BTB/POZ (broad-complex, tramtrack, and brica-brac/pox virus and zinc finger) and an ankyrin repeat domain (ARD).

Several npr1 alleles affect conserved amino acids within the ARD, suggesting that this domain is important for NPR1 function. Although one mutation (npr1-2) maps to the BTB/POZ, it is within a nonconserved region of the domain (unpublished observation) and thus should not be used as evidence that the NPR1 POZ/BTB is required for disease resistance.

NPR1 contains no well-recognized DNA-binding motifs and cannot bind to the as-1 element *in vitro*. This suggests that NPR1 does not function as a sequence-specific TF; however, its ability to bind DNA has not been rigorously assessed. Sequences encoding NPR1 or NPR1-related proteins can be identified in many higher plants, suggesting that the NPR1 function is well conserved.

In support of this notion, VIGS analysis has demonstrated that a tobacco homolog of NPR1 is required for resistance to TMV and a tomato homolog is required for resistance to *P. syringae* (*avrPto*).

npr1 Suppressors

In efforts to identify additional genes implicated in mediating defense responses, several groups have screened for mutations that restore *PR* gene expression or disease resistance in *npr1* backgrounds (i.e., genetic suppressors). Several suppressors appear to have activated defense responses constitutively; even in the absence of pathogen challenge, they contain elevated levels of SA, express *PR* genes, and display enhanced basal resistance to pathogens, typically *P. syringae* and *P. parasitica*.

Many of these suppressors also display dwarfism and spontaneous disease-like lesions. Several other mutants displaying these properties were also recovered in unrelated genetic screens. The observed phenotypes were subsequently shown to be largely independent of NPR1, prompting some researchers to classify them as *npr1* suppressors. Conversely, others have argued that any mutant displaying elevated levels of SA should not be considered as a true suppressor of *npr1*.

In fact, two suppressors of this class contain mutations in putative R-genes, and the phenotypes likely result from the constitutive activation of R-gene signaling. A recessive mutation that restores SA-inducible PR gene expression and disease resistance in the *npr1-1* background is *suppressor of npr1-1 inducible1* (*sni1*). Although plants are dwarfed, they contain wild-type levels of SA, express only low constitutive levels of PR genes, and do not produce any spontaneous disease-like lesions.

Neither *sni1 npr1-1* nor *sni1 NPR1* plants are more resistant than the wildtype in the absence of INA. SNI1 encodes a novel leucine-rich nuclear protein that likely acts as a negative regulator of SAR. It was suggested that the probable role of NPR1 in SAR is to remove SNI1 repression. Currently, no evidence suggests that NPR1 and SNI1 physically interact. Increased resistance to virulent *P. syringae* and *P. parasitica* in the *suppressor* of nim1 (*son1*) nim1-1 double mutant is independent of SA and not correlated with increased levels of PR genes. Accordingly, it has been proposed to define a novel type of SAR-independent resistance (SIR).

Several PR genes are constitutively activated in the son1 *NIM1* (*i.e., NPR1*) background, indicating that *SON1* also participates in SAR, possibly upstream of or at NPR1. *SON1* encodes an Fbox protein. Given that the son1 mutation is recessive, SON1 is likely a negative regulator of SIR and SAR. It has been proposed that SON1 may target specific positive regulators of defense responses for degradation by the ubiquitin/proteosome pathway.

Other Genetic Interactions

Genetic interactions between *NPR1* and genes encoding TFs were described in other section of this chapter. The current section summarizes findings that link NPR1 to other regulators of plant defense responses.

SA-Dependent Signaling

The recessive *enhanced disease resistance1* (*edr1*) mutation leads to increased resistance against *P. syringae* and *E. cichoracearum* without constitutive activation of *PR* genes. *EDR1* codes for a MAP kinase kinase kinase. This enhanced resistance is lost in the *edr1* npr1 double mutant, suggesting that EDR1 is part of a MAPK cascade operating upstream of NPR1 that negatively regulates plant defense responses.

MAPK4 is a MAPK that also acts as a negative regulator of SAdependent defense responses; however, the *mapk4* mutant phenotype is not attenuated in the npr1 background, suggesting that it is part of a MAPK cascade that acts independently or downstream of NPR1. Overexpression of two soybean calmodulin isoforms, GmCaM4-5, *in Arabidopsis* results in constitutive expression of SA-inducible PR genes, spontaneous microHR, and enhanced resistance to virulent *P. syringae.*

Constitutive expression of PR genes is lost when *GmCaM4-5* are introduced into an npr1 mutant background, suggesting that it is mediated by NPR1. NPR1 is also partly required for the enhanced resistance to *E. cichoracearum* observed in the *powdery mildew resistant4* (*pmr4*) mutant. Pathogen-induced expression of *SA-INDUCTION DEFICIENT2* (*SID2*), which encodes ISOCHORISMATE SYNTHASE1, a key enzyme in SA synthesis, is elevated in the npr1 mutant.

In contrast, SA-dependent expression of *PAD4* is compromised in the npr1 mutant, although *P.* syringae-induced expression of this gene is not affected. These provide further evidence that NPR1 acts downstream of SA and indicate that its role as a negative regulator of SA metabolism may be achieved, at least in part, at the level of *SID2* expression or through a signal amplification loop involving PAD4. The double mutant between npr1 and the gain-of-function mutation *accelerated cell death6* (*acd6*) displayed dramatic alterations in cell enlargement and division resulting in production of abnormal growths.

Subsequent detailed analysis of the npr1 single mutant revealed reduced cell numbers with high ploidy levels. Thus, in addition to a role in mediating SA defense responses, NPR1 is also required for SA-dependent control of cell growth. Such effects were not apparent upon casual observation of the npr1 mutants. The requirement of NPR1 for runaway cell death in the lsd1 mutant has already been discussed. HR-associated cell death is also increased in the npr1 single mutant in response to infection with avirulent races of *P. syringae*.

In contrast, no cell death or hydrogen peroxide accumulation is detectable in the *nonrace-specific disease* resistance1 (ndr1) mutant following challenge with *P. syringae avrRpt2*. Genetic analyses demonstrated that ndr1 is epistatic to npr1 with respect to both these parameters. However, *NDR1* and *NPR1* have an additive effect on SAR and *PR-1* expression.

JA/ET-Dependent Signaling

Epistasis analysis has revealed links between NPR1 and the ET signaling pathways. Resistance to *B. cinerea* is not compromised in the npr1-1 mutant and is moderately reduced in *ein2*, a mutant affected in the ET pathway; however, the npr1-1 *ein2* double mutant is considerably more susceptible to this necrotroph.

Furthermore, breakdown of R-gene-mediated resistance to *P. syringae* (*avrRpt2*) is more severe in the double mutant than in either single mutant alone. Simultaneous application of SA with JA can inhibit JA-dependent responses in wild-type plants.

This inhibition is alleviated in the npr1 mutant, suggesting that NPR1 acts as a negative regulator of JA-dependent signaling. Interestingly, this function of NPR1, unlike its role in SA-mediated signaling, does not require localization to the nucleus.

Biochemical Function of NPR1

The protein sequence of NPR1 provided few clues as to its function, other than it likely interacts with protein partners through its POZ/BTB or ARD. Using the yeast two-hybrid system, several groups established that NPR1 interacts with various TGA factors, including the *Arabidopsis* TGA2, TGA3, TGA5, TGA6, and TGA7. As described in other section of this chapter, TGA1 and TGA4 are capable of interacting with NPR1 only after reduction or mutation of key cysteines.

The *Arabidopsis* NPR1 was also used as bait to screen a rice cDNA library and found to interact with three TGA factors related to TGA2 and a putative homolog of the maize liguleless. Screening of a tomato library with a putative ortholog of NPR1 from this species identified a group II TGA factor named NIF1 (NPR1 interacting factor1).

The tobacco group II factors TGA2.1 and TGA2.2 were also shown to interact with *Arabidopsis* NPR1 in directed two-hybrid tests. With the exception of *nim1-4* (Chern et al. and our unpublished observations) all *NPR1* mutant alleles that compromise disease resistance encode proteins that fail to interact with TGA factors in the yeast two-hybrid system.

Where tested, these mutants also fail to interact with TGA factors *in vitro*. This suggests that the ability to interact with TGA factors *in vivo* is important for NPR1 function and establishes TGA factors as downstream components of the NPR1 signaling pathway. The NPR1–TGA2 interaction observed in yeast and *in vitro* has been confirmed in plant cells. Importantly, these studies have revealed that the interaction is stimulated by SA.

Visualization of the interaction in protoplasts using a proteincomplementation assay revealed weak diffuse interaction throughout the cell under uninduced conditions, but intense nuclear interaction following SA treatment. This pattern is consistent with reports that TGA2 and NPR1 are predominantly localized in the nucleus following SA treatment.

In fact, nuclear localization of NPR1 was shown to be required for *PR-1* expression. Several lines of evidence indicate that interaction with NPR1 is important for TGA factor function *in vivo:*

NPR1 stimulates the DNA-binding properties of interacting TGA factors *in vitro,* including the reduced form of TGA1.

Protein extracts prepared from wild-type transgenic plants expressing a chimeric TGA2:GAL4 DNA-binding domain (DB) protein bound a probe containing GAL4 binding sites substantially better than similar extracts prepared from *npr1* mutant plants expressing the chimeric protein.

Activation of a reporter gene under the control of a promoter containing GAL4-binding sites was only detected in wild-type transgenic plants expressing the TGA2:GAL4 DB fusion; no expression was detected in the *npr1* background. The DNA-binding enhancement and transactivation properties of the TGA2:GAL4 DB chimeric protein observed in the wildtype plants depended on SA.

The ability of TGA2 and TGA3 to bind to the *PR-1* promoter, as measured by ChIP, depended on functional NPR1 and SA.

Although NPR1 clearly enhances the binding of interacting TGA factors by EMSA, it does not alter the migration of the protein–DNA complex. Furthermore, supershift experiments using antibodies against NPR1 antibodies indicate that NPR1 is not present in these complexes. This suggests that NPR1 stimulates the DNA-binding activity of TGA factors without binding stably to the TGA–DNA complex.

It is possible that the NPR1–TGA interaction is insufficiently robust to withstand the polyacrylamide gel electrophoresis or that, upon binding DNA, the TGA factors release NPR1. Although there are no reports of direct interactions between NPR1 and WRKY factors, several lines of evidence suggest a role for this group of TFs as regulators of *NPR1* expression and potential mediators of the NPR1-signaling pathway.

First, many genes differentially regulated in the *npr1* mutant contain W-boxes in their promoters. Second, the expression of numerous *WRKY* genes depends on NPR1 function. Finally, WRKY factors bind to W-boxes in the promoter of the *NPR1* gene, and mutation of these elements compromises *NPR1* expression.

Thus, NPR1 appears to act as a novel cofactor to regulate PR-gene expression by modulating the activity of TGA factors and possibly other groups of TFs in response to pathogen challenge. However, the observation that cross-talk between the SA and JA signaling pathways does not require nuclear localization of NPR1 suggests that this versatile regulator may have additional modes of action.

Post-Translational Regulation of NPR1

Yeast two-hybrid screens also identified a novel group of four *Arabidopsis* proteins called NIMINs (*NIM1* INTERACTORs). The interaction of NPR1 with NIMIN1 has been confirmed in plants. Overall, NIMIN proteins share limited sequence identity (14 to 44%); however, they contain short stretches of high similarity.

For example, NIMIN1 and NIMIN3 possess stretches of acidic amino acids that may serve as effector domains, and NIMIN1 and NIMIN2 each contain a cluster of basic amino acids that may represent a nuclear localization signal.

All NIMIN proteins contain a motif similar to the EAR domain necessary for transrepression of ERF proteins and are nuclear localized. *NIMIN* genes are transiently expressed following treatment with SA. Overexpression of NIMIN1 in *Arabidopsis* reduced PR gene expression in response to SA and avirulent *P. syringae* and compromised SAR against virulent *P. syringae*. These phenotypes are reminiscent of the *npr1* mutation; however, unlike that mutant, NIMIN1 overexpressing plants were also compromised in R-genemediated resistance against *P. syringae* (*avrRpt2*).

Overexpression of a NIMIN1 mutant protein unable to interact with NPR1 had no effect on PR gene expression or disease resistance, indicating that NIMIN1 function is mediated through NPR1. Mutation and RNAi suppression of *NIMIN1* resulted in increased expression of *PR* genes but had no measurable effects on resistance to *P. syringae*.

It was proposed that NIMIN1 may regulate only a subset of NPR1-dependent defense genes and that derepression of this specific gene set in the nimin1 mutant and RNAi lines is insufficient to confer disease resistance.

It is also possible that the closest relative of NIMIN1 (NIMIN1b) may partly compensate for loss of NIMIN1 function under some conditions. Furthermore, one cannot discount the possibility that the phenotypes observed in the plants overexpressing NIMIN1 may be related to squelching, with the vast excesses of NIMIN1 protein binding all available NPR1 at the expense of other NIMIN proteins or interacting partners.

Overall, functional analysis of NINIM1 indicates that it is a negative regulator of NPR1, possibly involved in modulating the amplitude of certain defense responses that may be relevant to plant fitness. The fact that NPR1 is constitutively expressed but changes subcellular localization following pathogen infection or SA treatment suggests that it is regulated at the post-translational level.

NPR1 shares limited sequence similarity with the mammalian transcriptional regulator IK-B, including conserved N-terminal lysines and serines potentially involved in ubiquitination and phosphorylation events, respectively.

However, there is no evidence to support the notion that NPR1 is regulated by either of these mechanisms. In contrast, some evidence does support post-translational regulation by redox changes of conserved cysteines. Using nonreducing SDS-PAGE to monitor the presence of disulfide bonds, the mobility of a protein fusion between NPR1 and the green fluorescent protein (GFP) was found to change dramatically in response to INA.

In uninduced samples, a large cytosolic complex was detected, consistent with an NPR1 oligomer held together by intermolecular disulfide bridges. Following INA treatment, a band corresponding to the monomeric size of NPR1-GFP gradually accumulated; the fusion protein was detected in the nucleus and *PR-1* expression activated. Accordingly, it was proposed that INA triggered the reduction of NPR1 disulfide bridges, yielding biologically active monomers. To identify residues affected by changes in redox status, NPR1-GFP fusions containing site-directed mutations at each of ten conserved NPR1 cysteines were expressed in transgenic plants. Mutation of two cysteines (C82 and C216) resulted in the constitutive expression of monomeric, nuclear NPR1-GFP as well as *PR-1* expression in the absence of INA.

Because altered residues would no longer be capable of forming disulfide bridges, they were predicted to mimic the reduced state of the cysteine residues. Together these results suggest that reduction of C82 and C216 is critical for NPR1 activation.

Consistent with this notion, monitoring the levels and ratio of reduced to oxidized glutathione following INA treatment revealed that after an initial oxidative phase, the cellular redox status became more reducing.

Together with those obtained with TGA1, these results indicate that properties of proteins mediating SA-dependent defense responses are regulated by SA-mediated reduction of key cysteine residues. Currently, the enzymes responsible for modulating the redox changes of TGA1 and NPR1 cysteines or how these changes specifically affect protein structure and function is not known.

ANALYSIS OF TRANSGENIC PLANTS OVEREXPRESSING NPR1

Arabidopsis plants expressing higher levels of NPR1 protein have increased basal resistance to *P. syringae, P. parasitica,* and *E. cichoracearum.* In one study, transgenic plants also displayed enhanced SAR following treatment with low levels of a chemical activator (benzothiadiazole; BTH) that did not elicit a response in untransformed plants. Most transgenic lines did not constitutively express SA-inducible *PR* genes.

In one study, increased resistance was correlated with stronger, rather than faster, *PR* gene expression; in the other study, resistance was correlated with the speed, rather that the level, of *PR* gene expression. A direct correlation between levels of NPR1 and disease resistance was reported in only one of the two studies. The *Arabidopsis NPR1* gene has also been overexpressed in two heterologous hosts.

Expression in tomato resulted in substantial resistance against virulent strains of *P. syringae* and *F. oxysporum;* moderate resistance to *Xanthamonas campestris, Ralstonia solanacearum,* and *Stemphylium solani;* and no enhanced resistance to *P. infestans,* cucumber mosaic virus, and tomato yellow leaf curl virus.

Levels of resistance to *P. syringae* and *F. oxysporum* were reported to be comparable, but not as complete, as those conferred by *R-genes. In* general, levels of NPR1 were correlated with the effectiveness of disease resistance; however, several exceptions were noted, leading the authors to propose that resistance may require a threshold level of NPR1 expression.

There appeared to be no correlation between the levels of *six PR* genes tested with levels of NPR1 or disease resistance. Expression of *Arabidopsis* NPR1 in rice resulted in increased resistance to virulent *Xanthamonas oryzae.* These results suggest that NPR1 function and signaling are conserved among dicots, monocots, and, in particular, rice, which has very high constitutive levels of endogenous SA.

Resistance to *X. oryzae* conferred by NPR1 was not as effective as *R-gene* resistance; however, substantial reduction of pathogen growth was observed in the leaf central vein, which limited bacterial spread and enhanced survival. As suggested in tomato, it appeared that a threshold level of NPR1 was required to confer resistance in rice.

Closer examination of the rice transgenics under different growth conditions revealed that overexpression of NPR1 can trigger the development of spontaneous disease-like lesions. This phenotype was correlated with the accumulation of hydrogen peroxide and could be potentiated by exposure to SA or BTH.

The transgenic plants were found to contain lower levels of SA—once again providing a link between NPR1 and regulation of SA metabolism. Neither the production of lesions nor changes in SA levels were observed in dicotyledonous plants overexpressing NPR1.

NPR1-RELATED GENES

Completion of genome sequence revealed that *Arabidopsis* encodes six NPR1-related proteins that can be separated into three discrete groups. All of these proteins contain the POZ/BTB and ARD domains implicated in mediating protein–protein interactions. Every amino acid affected by known *npr1* mutations that compromise PR gene expression and disease resistance is conserved within the family.

The genes encoding the three family members most closely related to NPR1 (At4g26120, At5g45110, At4g19660) also contain introns at the same positions as NPR1. *Three Arabidopsis* NPR1-related genes have recently been functionally analyzed. Of these, the product of *NPR4* (A4g19660) is the most closely related to NPR1, sharing 36% identity.

Similar to NPR1, NPR4 is localized in the nucleus and can interact with the same spectrum of TGA factors. *NPR4* transcript is induced by SA and pathogen challenge and is rapidly repressed by JA. Mutation of *NPR4* results in a modest increase in susceptibility to virulent *P. syringae* and *E. cichoracearum* but not *P. parasitica*. However, partial breakdown in R-gene resistance against some races of *P. parasitica* was reported. PR gene expression was only marginally reduced in the *npr4* mutant.

Although these results implicate NPR4 in regulation of PR genes and resistance against biotrophic pathogens, they also suggest that its role is not as prominent as that of NPR1. It was suggested that NPR1 and NPR4 may have distinct roles in modulating cross-talk between SA- and JA-dependent defense signaling. *BLADE ON PETIOLE* (BOP) 1 and 2 are the two most distantly related members of the NPR1-gene family. These genes have been implicated in regulation of plant development and their contribution to defense responses is unclear.

TO CROP IMPROVEMENT

Thanks in large part to use of model systems, our knowledge of TFs involved in regulating plant defense responses has increased very rapidly in recent years. It is clear that numerous TFs, encoded by different gene families, are required to ensure the highly coordinated expression of defense genes in response to pathogen challenge.

Many TF genes are highly regulated at the transcription level and the products they encode can act as positive or as negative regulators of gene expression. Thus, in addition to the activation of positive factors, the elimination of negative regulators—at the transcriptional or post-translational level—is required for timely activation of plant defense responses. Highly related TFs can be recognized within gene families.

The roles of these proteins appear to overlap substantially, making assignment of function difficult. Overall, differences in DNA-binding preferences, gene expression patterns, transactivation properties, and, probably, post-translational regulation indicate that members of a TF family have diversified to fulfill unique roles.

Although interactions between members of a family and between different families of TFs are likely critical for controlling defense gene expression, they remain poorly understood. The same is true of the post-translational modifications that alter the functionality of TFs following pathogen challenge. Even though our understanding of how plant TFs regulate defense responses is far from complete, several studies have revealed that changing expression levels of individual TFs or cofactors can have a profound effect on disease resistance.

Given that plant defense signaling pathways are broadly conserved across plant species, TFs found to be effective at enhancing resistance against specific pathogens in *Arabidopsis*, or related proteins from crop plants, may also function against the same or similar microbes in the crop species. It is noteworthy that the vast majority of studies in Arabidopsis were performed under controlled laboratory conditions.

It will be important to determine whether overexpression of TFs in crop plants will confer commercially relevant levels of disease resistance under field conditions. A recent study indicated that Arabidopsis mutants constitutively expressing defense responses had reduced fitness, although NPR1 overexpressors were unaffected. As discussed in other section of this chapter, many plants overexpressing TFs implicated in mediating defense responses also display developmental abnormalities.

Obviously, these undesirable side effects must be minimized or eliminated before any commercialization can be considered. A better understanding of where the TF lies within a transcription cascade may allow targeting of proteins that have minimal impact on development. Manipulation of interactions with other proteins or of post-translational modifications may also alleviate negative side effects, as would use of promoters that offer better control of transgene expression.

9

Chapter

MOLECULAR BIOLOGY AND GENETIC ENGINEERING

Polyamines (PAs) are naturally occurring polycationic aliphatic amines, which due to their ubiquity and versatility are involved in the regulation of various cellular and molecular processes. They are positively charged compounds with their charge distributed along the molecule. The common PAs, spermidine (SPD) and spermine (SPM) and their diamine precursor putrescine (PUT) play a critical role in the normal functioning of all cells.

They are involved in the cellular functioning both at the molecular and physiological levels due to their association with various macromolecules (DNA, RNA and proteins) and membranes as well as their high concentration in the cytosol thus behaving as osmolytes. The role of PAs is much better studied in animal systems than plants, though they have been suggested to have a role as new plant growth regulators either by mediating the plant hormone effects or independently signalling other responses.

PAs exist in three forms in the cell, viz. as free cations, covalently bound to low molecular weight phenolic compounds like hydroxycinnamic acids (conjugated form of PAs) and bound to marcomolecules or membranes (bound form of PAs). Though the major form is the free cationic form of PAs, there are instances when the amounts of conjugated form exceed the free form and these are known to be critical in certain physiological processes including seed germination, flower development, defence responses and stress reactions.

Besides PUT, SPD and SPM, there are certain unusual PAs found in nature, e.g. thermo-SPM which have been detected in bacteria residing in hot springs and they seem to be important in protecting the enzymes from heat denaturation and aminobutylhomo-SPD found in fast growing cells of root nodule bacteria *Rhizobium*. NorSPD and norSPM are found in thermophilic red algae, brown algae, and *Chlamydomonas, Nitella and Chlorella.*

Similarly, some unusual PAs have been reported in plants, homo-SPD was first detected in sandalwood and also in mosses and ferns. In leguminous plants, other unusual PAs like canavalmine, homoagmatine, aminopropylcanavalmine and aminobutylcanavalmine have been detected. NorSPD and NorSPM have been detected in alfalfa grown under drought conditions and have been postulated to play a protective role under stress conditions.

As a matter of fact it has been suggested that PA distribution, especially of SPM, may serve as a phylogenetic marker. The study of plant PAs has come a long way since the first report by Bagni regarding the stimulatory effect of a PA (PUT) on growth of *Helianthus tuberosus* explants. Since then, PAs have been demonstrated to be associated with regulation of somatic embryogenesis, root and shoot formation, flower and fruit development, stress responses and senescence.

In fact, PAs may serve as 'biomarkers' for *in vitro* morphogenetic potential including plant regeneration via somatic embryogenesis. The multifaceted functions of PAs as well as the variations in their levels in response to changes in the physiological state, point towards their role as possible second messengers, though their high titres do not support the view.

Various studies have been conducted to investigate the involvement of PAs in cell functioning, using mutants of PA biosynthetic genes and specific substrate-based inhibitors of PAs. Though much information could be generated regarding the involvement and possible mechanisms of action, no clear picture of their functioning emerged.

Hence, transgenic plants expressing PA biosynthetic genes in constitutive and regulated manner were generated, with an aim to answer some of the queries regarding the functioning and role of PAs. This article deals with the molecular biology of PAs in plants, with special reference to transgenic plants expressing PA biosynthesis genes.

POLYAMINE METABOLISM

The diamine PUT is universally derived from ornithine by the rate limiting enzyme ornithine decarboxylase (ODC). Plants, bacteria and some fungi have an alternate pathway for PUT synthesis from arginine by arginine decarboxylase (ADC). In plants, ADC forms the major pathway for PA biosynthesis.

The PUT, hence derived from either of the two pathways is converted into higher PAs, the triamine SPD and the tetraamine SPM by the addition of aminopropyl groups obtained from decarboxylated S-adenosylmethionine (dcSAM). The dcSAM is formed by the decarboxylation of SAM by SAMDC, SAM is in turn synthesized from methionine by SAM synthase.

SAM is a major methyl donor in the cellular metabolism and also forms a part of ethylene biosynthesis. SAM is converted to 1-aminocyclopropane-1-carboxylic acid (ACC) by ACC synthase, which is converted to ethylene by ACC oxidase. The catabolism of PAs involves the enzymes diamine oxidase (DAO) and polyamine oxidase (PAO).

DAO preferentially acts on diamines (PUT, cadaverine) to form pyrroline, ammonia (NH_3) and

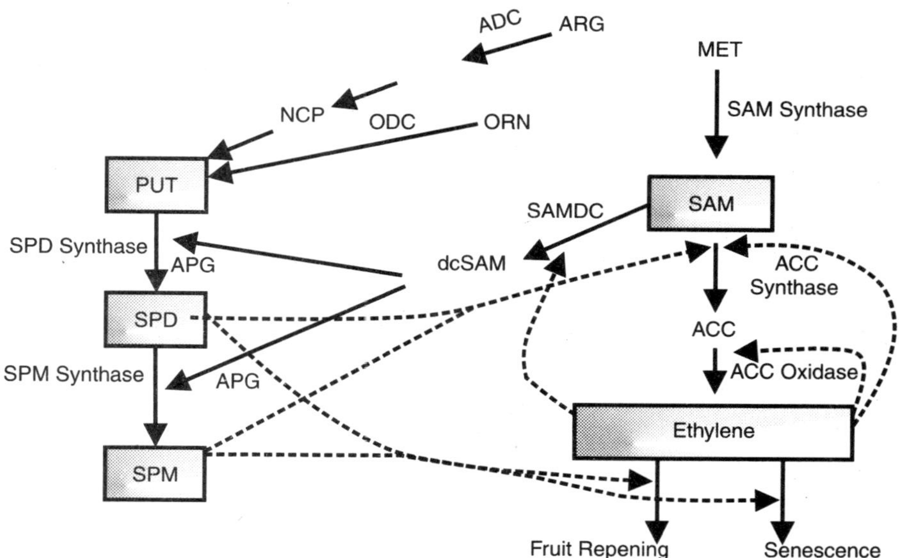

Figure 9.1: Metabolic and functional inter-relationships between polyamine and ethylene metabolism. Dotted lines show inhibitory effecs of respective metabolism on the other, whereas dashed lines depict stimulatory effects. ARG, arginine; ORN, Ornithine; MET, Methionine, PUT, Putrescine; SPD, Spermidine; SPM, Sperimine; ODC, Ornithine decarboxylase; ADC, Arginine decarboxylase; SAM, S-Adenosyl methionine; SAMDC, SAM decarboxylase; APG, Aminopropyl group, ACC, Aminocyclopropylcarboxylic acid; Agm, Agmatine; NCP, N-Carbomyl putrescine.

hydrogen peroxide (H_2O_2), though it can break down SPD to aminopropyl pyrroline, NH_3 and H_2O_2. PAO, on the other hand, breaks down SPD and SPM to pyrroline and aminopyrroline respectively, and diamine propane (DAP) and H_2O_2.

Pyrroline formed due to the activity of DAO and PAO is further converted to ψ -aminobutyric acid (GABA) by a nicotinic acid diamine (NAD) dependent dehydrogenase. Besides oxidases there might exist alternate pathways for PA catabolism as there are many plants in which oxidases have not been detected. Alternative diversion of the PAs into other metabolic pathways also plays a role in the regulation and the dynamics of the PA metabolism.

PUT forms the precursor for pyrrolidine ring of the nicotine and tropane alkaloids. PUT is converted to N-methyl PUT by PUT methyl transferase (PMT) which then forms the pyrrolidine ring of nicotine and other tropane alkaloids.

ROLE OF POLYAMINES IN BIOLOGICAL PROCESSES

Most of the PA functions can be attributed to their polycationic nature and the distribution of the charge along the molecule, which allows them to bind to a variety of molecules, including nucleic acids, protein and cell membranes in the cell and regulate their functions.

A very crucial binding of PAs is with membrane phospholipids thus stabilizing them and reducing chlorophyll loss when bound to the thylakoid membrane, as well as preventing the lowering of membrane potential and the Ca^{2+}, PO_4^{-3} fluxes of mitochondrial membrane under saline stress. Similarly, PAs have been found associated with cell wall components like lignin and pectins and have been implicated in maintaining cell wall characteristics by strengthening the links between cell wall components.

PAs also play a role in cell wall expansion and are part of modulators involved in host-pathogen interactions. The binding of SPD to the plasma membrane proteins in zucchini hypocotyls has been characterized. SPD was found to have a specific binding to a 44 and a 66 kDa protein.The binding of PAs with DNA is also known to be responsible for playing some part in the regulation of synthesis and function of DNA, including gene expression.

It had been proposed that PAs affect growth by interacting with DNA and SPM plays a part in B to Z DNA transitions. SPM has been reported to stabilize the triplex DNA formation and aggregation. PAs have also been reported to stimulate DNA, RNA and protein synthesis.

They are involved in joining of okazaki fragments and the depletion of PAs leads to accumulation of short DNA pieces. As a matter of fact *odc has* been suggested to be a proto-oncogene and its over expression leads to cell transformation. PAs are also known to bind with RNA molecules, and protect them from RNases. SPD and SPM are known to stimulate the reading of amber mutations and play an active role in the expression of specific genes.

The antisenescence effects of PAs are in part due to the inhibition of ribonuclease synthesis and activity by PAs. PAs also inhibit the protease activity, thereby delaying the degradative processes initiated during stress or senescence. PAs regulate their own biosynthesis by inducing a ribosomal frame shifting in the translation of ODC antizyme. Besides, being involved at the DNA and RNA levels, PAs are also known to regulate protein synthesis and activity. SPM has been reported to have a specific role in activity of cyclic AMP-independent caesin kinase.

A branched quarternary PA, tetrakis β-aminopropyl) ammonium along with SPM has been reported to support protein synthesis at high temperature in a thermophilic bacteria. SPD stimulates protein synthesis in chloroplast especially in light. PUT, SPD, SPM and cadaverine enhanced the phosphorylation of several plasma membrane proteins in tobacco, cucumber *and Arabidopsis.* PUT has been reported to increase the phosphorylation of many soluble proteins, unlike SPD and SPM, which decreased the phosphorylation of soluble proteins.

PAs are known to regulate cell division and also prolong the cell division phase by inhibiting the synthesis of phenylpropanoids. The conjugation of PAs with phenolics regulated the free PA levels and therefore led to the cessation of cell division. The critical role played by PAs in growth and development along with their role in protein phosphorylation/dephosphorylation and the binding of PAs (SPD) to specific protein in the thinlayer of tobacco together point towards the possibility of these being considered as signal transduction molecules.

Though the high PA titres in the cell are quite unlike secondary messengers, which increase rapidly, and transiently in response to stimuli. Therefore, this aspect of PAs warrants more attention to be able to pinpoint their role and mechanism of involvement, if any, in signal transduction. The study of physiology, biochemistry and genetics of the PA metabolism was initiated with the generation of mutants of PA metabolism in *E. coli,* yeast and plants.

The *E. coli* mutants defective in their PUT synthesis were the first to be isolated, these preferentially used the ODC pathway in the absence of any PA supplements and the ADC pathway in the presence of arginine. The substitution of PUT by its analogs which could not be converted into the higher amines could not restore the growth of the mutants, whereas the inclusion of SPD analogs was helpful, these results demonstrated the critical requirement of SPD for growth.

The *E. coli* mutants defective in the SAMDC function were also isolated, but their growth rate was almost unaffected. Further, *an E. coli* strain deficient in ADC, ODC and SAMDC function was isolated which was able to grow at a reduced growth rate. This strain was utilized in elucidating the role of PAs in ribosomal complex formation and hence protein synthesis.

Mutant studies in *Saccharomyces cerevisiae* revealed the presence of a single biosynthetic pathway (ODC) for the synthesis of PUT as well as the absolute requirement of SPD and SPM for growth and sporulation in yeast. PA mutants have also been raised in the model plants, *Arabidopsis* and tobacco. The *Arabidopsis* mutants with lower levels of ADC and ODC activities had abnormal root, shoot and floral morphology.

Malmberg and McIndoo have isolated tobacco mutants with abnormal floral morphologies, including flowers with large non-functional stigma, anthers with nonviable pollen and ovary with most ovules turned into anthers. These mutants were deficient in the function of SAMDC and brought out the role of SPD and SPM in flower development.

The ODC mutants failed to flower, demonstrating the involvement of ODC in floral initiation. Tobacco mutants have been raised by activation T-DNA tagging, such that the regions adjacent to the insertional position over-expressed the gene, and these were selected on selective concentration of MGBG.

The mutated plants showed altered phenotypes, abnormal floral morphology, male sterility and parthenocarpy, with higher SAMDC activity and SPD levels. The study of PAs pertains much to the availability of specific, irreversible, substrate- or product-based inhibitors of its biosynthetic enzymes. The substrate analogue of ODC, a-difluoromethylornithine (DFMO) was the first inhibitor to be synthesized. Similarly adifluoromethylarginine (DFMA) has been used as a potent inhibitor of ADC.

Another set of similar substrate-based analogous for ODC and ADC were monofluoromethylornithine (MFMO) and monofluoromethylarginine (MFMA), respectively. These have been reported to be much more potent than DFMO and DFMA. The substrate-based inhibitor for lysine decarboxylase, a-difluoromethyllysine (DFML) is also available.

A very potent inhibitor of the enzyme SAMDC is methylgloxyl *bis* (guanylhydrazone) (MGBG), though it has been put to limited use due to its non-specific effects on the respiratory enzymes. The inhibitors have provided insight into the dynamics of inter-conversion and regulation of the levels of PAs in the cell. Yet no clear picture of their mechanism of action or extent, period or stage of involvement emerges.

The polycationic nature of PAs allows them to interact with various molecules whose functioning is modulated and regulated by them. PAs also have a role in free radical scavenging due to their polycationic nature. Although there are many other suggested possible mechanisms regarding the functions of PAs, the exact role, the extent of involvement and the mechanism of action of PAs is as yet not very clearly understood. However, the development of transgenics expressing PA

Table 9.1: The polyamine metabolicgenes cloned from different organisms.

PA metabolic gene	Source
odc	*E. coli, Trypanosoma, Leishmania, Datura,* tomato, mouse, yeast and human
adc	Oat, pea, soybean and tomato
samdc	Potato, *Arabidopsis,* Spinach, *Catharanthus, Tritordeum* and *Pharbitis nil*
spd syn	*Nicotiana sylvestris, Hyoscyamus niger* and *Arabidopsis spm syn*
	Human
homo-spd syn	*Acetobacter, Senecia vernalis* and *Eupatorium cannavulgaris dao*
	Lentil and pea
pao	Maize

biosynthetic genes driven by constitutive promoters offered a good opportunity for the study of the PA functions.

Since PAs are fundamental to the process of morphogenesis, the generation of such transgenics presented a problem. Also the transgenics recovered had abnormal phenotypes. Inducible promoters were applied to overcome these problems but ambiguous effect of the inducer on the PA metabolism and the problem of sustained induction remained as hurdles.

The transgenics raised have been used to study the role of PAs in plant development and also the dynamics of their metabolism, but detailed studies were not conducted on the plant development and stress responses of the transgenics over-expressing PA biosynthesis genes as well as their response during *in vitro* morphogenesis. Therefore, much work is needed on the genetic manipulation of PAs in plants to clearly demonstrate the role of PAs in a variety of cellular and molecular processes.

CLONING OF POLYAMINE METABOLIC GENES

The PA biosynthetic genes were first isolated from animal systems, yeast and bacteria. The *odc* gene has been cloned from human, rat, mouse, holstein, *Trypanosoma, Leishmania,* yeast, *Neurospora* and *E. coli.* Recently PA biosynthetic genes have been cloned from plants too. The *odc* gene has been cloned from *Datura,* tobacco and tomato. The *adc* gene has been cloned from oat, tomato, *pea, Arabidopsis* and soybean.

The *samdc* gene has been cloned from *Arabidopsis, Datura,* potato, spinach, *Catharanthus roseus, Tritordeum, Pharbitis nil,* tomato, tobacco, rice and also from human genome. The *spd syn* gene has been cloned from *N. sylvestris, Hyocyamus niger and Arabidopsis.* Recently *spm syn* gene has been cloned from human genome. The genes coding for the enzymes involved in formation of conjugated PAs have also been cloned. The gene for the enzyme homo-SPD synthase has been cloned from bacteria (*Acetobacter*), *Senecio vernalis and Eupatorium cannavulgaris.* Besides, the biosynthetic enzyme, the gene for the catabolic enzyme PAO has been cloned from maize and DAO from lentil and pea. The cloned PA metabolic genes are summarized in Table elsewhere in this chapter.

The plant decarboxylases (both the ADC and ODC) belong to the group IV of decarboxylases and have the 23 amino acids which have been shown to be critical for the activity, conserved in them. There exists a strong homology between the oat, tomato and soybean *adc* genes, further their catalytic motifs are more than 75% identical.

The genes for related pathways like, homo-SPD synthase and PUT-N-methyltransferase seem to be evolved from the genes of the basic PA metabolism. There are evidences for the presence of more than one copy of *adc* or *odc* gene in the plant.

TRANSGENIC PLANTS EXPRESSING POLYAMINE METABOLIC GENES

Transgenic plants expressing PA biosynthetic genes were generated to gain better understanding of the PA metabolism, reconfirm the effects of the modulation of PA titres caused by the inhibitors at the molecular level and also to overcome the limitations of the inhibitor-based experiments.

The transgenic approach was highly specific to the target gene and moreover it provided a tool for manipulating the metabolic flux with the persistant shift in the PA metabolism. Even though some of the plant PA biosynthesis genes have been isolated and characterised, most transgenics have been raised using genes from heterologous source as these were the first to be isolated.

In most of the transgenics generated, CaMV35S promoter has been used to drive the transgene, though tetracycline (tet)-inducible promoter has also been used in cases where extreme deleterious effects of the transgene were expected. Some of the transgenic plants expressing PA biosynthesis genes are listed in Table elsewhere in this chapter.

The first report of the introduction of yeast *odc* gene was in root cultures of tobacco using *Agrobacterium rhizogenes*. The study was aimed to increase the nicotine content of the culture as PUT is a precursor for nicotine. Hence, over-production of PUT was attempted by using a double enhancer sequence containing promoter but only a 3-fold increase in ODC activity and a 2-fold increase in nicotine was observed.

This was suggested to have been the result of a tight regulation of nicotine biosynthesis or the activity of the enzyme PUT methyl transferase becoming limiting. PUT might be incorporated into many other secondary metabolic pathways, like alkaloids, which are present in significant amounts in Solanaceous plants and also conjugation of PUT could have been another pathway for the PUT synthesised due to the over-expression of the *odc* gene.

No significant increase in SPD and SPM levels was observed as their biogenesis is also precisely regulated. No plants were regenerated from the transformed root cultures. The tobacco transgenic plants expressing the mouse *odc* gene were the first PA transgenic plants raised by DeScenzo and Minocha. Two constructs were used for the transformation of tobacco, one having complete coding sequence of *odc* and the other in which 350 bp of 3' end were removed. The truncated gene produced a functional peptide 37 amino acids less at the C terminus and an increased half-life. The enzyme activity when checked at the pH optimum for mouse ODC was much higher in transgenics compared to endogenous plant ODC activity in the controls, whereas at the pH optimum for the plant ODC, no significant change was observed.

PUT was found to be 2–3 fold higher in leaves and 4–12 fold higher in callus, though no significant increase was observed in SPD and SPM content as the amounts of SAM were suggested to be limiting. The transgenics having high PUT titres were stunted with wrinkled leaves and reduced stamens. Carrot cell lines are known to have no detectable ODC activity, only the ADC pathway is functional in them.

Carrot cell lines were transformed with mouse *odc* gene driven by CaMV35S promoter and the effect of the high PUT titres on somatic embryogenesis was studied. The transformed cell lines showed improved somatic embryogenesis which could be correlated to higher PUT amounts. The somatic embryos were formed even in the presence of DFMA which inhibited the carrot ADC, therefore all the PA requirements of the embryos were fulfilled by the introduced mouse *odc* gene.

Exogenous addition of PUT was not found to be helpful thus suggesting that a fast turnover of PUT is also essential besides the high concentration for somatic embryogenesis. These transformed carrot cell lines were used to study the shift in metabolic flux as compared to the control cell lines. ^{14}C labelled arginine, ornithine, methionine or PUT was fed to the cell cultures

Table 9.2: Transgenic plants expressing polyaminemetabolic genes

Gene	*Gene source*	*Transgenic plant*
odc	Yeast	Tobacco root cultures
	Mouse	Tobacco
odc	Mouse	Carrot
	Mouse	Rice
	Mouse (antisense)	Rice, Tobacco
adc	Oat	Tobacco
	Oat	Rice
samdc	Human	Tobacco
	Potato (sense)	Potato
	Potato (antisense	Potato
samdc-odc	Human-Mouse	Tobacco
dao	Pea (sense)	Pea
	Pea (antisense)	Pea

and amount of label incorporated in different PAs and their fractions was analyzed. ^{14}C labelled PUT was much higher in trasgenic cell lines when 14C ornithine was given as substrate and there was no difference in labelled PAs when ^{14}C arginine was fed to the cultures. In correlation the conversion of ^{14}C-methionine to ethylene was much lower in transgenics due to a shift in the dynamics towards PA metabolism as more of PUT was available to be converted to SPD and SPM. Alterations in the PA levels during *Agrobacterium-mediated* genetic transformation with a reporter gene (*gus*) and mouse *odc* gene were found to affect the regeneration potential of indica rice.

It has been suggested that the modulation of PA metabolism may be used to improve the regeneration from transformed calli in rice and other crops. In a recent study, it was demonstrated that over-expression of human *odc* gene in transgenic rice plants alters PA pools in a tissue-specific manner. It was suggested that ODC rather than ADC is responsible for the regulation of PUT synthesis in plants.

In these transgenics, significant changes in the levels of all three major PAs were observed in seeds and also in vegetative tissues (leaves and roots) as compared to *oat odc* transgenics, wherein PUT and SPM levels were higher in seeds only. Transgenic tobacco plants over-expressing human *samdc* gene driven by CaMV35S promoter were generated by Noh and Minocha.

These plants were found to have 2–6 fold higher SPD than the untransformed controls, there was an increase in SPM too though PUT decreased. Since high amounts of SPD is cytotoxic, the regenerants obtained might have been moderate accumulators of SPD as the increase in SPD and SPM was not comparable to the dramatic decrease in PUT.

The cytotoxicity of the high amounts of SPD did not allow the regeneration of any plants overexpressing potato *samdc* in potato, hence a tet-inducible promoter was used to drive *samdc*. The sense *samdc* plants showed 7-fold increase in SPD, 3-fold in SPM and a decrease in PUT on tetinduction. Similarly potato plants expressing antisense *samdc* gene driven by both 35S promoter and tet-inducible promoter were raised.

The antisense *samdc* plants were stunted, branched, necrotic with few small tubers; this was attributed to increase in ethylene levels caused by the channelling of SAM for the formation of ethylene due to down-regulation of SAMDC. A decrease in PUT levels was observed due to down-regulation of ODC and ADC by the elevated levels of ethylene.

Phenotypic abnormality, i.e. delay in flowering was also observed in case of tobacco plants transformed with *Agrobacterium rhizogenes* and was attributed to a delay in the appearance of conjugated PAs due to the decrease in ODC and ADC activities. Bhatnagar et al. studied the genetic manipulation of PA metabolism in poplar cells by introducing mouse *odc* gene.

It was observed that over-expression of the heterologous gene resulted in high levels of PUT and increase in PUT degradation in the transgenic cells. In continuation of the above study, they showed that there was an increased turnover of PUT as well as its conversion to SPD as compared to the untransformed cells, the increase being proportionate to the cellular content of this diamine.

Furthermore, the increase in PUT catabolism in the transgenic cells did not result in any major changes in the activity of DAO or the half-life of PUT. The fact that elevated levels of PAs were detrimental for plant regeneration and were cytotoxic, led to the use of tet-inducible promoter for driving *oat adc* gene introduced into tobacco.

The PUT levels were increased by 16-46% on tet-induction and more significant increase was seen in the conjugated and bound fractions of PAs. A prolonged induction of the transgene at an early stage of development led to plant growth inhibition, necrotic, wrinkled leaves, but no such effects were seen in case of older plants on tet-induction of the transgene.

These results clearly brought forward the differential role of PAs at various developmental stages. Rice transgenics over-expressing the *oat adc* gene have also been raised. These transgenics showed a 4-7 fold increase in the activity of the ADC enzyme, along with a 4-fold increase in the PUT titres. The high PUT titres were found to be inhibitory to plant regeneration from the transformed calli.

The effect of the strength of the promotor driving the *adc* gene on the PA metabolism as well as the morphogenic capacity of the transformed calli has also been analyzed. In this study, *oat adc* gene under the strong maize ubiquitin promoter 1 (ubi-1) was introduced into rice but even then no significant change in PA levels was observed in seeds or in the vegetative tissues.

However, Noury et al. reported that only one specific transgenic line showed a significant increase in PUT and SPM levels in vegetative tissues and seeds. R1 generation rice transgenics expressing the *oat adc* gene driven by ABA responsive promoter were tested for their response to various environmental stresses.

Since PAs are known to play a role in stress responses, particularly the activity of the enzyme ADC is known to increase under stress along with increase in PUT levels, rice transgenics with *adc* gene were tested for their tolerance to abiotic stress (drought) and it was reported that no chlorophyll loss was observed after 8-days of drought as compared to the untransformed control plants. Tobacco transgenics over-expressing mouse *odc* gene affects cellular PAs *and in vitro* morphogenesis, and confers salt stress tolerance.

Further, it was seen that favourable changes in PAs titres and the optimum PUT: SPD ratio in transgenic lines showed better regeneration, and previously similar results were reported in indica rice genotypes. Transgenic pea plants with PA catabolic gene *dao* in sense and antisense orientation have also been generated in order to study the role of DAO in nodulation.

The sense plants showed an increased DAO activity and reduced PUT levels. It was observed that DAO activity was not involved in nodule formation, but may have a role in regulation of PA levels in host cells. Though the SPD synthase and the catabolic enzyme PAO genes have been isolated, no transgenic plants have been raised with them, neither are there any transgenics reported expressing the *adc and odc* genes in antisense orientation for studying the effects of long-term downregulation of these genes on plant development.

Transgenic plants of tobacco and/or rice were also generated in our laboratory with *oat adc,* human *samdc and spd syn,* and the transgenics have been analysed for the effects on cellular PA concentrations, PA biosynthetic enzyme activities, plant development and abiotic stress responses.

The PA titres in *adc and samdc* tobacco transgenics were also analysed. In case of *adc* transgenics a significant increase in the PUT and SPD levels with no apparent changes in SPM was observed. The increase in the PA levels was comparable to the concurrent increase in the activity of ADC as well as SAMDC. The ODC activity was found to be decreased in these transgenics.

The activity of the PUT catabolic enzyme DAO was measured, and a higher activity was found in all the transgenic lines tested, suggesting that the increase in DAO activity may be important to maintain optimum PA levels in the cell. In case of SAMDC transgenics, in addition to an increase in SPD and SPM levels, there was a significant increase in PUT levels, which might be due to the re-conversion of SPD to PUT via acetyl-SPD or amino butyraldehyde.

These transgenics have exhibited very high SAMDC activity, which was accompanied by higher DAO activity. They also showed marginal increase in ODC activity. Tobacco transgenics with *adc and samdc* genes were also tested for stress responses. They showed increased tolerance to salinity (250 mM NaCl) and PEG (10%) mediated drought. Interestingly, these transgenics also showed enhanced resistance against fungal (caused by *Verticillium dahliae and Fusarium oxysporum*) and bacterial (caused by *Ralstonia solanecearum*) wilts.

Some of the above single PA transgenics (e.g. *odc* transgenics) exhibited morphological aberrations like stunted plants with wrinkled leaves, which might be due to altered PA levels and PUT: SPD ratios. This problem may be overcome by up-gradation of the entire PA pathway in transgenic plants by the simultaneous introduction of the PUT synthesis gene (i.e. *odc*) and SPD synthesis gene (i.e. *samdc* or *spd syn*) by co-transformation. Indeed, such an attempt has been made in our laboratory, and double transgenic tobacco plants were produced with mouse *odc and* human *samdc* genes.

It was observed that the double transgenic tobacco plants were normal and did not show any morphological abnormalities. Further, regeneration was also better from the transformed leaf explants with both *odc and samdc* genes as compared to explants from single transformants. These results further substantiate the role of PUT : SPD ratio in *in vitro* plant regeneration.

Further, stress assays with double tobacco transgenics revealed increased tolerance to salinity and bacterial wilt. Since most of the phytopathogenic fungi have only the ODC pathway for the synthesis of PAs, a novel method for the control of fungal plant infections by the selective inhibition of the fungal ODC by using its specific inhibitor (DFMO) has been reported.

The selective inhibition of the fungal ODC might also be achieved at the molecular level by the use of the antisense RNA technology. The endoparasitic fungi which infect the plant might take up the antisense *odc* transcripts produced in the transgenic plants along with the nutrients from the plant cells, thereby leading to the inhibition of the fungal ODC and growth.

The above hypothesis was tested for the antisense *odc* transformed tobacco plants for the control of fungal wilt caused by *Verticillium dahliae.* The transgenic lines tested showed increased resistance to fungal infection as compared to the untransformed control plant. However, some more studies would be needed to prove this hypothesis. There seemed to be distinctive roles played by either of the PA biosynthetic enzymes, with the ODC being critical in the early development of roots as the root development was affected during the regeneration of the antisense *odc* transformed rice and tobacco plants.

The high ODC activity or perhaps the high PUT titres were on the other hand inhibitory for the early growth and development probably because they might lead to an increase in DAO activity as PUT forms the substrate for the enzyme, which results in the production of H_2O_2, higher concentrations of which are cytotoxic.

Even though there existed distinctive roles for either of the two enzymes, it was observed in case of sense *odc* plants that the high titres of PUT could compensate for the decrease in the activity of the enzyme ADC, as was observed in the response of sense *odc* plants to abiotic stress.

CONCLUSION

The PA transgenics have provided a lot of information regarding the long-term effects of the shifts in the metabolism on plant growth and development, but they can be further used for gathering more information regarding the effects of the variations in the PA metabolism on other fundamental metabolic pathways in the cell. Such studies may provide an insight into the extent of involvement of each metabolic pathway in a specific response or phenotype.

Also, there seems to be distinctive roles played by either of the PA biosynthetic enzymes, as can be deduced from the variable phenotypes observed in the various transgenics; this suggests a possible role for the improvement of modulation of the specific responses.

The transgenics also have the advantage of being distinctive from each other due to the random integration of the transgene, which influences the expression of the transgene and hence the shifts in the physiology in each plant were different.

Therefore, an array of different transgenic lines each showing variability in the expression of transgene that were generated, could provide information regarding the physiological effects of variable gene expression.

A clearer picture would emerge for these transgenics with the analysis of the fate of the PAs accumulated in these plants, as the PA catabolic enzymes as well as their channelling into the other pathways like ethylene and tropane alkaloids etc. also play a role in the metabolic flux of the cell.

Besides the PA metabolism, the variations in the dynamics of other related metabolic pathways and their cumulative effect on the phenotype and physiological responses of the transgenics would give a better view of the cross-talk amongst pathways and its role in the functioning of the cell.

Also, the PA transgenics raised need to be tested for their tolerance to various stresses to elucidate the role played by the transgenics in stress responses and also the degree of tolerance imparted by them. These transgenics may also be used for studying various other biological processes, including senescence and fruit ripening.

The use of the PA biosynthesis genes to generate stress tolerant plants has a major hurdle of the plants having an abnormal phenotype. Therefore a better approach to the problem would be to up-grade the entire metabolism (by the introduction of the *odc/adc* gene in conjunction with SPD and SPM synthesis gene *samdc/spd* syn) instead of a singular step, so that the plants would possess a normal phenotype and yet be tolerant to stresses.

In fact, this approach was examined in our laboratory and proved to be correct. Further, it has been observed that the optimum PUT : SPD ratio is very important for normal plant growth and development. Therefore, the maintenance of a balance between PUT and SPD ratio appears to be very important for obtaining normal transgenics.

This may be achieved by transforming the plants with the genes for both PUT and SPD synthesis. We were able to produce a large number of normal tobacco transgenic plants with mouse *odc and* human *samdc* genes by co-transformation as well as step-wise transformation. Such an approach would be quite useful in producing normal transgenic plants with PA biosynthesis genes and they would also impart abiotic stress tolerance.

10

Chapter

TRANSGENIC CROPS

This review concentrates on recent advances in the production and use of transgenic crops and should be read in conjunction with previous reviews. It does not consider issues such as the possible impact of genetically modified (GM) crops on the environment or the safety of GM food, as these subjects have been adequately reviewed elsewhere.

During the preparation of this chapter, as well as consulting the usual sources of research publications, extensive use has been made of the freely available patent databases in the United States, Europe, World International Patent Organization and other international sites.

In addition, this review includes data from various sites providing information on the field tests of GM crops in the United States and elsewhere, as these often provide useful perspectives into future commercial products and international trends.

The following subheadings first summarizes the state of those GM crops already being commercialized and those in the development then describes some of the recent research and patent publications detailing advances in the underlying technology for GM crop production and some of the novel approaches to modifying both agronomic (input) and product quality (output) traits.

PRESENT STATUS OF GM CROPS

Table elsewhere in this chapter provides data on the global areas of GM crops grown over the last 2 yr. The estimated global area for 2003 was 67.7 million hectares or 167.2 million acres, grown by 7 million farmers in 18 countries. The increase in area between 2002 and 2003 was 15%, equivalent to 9 million hectares or 22 million acres. Since the introduction of GM crops in 1996, the average growth in area has been more than 10% per year, an overall increase of 40-fold; this ranks as one of the highest adoption rates for any crop technology.

Of the global area, almost all (99%) is found in six principal countries: the United States, Argentina, Canada, Brazil, China, and South Africa. The latter two countries have had the highest year-on-year growth, with 33% increases in their total areas of GM crops. Specifically, China showed a 40% increase in its area of cotton expressing the *Bacillus thuringiensis* (Bt) protein with this area comprising 58% of the total national cotton area of 48 million hectares.

An increasing number are grown in developing countries, accounting for more than one-third (increase from 27%) of the global crop area. Two countries, Brazil and Philippines grew approved transgenic crops for the first time in 2003. Of the global area of transgenic crops in 2003, herbicide tolerant soybean occupied 61%, accounting for 41.4 million hectares, with Bt corn second at 9.1 million hectares.

Table 10.1: Areas of GM Crops Grown Worldwide in 2002 and 2003

Country	*Area 2002 (million hectares)*	*Area 2003 (million hectares)*
United States	39.0	42.8
Argentina	13.5	13.9
Canada	3.5	4.4
Brazil	?	3.0
China	2.1	2.8
South Africa	0.3	0.4
Australia	0.1	0.1
India	<0.1	<0.1
Romania	<0.1	<0.1
Spain	<0.1	<0.1
Uruguay	<0.1	<0.1
Mexico	<0.1	<0.1
Bulgaria	<0.1	<0.1
Indonesia	<0.1	<0.1
Colombia	<0.1	<0.1
Honduras	<0.1	<0.1
Germany	<0.1	<0.1
Total	58.7	67.7

The other six crops, planted on 5% or less of the global transgenic crop area, include herbicide-tolerant canola, herbton, Bt cotton, Bt/herbicide-tolerant cotton, herbicide-tolerant corn. There is an increasing trend towards stacked genes (combinations of herbicide and insect resistance). An additional method by which to access the impact of GM crop is to estimate their having increased from $4.0 billion in 2002.

These values are basedon the sale price of GM seed plus any technology fees that apply.

Table 10.2: Selected Examples of Recent US Field Trial Applications for Product Quality Traits

Number	Crop	Applicant	Gene	Trait
04-083-09	Barley	Washington State	Lysozyme	Novel protein
04-029-02	Barley	Washington State	Glucanase	Heat stable enzyme
04-021-01	Corn	Pioneer	Amino polyolamine oxidase	Fumonisin degradat.
04-020-02	Alfalfa	Forage Genet. Inc	CBI[a]	Altered lignin
04-033-15	Wheat	ARS	Glutelin	Storage protein
03-345-05	Tomato	BHN Research	Sucrose phosphate	Altered ugar[s] synthetase
03-288-26	Fescue	Noble Foundat.	Cinn. alc. dehydrog.	Decreased lignin
03-268-07	Soybean	Univ. Nebraska	8-6 desaturase	Altered oil
03-253-05	Sweetgum	ArborGen	CBI	Decreased lignin
03-203-09	Eucalyptus	ArborGen	CBI	Decreased lignin
03-153-01	Pineapple	Univ. Hawaii	ACC oxidase	Fruit ripening
03-136-01	Paspalum	Univ. Florida	O-methyl transfer.	Decreased lignin
03-091-08	Tobacco	Vector Tobacco	Aquaporin Quino. phos. trans.	Reduced nicotine
03-090-11	Lettuce	Harris Moran	Trans. int. fact.	Reduced senescence
03-073-07	Apple	Cameron Nursery	?	Brown spot resistance Polyphenol oxidase
03-057-02	Potato	Univ. Idaho	CBI	Reduced bruising
03-052-07	Petunia	Scotts	CBI	Altered colour
03-035-10	Tobacco	Virginia Tech.	Gulono-lactone oxidase	Increased vitamin C
03-034-05	Wheat	Montana State	Purindoline	Better breadmaking

Within Europe, the most comprehensive recent survey is probably that prepared by the European Science and Technology Observatory.

Fourteen crops have been approved for commercialisation to date; these are maize, oilseed rape, carnation, chicory, soybean, and tobacco further authorizations have been granted since October 1998, 13 applications were pending approval under the old Directive 90/220/EEC and 19 applications have been submitted under the new Directive 2001/18/EC. Summaries of field trial data for the United States and Europe are provided in Figures elsewhere in this chapter.

The overall total for the United States is now over 9300 (from 227 institions) since 1987, with an average annual total of about 1000. In the United States, two companies (Monsanto and DuPont/Pioneer) account for about 50% of all field trials, with almost two-thirds of the trials on maize, potato, and soybean.

Analysis of the data for 2003 shows that about 32% of these trials involved herbicide tolerance, 22% product quality traits (starch, sugar, proteins, etc.), 19% insect resistance, and 12% agronomic traits, with most of the remainder covering resistance to fungi, viruses, bacteria, and nematodes. In the European Union, by contrast, the number peaked at 250 in 1998 and has declined by 80% since then, because of the 1999 decision to block any

European Union new commercial release of GMOs and the general opinion of the European Union public. Further details of the present position within the United Kingdom can be obtained from the DEFRA websites on GM crop farm scale evaluations the Royal Society pages on GM plants the Friends of the Earth report on farm scale evaluations, and the Food Standards Agency public forum on GM food.

REGENERATION AND TRANSFORMATION TECHNIQUES

The most commonly used transformation technologies are those involving either particle bombardment or *Agrobacterium* and these are not considered in detail. A noticeable recent trend is the development of efficient *Agrobacterium*-mediated methods for cereals and other crops previously considered recalcitrant.

Among the associated recent improvements in underlying regeneration technology for important crops is that of wheat, in which it is claimed that the addition of copper at a concentration range from 50 to 300 μM and a growth hormone at a concentration from 0.05 to 10 mg/L gives greatly im roved results.

Such culture medium allows reliable high-frequency regeneration from a range of current elite wheat germplasm leading to up to 12-fold increases in the production of fertile transgenic plants. The latest proposed method for soybean transformation comprises *Agrobacterium* treatment of an explant (particularly after pretreatment with high doses of cytokinin), transferring embryonic axes explants of the mature seeds incubated on wet filter papers in the presence of at least one phenol compound, to induce *vir* genes, and incubation in the dark at 20-25°C for at least 24 h.

After incubation, the explants are transferred to a medium to develop shoots from explants, control *Agrobacterium* growth, and after shoot elongation, separated shoots—with or without roots—are either transferred to soil or treated with at least 1 mg/L Indole butyric acid (IBA) before transplantation.

Other recent improvements include novel methods for *Eucalyptus* and *Tagetes* , the latter involving *Agrobacterium-mediated* transformation of cell cultures. For certain less amenable crop species (e.g., maize) only particular genotypes are easy to transform. In these cases, it may be prefera hybrid between the target genotype and a transformation competent line of the same species or of another closely related species.

The gene of interest can then be introgressed into the genetic line from which the original recipient parent was derived or into other genetic lines. With the increasing regulatory pressure to avoid antibiotic selectable marker systems there has been a search for alternatives, the latest being a development of positive selection methods that involve conferring on cells the ability to metabolize certain compounds, preferably arabitol, ribitol, raffinose, sucrose, mannitol, or combinations of these compounds. Anothe developed in potato is a strategy that relies on the transformation of tissue explants or cells with *a Agrobacterium* strain and selection of transformed cellsor shoots after polymerase chain reaction (PCR) analysis.

For example, incubation of explants with *Agrobacterium* strain AGL0 resulted in transformed shoots at an efficiency of 1–5%, depending on the genotype used. Because this system does not require genetic segregation or site-specific DNA deletion systems to remove marker genes, it may provide a reliable and efficient tool for generating transgenic plants for commercial use, especially in vegetatively propagated species such as potato and cassava.

Another variation (19) designed to avoid the possibility of transfer of non transfer (T)-DNA sequences is one in which the construct includes non-T-DNA sequence comprising a lethal gene. Selection then allows the identification of plant cells that contain the T-DNA sequence and not the lethal gene.

A final alternative (20) is provided by a system that enhances the selection of transgenic plants having two T-DNA molecules integrated into the genome at different physical and genetic loci. The constructs comprise novel arrangements of T-DNA molecules containing genes of interest, positive selectable marker genes, and conditional lethal genes.

In this system, first the transgenic plant and independent transgene loci are identified; subsequently, the selectable marker genes and introduced genes of interest can be segregated in the progeny. As well as the "traditional" process of introducing genes into GM crops, there is increasing interest in the selected inactivation of endogenous genes or targeted integration. For example, one such method involves the use of a combination of target sites for two site-specific recombinases and expression of a chimeric recombinase with dual target site specificity.

As well as the research described elsewhere in this chapter concerning the introduction of specific coding regions, there are also interesting developments in the design of specific novel regulatory constructs (e.g., those comprising zinc finger transcription factors, and also in the identification of novel promoter sequences such as those targeting expression to the seed or caryopsis, and those that can be regulated by the application of exogenous chemicals.

Chloroplast transformation has also become a meth academic and commercial interest in recent years, partly because of the ability of these organelles to accumulate introduced proteins at very high yield but also because of the theoretical ecological advantages of reduced transfer of the transgene via pollen dispersal. Another associated technology concerns a method of modifying mitochondrial-encoded traits in plants by physically inserting this organelle by spraying a filtrate under pressure onto target cells.

INPUT TRAITS

The priority for most plant breeders is maximising yield, and improvements in this character has been claimed in several recent patents involving the plastid expression of cyanobacterial fructose-1,6-bisphosphatase/sedoheptulose- 1,7 bisphosphatase or other enzymes, and the use of the elf transcription factor. General approaches have been reviewed previously.

Agronomic Traits

The first transgenic products were those that contained tolerance to various herbicides, and these make up the majority of the GM acreage. However, there is still further extension and fine-tuning of these technologies either to improve efficacy or extend the technology to other species. One example of this latter type is the project designed to improve resistance to broomrape (*Orobanche spp.*), parasitic weeds that are major constraints to vegetable crop production in the Mediterranean basin and in localized areas in India, China, and the United States.

Transgenic target-site herbicide resistance (e.g., to acetolactate synthase [ALS] inhibitors) allows for movement of unmetabolized herbicide through the crop to the photosynthate sink in the parasite, as well as through the soil. The authors report the successful engineering of a mutant *ALS* gene into carrot, allowing control of broomrape by imazapyr, an imidazolinone *ALS* inhibitor.

Apart from the more common strategies for herbicide tolerance, another novel approach consists of introducing genes encoding the enzymes of the complete mevalonate pathway. Regarding insect resistance, there is also continuing development of strategies based on Bt.

For example, there is recent claim δ-endotoxin, designated CryET29, that exhibits insecticidal activity against siphonapteran insects, including larvae of the cat flea (*Ctenocephalides felis*), as well as against coleopteran insects, including the southern corn rootworm (*Diabrotica undecimpunctata*), *Western corn rootworm* (*D. virgifera*), *Colorado* potato beetle (*Leptinotarsa decemlineata*), Japanese beetle (*Popillia japonica*), and red flour beetle (*Tribolium castaneum*).

Such research needs to be assessed against a background suggesting that long-t suppression after deployment of Bt crops (e.g., Bt cotton resistant to the major pest, pink bollworm *Pectinophora gossypiella*) may contribute to reducing the overall need for insecticide sprays.

Abiotic Stress

There are many, varied transgenic approaches to improving the resistance of a crop to abiotic stress. Among the recent suggestions are the use of a novel heat-shock protein with high homology to chloroplast elongation factor EF Tu; protein phosphatase stress-related polypeptides; an *Arabidopsis* Na^+/H^+ exchanger polypeptide that allows crops to survive in soil with high salt levels; and a farnesyl transferase that improves the tolerance of plants to environmental stresses and senescence (expression of inhibitors of these enzymes enhance drought tolerance, improve resistance to senescence and modify growth habit).

In this context, the promoter of the Wcs120 gene, which encodes a highly abundant protein induced during cold acclimation of wheat, has been proposed to drive the expression of genes needed for low-temperature tolerance in sensitive species.

Sterility and Flowering

A recently described example of modifying these characteristics involves manipulation of the first example of a phytochrome-regulated transcription factor; this protein, designated CCA1, binds to the promoter region of a chlorophyll binding protein gene of *Arabidopsis.*

When CCA1 is overexpressed, the normal circadian rhythms of the plant are disrupted, and the transgenic plants take a significantly longer time to reach flowering, even in the presence of day length conditions that normally induce flowering.

Thus, this method may represent a valuable means of extending vegetative growth and delaying flowering. In a similar example, a late-flowering transgenic radish has been produced by the expression of an antisense GIGANTEA (*GI*) gene fragment.

This study provides evidence that downregulation of the *GI* gene by cosuppression could delay bolting in a cold-sensitive long-day (LD) plant. Production of late flowering germplasm of radish may allow this important crop to be cultivated over an extended period and also provide further food to the famine countries of Southeast Asia.

Modification of flowering time or flowering period is also reported as a consequence of altering the levels of a specific transcription factor. Induction of floral sterility is claimed to result from expression of selected floral homeotic genes from poplar- or biotin-binding compounds, and female sterility has been induced by expression of a deacetylase gene under the control of a tissue-specific promoter that is specifically active in the female organs; these organs are killed by treatment with N-acetyl phosphinothricin. Other examples of altered phenotype include methods for inducing dwarfism and reduced senescence linked to expression of deoxyhypusine synthase.

OUTPUT TRAITS

These have been recently reviewed, and selected trials of such transgenics are included in Table elsewhere in this chapter.

Seed Quality

Apart from general attempts to modify seed size, seed quality can also be modified. For example, chlorophyll reduction in the seed of *Brassica* can be achieved by downregulating its synthesis.

To achieve this, expression of an antisense glutamate 1-semialdehyde aminotransferase gene (*gsa*), directed by a *Brassica* napin promoter, was targeted specifically to the embryo of the developing seed of oilseed rape.

These transgenic lines have provided useful materials for the development of a low chlorophyll seed variety of this crop. Another novel method of improving processing grain in crops such as corn and soybeans involves utilizing thioredoxin and/or thioredoxin reductase to enhance extractability and recovery of starch and protein. Other methods of reducing fungal toxin content of seed to improve palatability have also been developed.

Protein

The nutritive value of storage organs for human and/or an was achieved by transferring a gene that encodes a sulfur-rich protein, such as sunflower seed albumin (SSA) containing 16% methionine and 8% cysteine, placed under the control of a promoter that confers storage organ-specific expression.

It was discovered that, in addition to the expected changes in sulfur-rich protein content of seeds, the overall composition of the seed was altered unexpectedly so as to produce a dramatic improvement in many different, unrelated nutritive parameters.

In particular, the process produced an increase in the total protein content (e.g., in rice, peas and chickpeas), altered fiber composition (lupins and peas), modified oil content and composition (lupins), altered starch content (peas), and a decrease in the content of endogenous antinutritional factors (peas and chickpeas). Two related studies describe different approaches to improving the amounts of free amino acids in plants.

In the first, Kisaka described the introduction of a glutamate dehydrogenase gene as a means of increasing the amount of at least one of the following glutamic acid, asparagine, aspartic acid, serine, threonine, alanine, and histidine. This process is also claimed to increase potato yield. In a related study on maize, a prolamin box binding factor peptide, or a subunit thereof, was used to increase the amount of a preselected amino acid, such as lysine and/or methionine, in the seed.

This beneficial alteration occurred without substantially altering the total protein content of the seed, which might be deleterious to other agronomic characteristics of the transgenic plant. Several plant proteins, particularly those from seeds, have proven or possible allergenic potential, and various transgenic approaches have been made to reduce their amount.

Among these is one study on hypoallergenic soybean with reduced amounts of the major soybean allergen, the vacuolar protein known as P34, as well as other allergens.

Soybean protein products made from these hypoallergenic soybeans should be substantially free of P34, as well as a series of other minor allergens such as various glycinins and conglycinins.

Carbohydrate

One proposed method of producing modified starch relate gene encoding a plastid protein referred to as an R1 protein. It is probable that this protein exists in the plastids in a form bound to the starch granules as well as in a soluble form, and that this protein is involved in starch phosphorylation.

Similar cases include those claiming the use of reserve polysaccharide biosynthetic enzymes, such as glycogen biosynthetic enzymes, glycogen synthase, and/or ADP-glucose pyrophosphorylase; pea plastidial phospho glucomutase for altering the sucrose and starch content of plants; modified sucrose binding proteins with enhanced sucrose uptake activity; and a method of decreasing the oil content of seeds by expression of ADP-glucose pyrophosphorylase. Among studies on specific sugars rather than total carbohydrate is one involving palatinose (isomaltulose, 6-O-(-D-glucopyranosyl-D tural isomer of sucrose with very similar physicochemical properties.

Owing to its noncariogenicity and low calorific value, it is an ideal sugar substitute for use in food production. Usually, palatinose is produced on an industrial scale from sucrose by an enzymatic rearrangement using immobilized bacterial cells, but production in potato has now been achieved by use of a chimeric sucrose isomerase gene from *Erwinia rhapontici* under control of a tuber-specific promoter.

Despite the soluble carbohydrates being altered within the tubers, growth of transgenic plants was indistinguishable from wild-type plants.

Oils and Fatty Acids

Modification of these compounds has been reviewed recently. cDNAs for a wide variety of unusual fatty acid biosynthetic enzymes have been identified, particularly through the use of expressed sequence tags. Amongs the GM crops undergoing field tests in the United States in the las expressing A-6 desaturase (from *Borago officinalis*), A-12 saturase (from soybean), palmitoyl thioesterase (from soybean), lysophosphatidate acyltransferase (from *Saccahromyces cerevisiae*), and stearoyl ACP desaturase (from *Rattus norvegicus*)

Among the most comprehensive generic applications is one from Pioneer covering GM soybean-expressing genes (not described) from Euphorbia lagascae, Isochrysis galabana, Mortierella alpina, Parthenium *argentatum, Saprolegnia diclina, Schizochytrium aggregatum, Thraustochytrium aureum*, and *Veronia galamensis.* One specialized example involves exploitation of long chain polyunsaturated fatty acids (LCPUFAs) such as arachidonic acid (ARA, *20:4n-6*) and docosahexaenoic acid (DHA, *22:6n-3*).

These compounds are highly concentrated in the phospholipid bilayer of biologically active brain and retinal neural membranes and are important in phototransduction (retina) and neuronal function (brain). They are present in large quantities in human milk, and it has been recommended that plant-derived versions could be added to infant food.

Other Food Components

Isoflavones are compounds with claimed health benefits, and there is much interest in modifying the expression of genes encoding enzymes in their biosyn thetic pathway; these enzymes include chalcone isomerase, isoflavone reductase, and vestitone reductase. Similarly, several eukaryotic genes encoding c-cyclase, isopentenyl pyrophosphate isomerase, and (-carotene hydroxylase have been used as a means of increasing the production of novel and rare carotenoids.

Among these is zeaxanthin, an important dietary carotenoid, although its abundance in food is low. To provide a better supply of this compound in a staple crop, two different potato varieties were genetically modified by transformation with sense and antisense constructs encoding zeaxanthin epoxidase; zeaxanthin conversion to violaxanthin was inhibited.

Both approaches (antisense and cosuppression) yielded potato tubers with higher levels of zeaxanthin (between 4- and 130-fold). Among the various efforts to modify food quality is one involving flavour improvement by increased expression of a lipoxygenase transgenic grapes with optimal levels of the enzyme.

Other recent examples of plants with modified vitamins include the field testing (US field trial APHIS no. 03-035-10n) of a tobacco line expressing L-gulono-γ-lactone oxidase from *Rattus* norvegic*us* and containing increased amounts of vitamin C. Similarly, a barley gene encoding 4-

hydroxy phenylpyruvate dioxygenase was overexpressed in tobacco plants under control of the 35S promoter with the aim of enhancing the vitamin E content. Seeds from transgenic lines had up to twofold enhanced levels of this vitamin without any change in the ratio of γ-tocopherol and γ-tocotrienol.

A particularly specialized example of this type relates to the use of transgenic plants for the expression of vitamin B_{12} (cobalamin) binding proteins. Such recombinant proteins can be used in analytical tests and in the treatment of vitamin B12 deficiency.

Industrial Products

1. Lignin. Expression of several of the enzymes that comprise the biosynthetic pathway for lignin have been altered in efforts to modify lignin, either as a means of improving paper-making quality or digestibility for animals.

   A recent example of the latter approach is provided by a study on alfalfa in which reduction in caffeic acid 3-O-methyltransferase and caffeoyl CoA 3-O-methyltransferase led to a dramatic decrease in lignin content (approx 20%) and modest increase in cellulose (approx 10%). These compositional changes potentially allow enhanced use of alfalfa as a major forage crop by increasing the digestibility of its stem fraction.

2. Plastics. Polyhydroxyalkanoates (PHAs) and polyhydroxybutyrates (PHBs), polyesters of hydroxyacids naturally synthesized in bacteria as a carbon reserve, have properties of biodegradable thermoplastics and elastomers and their synthesis in crop plants is seen as an attractive system for the sustained production of large amounts of polymers at

Table 10.3: Predicted Areas of GM Crops in the European Union (EU) up to 2013.

Crop/trait	*Commercially available in EU*	*% EU area planted 2008*	*% EU area planted 2013*
Insect-resistant maize	2005–2007	10	25-30
Herbicide-tolererant maize	2005–2007	10	35-45
Herbicide-tolererant oilseed rape	2006–2008	0–5	20-30
Herbicide-tolererant sugar beet	2006–2008	5–10	40-50
Insect-resistant cotton	2006–2008	5–10	40-50
Herbicide-tolererant cotton	2006–2008	5–10	40-50
Herbicide-tolererant wheat	2008–2011	0	15-25
Herbicide-tolererant soybeans	2007–2009	0–10	30-40
Herbicide-tolererant rice	2007–2009	0–5	30-40
Nematode- and fungus-resistant potatoes	2010–2012	0	5-10
Fungus-resistant oilseed rape	2010–2012	0	5-10

low cost . Various PHAs and PHBs having different physical properties have now been synthesized in the cytoplasm, plastid, or peroxisome of numerous transgenic plants, including *Arabidopsis,* rape, corn, and hairy roots of sugar beet. This latter study is the first example of plastidic PHB production in roots of a carbohydrate-storing crop plant.

3. Cellulose. Modifying the amount of this compound in plants is now feasible following an understanding of its biosynthetic pathway.

Enzymes

A bacterial thermostable cellulase, the endo-1,4-(-D-glucanase El from *Acidothermus cellulolyticus,* has been expressed in chloroplasts, and an active enzyme recovered both in vitro and in vivo.

Plant-Based Pharmaceuticals

Much investment has been committed recently to this subject; a wide range of different production systems have been proposed and are at various stages of commercial development. These include expression in chloroplasts, oil bodies, and those using specific species for production, for example, maize, rice, and barley.

This latter system is being used for production of lacto-ferrin, lysozyme, a1-antitrypsin, fibrinogen, and thioredoxin. Other companies are specializing in noncrop species such as the water plant Lemna, being tested in a joint venture with Bayer for the production of human plasminogen, or the moss Physcomitrella.

Other specialized products include antibodies such as that active against one of the bacteria responsible for dental caries. Other less developed projects in this area include one on the humanizing of plant cDNAs. Another potentially significant advance is the production of enzymatically active recombinant human and animal lysosomal enzymes in plants that has been accomplished by construction and expression of recombinant constructs encoding human glucocerebrosidase and a-galactosidase sequences.

Various novel extraction technologies have also been developed for extraction of enzymes produced in this manner. In a recent regulatory change, a problem with "contamination" by corn expressing pharmaceutical proteins has led to stricter isolation distances for such transgenics growing in the field.

Vaccines

The prospects for plant-derived vaccines, sometimes cines," have been well reviewed. Recent examples include the production of transgenic carrots expressing an immunodominant antigen, hemagglutinin (H) glycoprotein, of the measles virus.

THE FUTURE

It has been predicted that the global area and the number of farmers planting GM crops will continue to grow in 2004 in the six principal countries already growing GM crops. Among the other 12 countries growing such crops, India, is expected to increase its Bt cotton significantly and one or more new countries will also grow GM crops for the first time.

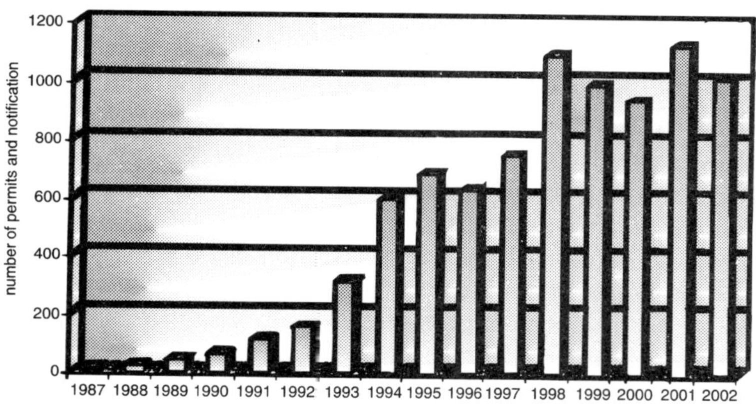

Figiure 10.1: Total number of US field trial permits and notifications approved by year.

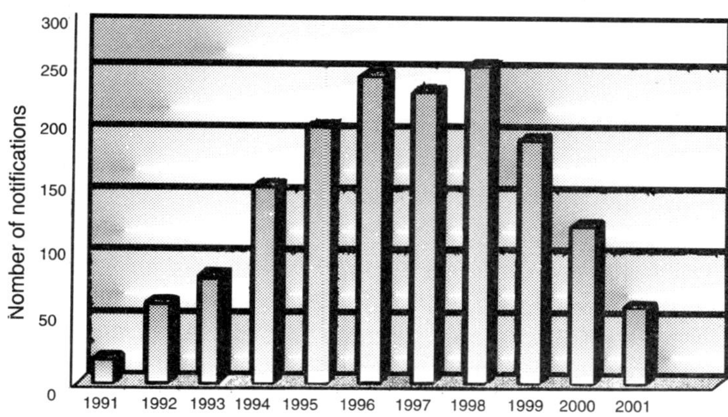

Figure 10.2: Annual number of EU field trial notifications.

The three most populous countries in Asia, namely China, India, and Indonesia, with 2.5 billion people, are all now growing GM crops commercially. Looking further into the future, detailed predictions of the prospects for commercialization in Europe have recently been published and are summarized in Table elsewhere in this chapter.

The key elements from this study are first that it is likely to be another 2–3 yr before GM seed is widely available to European Union producers f maize and possibly 3–4 yr for other crops such as oilseed rape and sugar beet; GM wheat is unlikely to be available until 2008–2010. Second, in a 5 yr time period the extent of GM crops is likely to be limited to no more than 10% of cultivation in some crops such as maize.

This largely reflects the time required to complete the various regulatory approvals and to introgress introduced genes into leading varieties, and the continued existence of anti-GM sentiment among some consumers. Thirdly, in a 10 yr time period, it is predicted that the acreage of GM crops will largely reflect the degree to which specific pests and weeds (which are targeted by GM traits) are considered to be a problem for farmers.

Thus, takeup of insect resistant and herbicide tolerant crops such as oilseed rape, maize, and sugarbeet will be concentrated in regions where pests and weeds are perceived to be causing significant crop losses and/or conventional control methods have been found to be of limited effectiveness (or are more expensive than the GM alternative).

However, in 2013, areas of GM wheat and potatoes will probably be more limited than for other crops, mainly because the traits available will be fairly new to the market. The potential impact of biotechnology in improving pest management in Europe is also the subject of a recent detailed case study analysis. Whether these predictions are fulfilled will depend no available novel genotypes, but rather on the uncertain combination of economic pressures, regularoty systems, biosafety assessments like those that operated under the Cartagena Protocol that came into force in September 2003, and public perception.

This is well-demonstrated in the United Kingdom where, despite an extensive debate on the future of GM crops, an assessment of various environmental impacts, and conditional permission from the government, commercial planting has recently (April 2004) been postponed.

COMPARATIVE DEVELOPMENT AND IMPACT OF TRANSGENIC PAPAYAS

Transgenic commodity crops with resistance to herbicide or lepidoptera insects have been widely adopted in North America by the United States and Canada, and in other countries as Argentina. However, the progress in commercialization of minor crops even within the United States has been rather slow.

The transgenic papaya with resistance to *Papaya ringspot virus* (PRSV) is one of these few minor crops that have been commercialized in the United States (Hawaii). Interestingly, because papaya is widely grown in the tropical and subtropical lowlands of the world and is severely infected by PRSV, there is widespread interest in using PRSV-resistant transgenic papaya for these areas outside of the United States.

Shortly after the Hawaiian transgenic papaya was developed and shown to be resistant, personnel from Jamaica and Venezuela launched efforts to develop transgenic papaya for their countries in concert with the Hawaii transgenic papaya effort.

As a result, PRSV-resistant transgenic papayas have been developed for Jamaica and Venezuela. This chapter covers the technical aspects of the transgenic papaya but focuses on describing the factors that affected the differential adoption of the transgenic papaya in Hawaii, Jamaica, and Venezuela. We feel that the approach in this chapter might provide insights that may help future efforts at transferring transgenic products to countries.

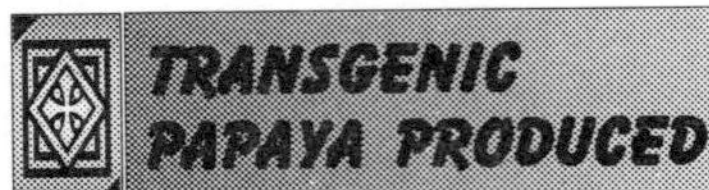

TRANSGENIC PAPAYA PRODUCED

Papayas are fast-growing and high-yielding fruit trees (actually the papaya "tree" is a gigantic herb) very popular in the lowlands of tropical and subtropical regions. Papayas are grown domestically or at industrial scale; 6 million Mt of papaya were produced worldwide in 2002.

However, papaya orchards can be severely attacked by PRSV, arguably the most important pathogen that attacks the plant. Since the first demonstration that the *coat protein* (*CP*) gene of a plant virus inserted into a plant genome could protect transgenic plants from the homologous virus, many plants have been engineered for viral re concept of pathogen-derived resistance .

PRSV belongs to the Potyviridae family of plant viruses. Potyviruses have a sense polarity RNA genome, which is translated into a polyprotein that, on sequential processing by different virus-encoded proteases, yields all viral proteins including the CP. CP-transgenic papayas resistant to PRSV were commercially released in Hawaii in 1998 and have helped to save the papaya industry.

The transgeniccultivars, "SunUp" and "Rainbow," contain the CP gene of PRSV HA 5-1, a mild PRSV mutant strain from Hawaii. "Rainbow" is a he, izygous (*CP+/–*) transgenic papaya as it is an F1 hybrid that was obtained by crossing homozygous (CP+/CP+) "SunUp" and nontransgenic "Kapoho" . "SunUp" was obtained by particle bombardment, and it contains only one *CP* insert in homozygous state.

Resistance to PRSV in transgenic "Rainbow" and "SunUp" has been the target of numerous studies that can be summarized as follows: the CP protein can be detected in transgenic tissues, although at low levels; there is only one transgenic insertion in which all transgenes present, *uidA* for G-glucuronidase, *CP* for the PRSV coat protein, and *nptII* for neomycin phosphotransferase, have suffered no duplications or rearrangements; nuclear run-on analysis has detected a high level of expression of the CP gene in the nucleus, but low levels in the cytoplasm.

The transgene messenger thus is degraded posttranscriptionally and resistance is RNA mediated. Because resistance is better achieved when the challenging virus shares a high degree of similarity to the transgene, resistance is said to be homology dependent. Resistance is also affected by gene dosage and the developmental stage at which the transgenic plants are challenged with the virus. Slight, but important differences exist between the CP transgene in "Rainbow" and the engineered CP transgene in the transgenic papaya from Jamaica and Venezuela.

The engineered PRSV HA 5-1 CP transgene in "Rainbow" is fused to the nucleotides that code for the first 16 amino acids (aa) of the *Cucumber mosaic virus* (CMV) *CP* gene. The engineered Jamaican and Venezuelan papayas includes only the PRSV CP gene sequence four aa downstream of the Q/S cleavage site "Rainbow" transgene contains the 5' UTR of CMV, whereas the other two possess the translational enhancer of the "white leaf strain" of CMV. In all cases, the CaMV 35S promoter drives the transgenes; the "Rainbow" is derived from the CaMV 35S DNA, whereas the other transgenes use the *nos* terminator.

However, more remarkable is the fact that the transgene in "Rainbow" was isolated from a mild mutant of PRSV HA, while in all other cases the transgenes were isolated from aggressive geographical isolates of the virus. Resistance in transgenic papayas transformed with the CP gene from Jamaica and Venezuela share with transgenic "Rainbow" the following features: transgene-derived resistance is RNA-mediated, homology-dependent and not related to the number of CP insertions.

Yet the range of resistance in the Jamaican and Venezuelan transgenic papayas is wider as compared to "Rainbow" because resistance can be observed even with isolates showing less

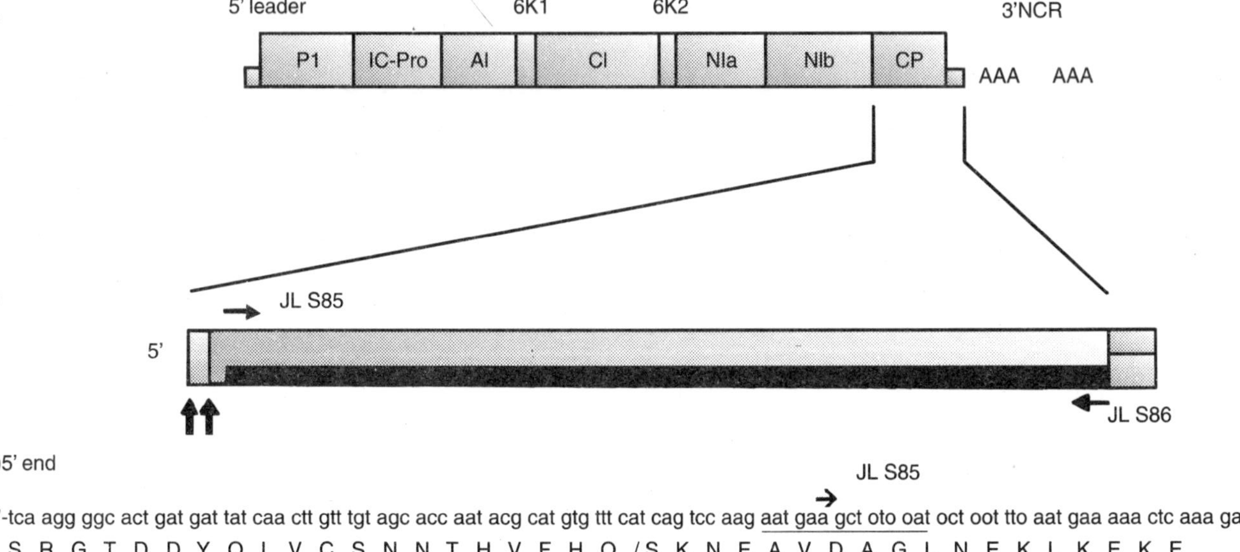

(i)5' end

JL S85

5'-tca agg ggc act gat gat tat caa ctt gtt tgt agc acc aat acg cat gtg ttt cat cag tcc aag aat gaa gct oto oat oct oot tto aat gaa aaa ctc aaa gag aag gaa

/ S R G T D D Y Q L V C S N N T H V F H Q /S K N E A V D A G L N E K L K E K E

(ii) 3' end

JL JL S86

5'-ggt atg cgc acc taa atacctgcgcgtgtgtgtgttgagtctgactcqaccctqtttcacc-3'

G M R N

Figure 10.3: (A) Genetic map of the Papaya ringspot virus genome showing all potential viral products after auto-proteolytic processing of the precursor polyprotein. (B) Schematic representation of the coat protein (CP) transgene with the leader sequence (white box) and the CP sequence as engineered in 'Rainbow' (dark box), with the corresponding location of primers JLS85 and JLS86 (horizontal arrows) used to engineer the PRS V CP gene from Jamaica and Venezuela (light box); (i) the 5' end CP coding region with its translated product showing the putative sites of proteolytic cleavage (/), and (ii) the 3' UTR that is followed by a terminator sequence.

than 90% similarity to the transgenes. However, "SunUp" shows a wide range of resistance to PRSV isolates, because of its homozygous *CP* state.

The untranslatable PRSV HA transgene is able to confer resistance to transformed papayas in the very same way the translatable transgene does. However, the nontranslatable Jamaica transgene was not able to confer resistance to transformed papaya plants the corresponding version of the untranslatable Venezuelan CP transgene has not been used for transformation.

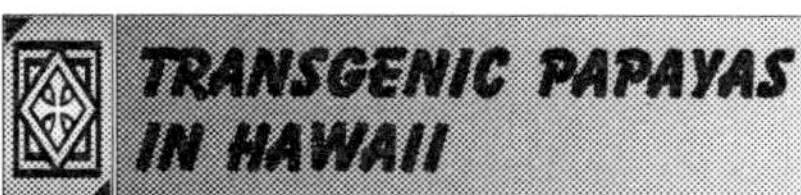

TRANSGENIC PAPAYAS IN HAWAII

The transgenic papaya story in Hawaii represents a case in which a transgenic product was introduced in a timely manner to stop further devastation of an industry by PRSV. The transgenic papayas developed for the papaya industry in Hawaii have been the subject of several recent reviews.

This section focuses primarily on nontechnical factors that led to the development of the transgenic papaya in Hawaii, its subsequent adoption and impact on the industry, and challenges facing Hawaii's papaya industry.

The Rationale for Developing Control Measures for PRSV in Hawaii

PRSV was discovered in the mid-1940s on Oahu Island, where Hawaii's papaya industry was located. However, by the mid-PRSV had caused severe damage to papaya grown on Oahu, pushing the industry to relocate to the Puna district of Hawaii Island in the late 1950s.

Why relocate to Puna? In addition to the absence of PRSV, Puna had lots of land available for leasing at a reasonable price, had an abundance of sunshine and rain, and the well-drained lava-based soil was suited for papaya.

Another very important reason was the grower-selected "Kapoho" variety, which had excellent flavor and shipping qualities and was well adapted to Puna. In fact, "Kapoho" generally does not grow well outside of the Puna area. By the 1970s, Puna was producing 95% of Hawaii's papaya crop.

However, the potential threat of PRSV existed because, by the 1970s, PRSV was in papaya growing in backyards in the town of Hilo, which was only 19 miles from Puna. Recognizing the consequences to the industry if PRSV got established in Puna, the Hawaii Department of Agriculture (HDOA) formed a small task force that was responsible for identifying and roguing PRSV-infected trees in Hilo and nearby areas.

Quarantine on movement of papaya seedlings into Puna was also put in place. Another major factor that contributed to the success of the Hawaii papaya story is that efforts to develop control measures were started early. In fact, research to breed papaya with tolerance to PRSV began in 1979, scientists at Cornell University and the University of Hawaii started research on cross-protection to control PRSV .

Although cross-protection did not reach the stage of large-scale commercial adoption in Hawaii, these research efforts made important contributions to the control of PRSV. For example, the now well-known PRSV HA isolate was purified and characterized, serology was used for rapid detection, the mild strain PRSV HA 5-1 was selected following nitrous acid treatment of the severe PRSV HA, and the *CP* gene of PRSV HA 5-1 was cloned and sequenced.

Development of Transgenic Papaya for Hawaii

Research on the development of transgenic papaya was started in the mid 1980s. As with the cross-protection work, the research was supported by modest levels of special USDA grants that were aimed at helping agriculture in Hawaii and the Pacific Basin. The research team had a proper balance in expertise and a strong desire to develop a practical control measure for PRSV in Hawaii.

The team consisted of Richard Manshardt, a horticulturist at the University of Hawaii; Maureen Fitch, a graduate student of Richard Manshardt, Jerry Slightom, a molecular biologist with whom D. Gonsalves had collaboration to develop virus resistance transgenic vegetables, and D. Gonsalves, a virologist and coauthor of this chapter. After unsuccessful attempts to develop transgenic papaya by the transformation of papaya leaf disks with Agrobacterium *n* tumefaciens, efforts to transform somatic embryogenic cultures via biolistics were started in fall of 1988.

The yellow-flesh cultivar "Kapoho" and the red-fleshed cultivar "Sunset" were targeted for transformation. "Kapoho" was the dominant cultivar, accounting for more than 90% of the commercial papaya grown in Hawaii. Transformation efficiency was relatively low and resulted in less than 20 transgenic papaya lines.

In 1991, clones of an R0 transgenic "Sunset" (designated 55-1) were resistant to inoculation by PRSV HA under greenhouse conditions. This line would be carried through to eventual commercialization. Line 55-1 was a female and thus would need to be crossed with nontransgenic papaya to obtain seeds for further characterization, which would take a year.

While the process of obtaining seeds under greenhouse conditions was being carried out, a permit to perform a field trial to test clones obtained from APHIS (Animal and Plant Health Inspection Service) in 1991. The plants were put in the field in June 1992.

The value of this trial cannot be overestimated: it served to confirm the resistance of line 55-1 under field conditions, showed that line 55-1 was of suitable horticultural characteristics, and served to provide material for the deregulation process, and for developing the red-flesh cultivars "SunUp" and yellow-flesh cultivar "Rainbow". "SunUp" is transgenic line 55-1 that is homozygous for the CP gene insert, and the yellow flesh "Rainbow" is an F1 hybrid of "SunUp" and nontransgenic "Kapoho".

The timely development of the transgenic virus-resistant papaya became evident when PRSV was identified in Puna in May 1992, just about the same time that the field trial was started on Oahu Island. The severity and speed of the virus spread in Hawaii has been reviewed (1) ; nevertheless, pictures of healthy papaya in Puna in 1992 (Fig. 2A) when the virus was discovered and a devastated field in 1994 illustrate the rapid damage that was occurring to the industry in Puna.

In brief, the Kapoho area of Puna was the first to become severely infected because it was close to Pahoa, the initial site of infection in Puna. Roguing infected trees to suppress the spread of PRSV in Puna were begun immediately by HDOA, but the virus spread could not be contained.

In late 1994, these efforts were abandoned, causing even faster spread of the virus because

Table 10.4: Fresh Papaya Production in Hawaii and in Puna District From 1992 to 2001.

Year	Fresh Papaya Production in Hawaii Total	Puna (× 1000 pounds)	
1992	55,800	53,010	PRSV enters puna
1993	58,200	55,290	
1994	56,200	55,525	
1995	41,900	39,215	
1996	37,800	34,195	
1997	35,700	27,810	
1998	35,600	26,750	Transgenic seeds released
1999	39,400	25,610	
2000	50,250	33,950	
2001	52,000	40,290	

of abandonment of infected trees by growers. By 1997, PRSV was widespread in Puna, making it nearly impossible for growers to economically grow papaya in many areas of Puna.

Yield of papaya in Puna dropped from 53 million pounds of fresh papaya in 1992 when PRSV was discovered in Puna to 26 million pounds in 1998. The industry was clearly in a crisis.

The growing of papaya in the Hamakua district of Hawaii Island and Oahu Island helped the state of Hawaii continue production but statewide yields of fresh papaya were only 35 million pounds in 1998 compared to 55 million pounds in 1992. A field trial was started on a farm in the Kapoho area of Puna in October 1995.

This trial, which was headed by Steve Ferreira of the University of Hawaii, was pivotal for several reasons. It allowed us to: (a) test the resistance of "SunUp" and "Rainbow" under severe virus pressure in Puna, (b) develop data on the horticultural characteristics of the cultivars, and (c) provide a demonstration site for growers and packers to observe the cultivars that they might produce and sell in the future.

To accommodate these aims, the field trial consisted of replicated plots, and a large solid block of "Rainbow," which was surrounded by several rows of nontransgenic "Sunrise." The nontransgenic "Sunset" is a sib selection of "Sunrise." The large "Rainbow block was to serve as a "simulated" commercial planting and the latter would provide data that could be analyzed statistically.

Even though "Rainbow"-type papaya had not been tested in Puna, it was used in the simulated commercial plantings because Hawaiian growers overwhelmingly prefer the yellow-flesh papaya, as exemplified by "Kapoho." Data from the field trial were collected for a period of 28 mo after planting. "SunUp" and "Rainbow" were resistant under heavy disease pressure, the performance of "SunUp" and especially "Rainbow" were acceptable to growers and packers, and the comparative effect of growing virus resistant and susceptible papaya plants in an area with heavy and continual virus pressure were dramatic.

For example, the annual yield of "Rainbow" in the solid block was about 2242 kg/ha of marketable fruit and yields were steady over the harvest period, and the susceptible "Sunrise" matrix annualized yields were 419 kg/ha initially and dropped to 56 kg/ha after 1.5 yr of harvest. Clearly, the data showed that "Rainbow" was an excellent substitute for '"Kapoho" as judged by its resistance, horticultural characteristics, and the opinions expressed by growers and packers.

Deregulation and Commercialization of Transgenic Papaya

The papaya crisis in Hawaii spurred the researchers and the industry to move aggressively toward deregulation and commercialization of the transgenic papaya. The details of the events have been described in a review, and thus are not repeated here. Instead, some circumstances and efforts will be described to illustrate the interrelations of the researchers with the industry. The Papaya Administrative Committee (PAC) represented the organized section of the papaya industry.

The PAC consisted of growers who voluntarily created a Marketing Order that was under the oversight of USDA. The PAC focused primarily on solving market problems such as quality regulations, promotion, and standardization of containers or packs.

Funds for these projects came through fee assessment on papaya that was sold through the packinghouses. The governmental agencies that oversee the regulatory aspects of transgenic crops in the United States are APHIS, Environmental Protection Agency (EPA), and the Food and Drug Administration (FDA). Petitions usually are developed and filed by private companies.

Because private companies were not involved in the papaya project, the task of moving the papaya through the regulatory processes fell in the hands of the researchers who lacked experience in this subject. Thus, they simply learned by doing.

These efforts were started in late 1995. The key steps that led to the deregulation of the transgenic papayas in Hawaii can be summarized as follows. APHIS, the US federal agency concerned with the potential risks of the papayas released on the environment, deregulated the transgenic plants in November 1996 after establishing that there were no risks associated with potential virus heteroencapsidation or plant weediness.

Because the CP gene confers virus resistance, it is considered a pesticide and subjected to EPA regulations. The PAC provided the filing fees that accompanied the petition to the EPA. It was argued that PRSV-infec fruits contain detectable levels of CP and have been consumed with no reported ill effects. For instance, large numbers of fruit from trees deliberately infected with the attenuated strain PRSV HA 5-1 were consumed when the technique of cross-protection was being used for papaya and the fruits eaten after they were infected.

Data also showed that the transgenic fruits contained less CP than the infected nontransgenic fruits. In August 1997, the EPA granted an exemption from tolerance levels of CP in transgenic line 55-1. Finally, the FDA assessed an application with data and statements on the product's safety to human health.

The information submitted included data on vitamin content, including vitamin C, presence of the *uidA* and nptII transgenes, levels of benzyl isothiocyanate (normally present in nontransgenic papayas) and more. FDA approval was granted in September 1997. What remained was to obtain the necessary licenses for commercialization of the transgenic papaya.

The PAC took on the tasks to obtain the licenses for use of intellectual property rights that were used to develop the transgenic papaya and to produce and distribute seeds of the transgenic papaya. The necessary licenses were obtained in April 1998, and seeds were distributed to growers on May 1, 1998, almost 6 yr to the day after PRSV was discovered in Puna.

Although seeds were distributed for free, prior to obtaining seeds the person had to register with the PAC, attend an educational session, and sign a material transfer and proprietary rights agreement. The latter states that the transgenic papaya seeds cannot be grown outside of Hawaii.

Thus, the papaya industry was ready to start the reclamation of the Puna areas that were devastated by PRSV and to plant newly cleared areas (Fig. 3A,B).

Adoption of the Transgenic Papaya in Hawaii

The Hawaiian efforts also presented a rather unique opportunity to critically measure the adoption and impact of a transgenic product. The geographical area to be surveyed was rather small and the efforts were undertaken by Carol Gonsalves, who was intimately familiar with the activities involved in developing the transgenic papaya and understood the nature of the people in the papaya industry papaya industry.

Following the distribution of transgenic seeds on May 1, 1998, a survey was conducted to examine the rate of adoption by papaya farmers in the Puna area.

This study was necessary to ascertain the usefulness of the new virus resistant papaya varieties for the farmers. It was an attempt to capture an early picture of farmer adoption. A farmer was considered to have adopted the technology based not merely on whether seeds were obtained but on whether the farmer had actually planted the seed in an effort to obtain fruit production.

Other pertinent issues were also addressed regarding farmers' opinions on the quality of the transgenic fruit and resistance to PRSV, prod attitudes toward the technology. In an effort to become familiar with the workings of the papaya industry, farmers and personnel from packing houses, the PAC, the University of Hawaii, HDOA, Hawaii County, the Hawaii Agricultural Statistics Service, and others were consulted one or more times.

Many farms in the Puna area were abandoned with infected trees left in the fields. Unfortunately, these trees provided a virus source that aphids could feed on and transmit the virus to other trees in the area. In some places, the environment had changed drastically, such that tall weeds had grown in the fields and the HDOA personnel were where some of the papaya fields had been. Some farms had been converted to pasture land. It was nearly impossible to find farmers working in their fields because many farms were abandoned, and as mentioned earlier, many farms ceased to exist.

Furthermore, most of the papaya farms were established on leased land, thus there was little chance of finding a farmer's home next to his papaya field. The interviews took place over numerous weeks from June through September 1999, concluding at 16 months following the release of transgenic seeds in May 1998.

Because the growers had decided to organize themselves as a Marketing Order of the USDA, all of the commercial papaya farmers were registered with the PAC. A total of 262 farmers were located on the island of Hawaii. Of these, 171 growers farmed in the Puna area.

Attempts were made to contact all of the Puna growers, either by telephone or by letter, and to do interviews. This represented a response rate of 54%. Ninety-two of these registered farmers were qualified to receive transgenic seeds because they had also watched a mandatory training video or educational session and signed a sublicensing contract. Farmer adoption of the transgenic papayas was astounding: 90% of the farmers had obtained seed, 76% adopted by planting their se already harvesting fruit.

Of a group of 93 respondents, 57 grew papayas in Kapoho, an area that had been hit early and completely by PRSV. The Kapoho farmers who were interviewed had a higher adoption rate, with 88% adopting the technology by planting the seed and 29% far enough along to be harvesting fruit.

Not only was there a high rate of adoption but it also occurred rapidly. Overall, 61% of the Puna farmers who received seeds had planted their seeds within 3 mo of obtaining them. Demographic data showed that most of the farmers were married males, average age 47 yr, primarily of Filipino ethnic heritage.

Their education ranged from elementary school through college in the United States or the Philippines. The PAC was instrumental in assisting farmers in the qualification process and later in seed production and distribution to the farmers. Farmers gave most of the credit to the PAC for disseminating information on "Rainbow."

But it is important to acknowledge that the University of Hawaii provided much support to the PAC by providing educational materials, research data, and field testing demonstrations for farmers. Personnel from Cornell University and the University of Hawaii were involved in the development and initial testing of the transgenic materials required for deregulation from the various governing bodies.

They also provided a tremendous support to the PAC at all stages of information dissemination and production of the transgenic materials that were released.

In general, the farmers appeared to be satisfied with the performance of the transgenic product. Farmers ranked resistance to PRSV as the greatest reason why they planted the transgenic "Rainbow" variety. When asked to rate the quality of "Rainbow" papaya (based on what they learned from others or experienced firsthand), farmers who commented noted that fruits possessed the desired sweetness and firmness and trees exhibited excellent resistance to PRSV.

Although the transgenic seeds were given free of charge to farmers, 86% of the farmers expressed a willingness to purchase transgenic seeds in the future provided that the cost is reasonable. In addition, if new virus-resistant transgenic varieties were introduced in the future, 88% said they would be interested in trying them, and 12% were not sure whether they would try them or not.

Concerning a question on labeling, farmer attitudes were optimistic, with a twist. In response to the question, "Do you think it's important to la"SunUp" so that the consumer knows that these are transgenic fruit?", 77% said yes, 12% said no, and 11% were not sure. The twist was that farmers were primarily thinking of whether to label their fruit "Rainbow" or "SunUp," and not whether they should label them according to whether they were GMO or not. Farmers felt that labeling by variety name would distinguish their fruit as a superior variety and thus would be a good marketing tool. Apart from one change in establishing the crop, there were no major changes to their farming practice.

Farmers began to establish and transfer transgenic seedlings to the field rather than using their routine practice of direct sowing. In the past, direct seeding of nontransgenic papayas was practiced by most farmers because they either produced the seed themselves, or purchased their seed at a nominal cost (in this survey, an average of $27 per acre when purchased).

In direct sowing, 15 or more seeds, sometimes a handful, are tossed into the planting hole to insure that one of them could be healthy plant of the hermaphrodite sex, which produces the typical papaya shape desired in the marketplace.

Although the transgenic seeds were distributed free of charge, farmers needed to be judicious in planting because the seed supply did not allow for direct seeding. This change from direct sowing to transplanting seedlings allowed farmers to plant fewer seed to avoid skips and the need to replant in case there was a problem with seed quality or low germination rates. Production costs of growing nontransgenic during the first year varied widely from farmer to farmer with the average cost during the first year at $2515 per acre.

Many people are curious about whether it costs more to grow transgenic papayas than the nontransgenic ones. Most respondents (66%) were not sure that there was a difference or made no comment on the matter, 25% felt there was no cost difference, whereas 2% stated that growing transgenics costs less and 8% said that it costs more.

On a global scale, and unlike large-scale farmers who benefit from genetically engineered corn, soy, or cotton, the papaya growers in Puna and hence, in Hawaii as a whole, were small-scale farmers.

Fully 95% of those who were farming in this study were growing from one to nearly 50 acres of papayas. Thus, this study shows that small-scale farmers can reap benefits from the fruits of agricultural biotechnology. Taken together, the transgenic papayas were well accepted in Hawaii because there was PRSV destruction in the papaya fields, and there was a solution: virus resistant "Rainbow" and "SunUp" varieties.

It is undeniable that agricultural bio-technology provided an excellent and useful transgenic product that performed well in the hands of Hawaiian farmers.

Impact of the Transgenic Papaya in Hawaii

Hawaii Agricultural Statistics Service published Hawaiian papaya production data showing a steady decline in Puna papaya production from 1992 when the PRSV virus was first found in the Puna growing region. At the time, Puna's production was 53 million pounds.

By 1998, the year the transgenic varieties were available to the farmers, production had dropped nearly 50%, to 26.750 million pounds. Production increased since then to 40.290 million pounds in 2001. Remarkably, in 2000, HASS reported that the varietal distribution of papaya-bearing acreage was 56% "Rainbow" on Hawaii Island and 54% statewide.

These figures are in concert with the high rates of farmer adoption reported in the survey. One impact of the transgenic papayas, which had never been grown before, as they help to revive an industry where plants that were once entirely susceptible to a killer virus are now resistant to the effects of the virus.

Another testimony to the beneficial effect of agricultural biotechnology is the report by HASS in 1998 compared to the one in 2001. Papaya has been able to maintain its rank as Hawaii's eighth most important farm products in a listing of the state's top 20 products.

Farm level revenue generated from papayas was $12.6 million in 1998 and grew to $14.6 million in 2001. Also, the papaya has maintained its spot as Hawaii's second most important fruit crop next to pineapple.

Some Challenges Facing Hawaii's Papaya Industry

Hawaii's papaya industry faces two challenges pertaining to its future: maintaining its share of the Japanese market and maintaining durability of resistance of the current transgenic papayas. A significant amount of Hawaii's papaya is shipped to Japan, where Hawaii's transgenic papaya is not yet permitted to be imported.

Thus, Hawaii needs to grow nontransgenic papaya to retain its share of the papaya market in Japan. Furthermore, measures must be taken to minimize the accidental contamination of transgenic papaya in papaya shipments that are exported to Japan. The transgenic papaya actually has helped growers to economically produce nontransgenic papaya in Puna.

Currently, a number of growers use transgenic papaya as a buffer to cut down virus pressure in their nontransgenic blocks. To accomplish this, nontransgenic blocks of papaya are surrounded by large plantings of transgenic papaya, and the nontransgenic blocks are constantly surveyed for virus-infected plants that are immediately removed.

Presumably, this practice works because the transgenic papaya provides physical isolation by increasing the distance that viruliferous aphids need to travel to contact nontransgenic plants, and viruliferous aphids that do feed on transgenic papaya before migrating to the nontransgenic papaya would lose their virus dosage. Naturally, the constant vigilance to identify and remove infected plants cuts down on the secondary spread of the virus in nontransgenic blocks.

A company that focuses almost exclusively on exporting papaya to Japan has successfully followed this practice for several years. The issue of developing practices that minimize the introduction of transgenic papaya into Japan has also been addressed cooperatively by Japan and Hawaii.

At the request of Japanese importers, HDOA adopted an Identity Preservation Protocol that growers and to which shippers must adhere to receive an Identification Preservation Protocol (IPP) certification letter from HDOA that accompanies the papaya shipment. This is a voluntary program. Papaya shipments with this certification are allowed to be distributed in Japan without delay, while Japanese officials perform spot testing to detect contaminating transgenic papaya in the shipment.

However, although papaya shipments without this certificate can be shipped to Japan, the shipments are not allowed to be distributed until the Japanese officials have completed their tests, which may take several days to a week. During this time, the quality of the papaya will decrease even before it is distributed for sale. Some significant features of the Identity Preservation Protocol are that the nontransgenic papaya must be harvested from papaya orchards that have been approved by HDOA.

To get approval, every tree in the proposed field must be initially tested by the GUS test and found negative, the trees must be separated by at least a 15-ft nonpapaya buffer zone, and new fields must be planted with non genetically modified organism (GMO) papaya seeds that have been produced in approved non-GMO fields.

These tests are monitored by the HDOA and conducted by the applicant who must submit detail records to the HDOA. Before final approval of field, the HDOA will randomly test one fruit from 1% of papaya trees in the field. If approved by the HDOA, fruit from these fields can be harvested. In addition, the applicant must submit the detailed protocols that the applicant will follow to minimize the chances of contamination of non-GMO papaya by GMO papaya.

This includes applicant on the random testing of papaya before they are packed for shipment. If the procedures are followed and tests are negative, then a letter from the HDOA will accompany the shipment stating that the shipment is in compliance with a properly conducted Identity Preservation Protocol. The above procedure represents a good faith effort by the HDOA and applicants to prevent the contamination of transgenic papaya in nontransgenic papaya shipments being exported to Japan.

It also illustrates meaningful collaboration between Japan and the HDOA to continue shipment of non transgenic papaya to Japan with the least delay once it arrives in Japan, and yet adhere to the policy that transgenic papaya will not commercially enter Japan until it is deregulated by the Japanese government.

These efforts, along with the effectiveness of the transgenic papaya in helping the economic production of nontransgenic papaya, have allowed Hawaii to maintain significant shipment of the latter to Japan. Obviously, deregulation of transgenic papaya in Japan will circumvent much of the concern of accidental introduction of transgenic papaya into Japan.

To this end, efforts to get the transgenic papaya in Japan were initiated by PAC soon after the transgenic papaya was commercialized in Hawaii. Again, the researchers took the lead in developing the petition. Approval of the transgenic papaya in Japan requires the approval of the Ministry of Agriculture Fisheries and Forestry (MAFF) and the Ministry of Health Labor and Welfare (MHLW).

The petition to the MAFF was approved in December of 2000. The petition process for approval by MHLW is still in progress. An initial petition was submitted to MHLW in April 2003. MHLW requested more information and this information are currently being generated by the researchers. Canada approved the importation of Hawaii's transgenic papaya in January 2003.

Finally, the issue on durability of resistance has to be considered. Studies have shown that "SunUp" papaya has broader resistance than "Rainbow" (18), but the reality is that "Rainbow" is the dominant transgenic papaya grown in Hawaii. So far, we have not observed breakdown "Rainbow" in Puna, and on Oahu. However, we need to be on guard for this possibility.

The occurrence of new strains due to recombination of PRSV strains in Puna with the CP transgene of "Rainbow" is remote. A more realistic danger is through the introduction of PRSV strains resulting from outside of Hawaii. We have shown that "Rainbow" or hemizygous 55-1 is susceptible to many strains of PRSV from outside of Hawaii.

These include strains from Guam and Taiwan, which are quite close to Hawaii. Technically "SunUp" should be resistant to many strains of PRSV that might be introduced into Hawaii. However, as noted previously, the red-fleshed "SunUp" is not the preferred cultivar in Hawaii. A potential solution is to develop transgenic "Kapoho" that is resistant to a wide range of strains.

This could be used as a stand-alone cultivar, or it could serve as a transgenic parent for creating a new type of "Rainbow" by crossing the transgenic "Kapoho" with "SunUp." This F1 hybrid should have the horticultural characteristics of current "Rainbow" but very likely have much wider resistance than the current "Rainbow," owing to increase in CP gene dosage.

We indeed have developed transgenic "Kapoho" that is resistant to range of PRSV strains (D. Gonsalves, *unpublished data*). However, the time frame to commercialize this transgenic "Kapoho" may be longer than the time it took to commercialize line 55-1. These circumstances point to the fact that we need to carefully guard against the introduction of PRSV strains into Hawaii and seek ways to develop new broader resistant cultivars by using the existing germplasm of line 55-1.

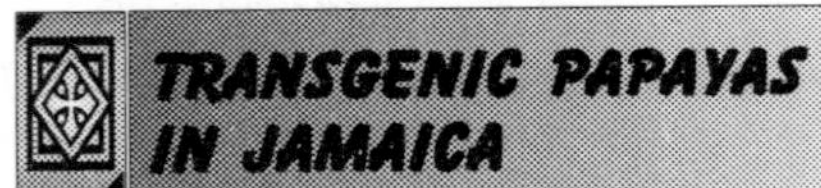

TRANSGENIC PAPAYAS IN JAMAICA

Jamaica is one of the few countries in the Caribbean that has maintained a consistent supply of papaya for its domestic and international markets. "Sunrise" solo selections originally from Hawaii are exported as fresh produce to the European Community, Canada, and the United States, whereas large fruited varieties, from Floridian and South American selections, are cultivated mainly for use in the local processing industries.

Early accounts document a virus disease of papaya, presumably PRSV, in Jamaica in 1929; however, the first epidemic occurred in 1989 in the traditional papaya growing areas of St. Thomas and St. Catherine. At this time, the papaya industry of the country was expanding and had grown from 0.1 ha, in 1986, to 60 ha. In an attempt to eradicate the disease, a reported 50,0 holdings were destroyed between July and November 1990.

Recultivation was halted for about 6 mo. Production and expansion of the industry continued thereafter until produce inspectors in 1994 observed the characteristic ring-spotting pattern on fruit destined for export. Field visits confirmed that the disease was prevalent in St. Mary, Clarendon, Manchester, and once again in the parishes of the initial outbreak, St. Thomas and St. Catherine.

Forty percent of papaya orchards have been devastated by PRSV (Jamaica Papaya Growers Association, *personal communication*). Shortly after the major outbreak of PRSV in 1994, a papaya-breeding program with collaborators at the University of the West Indies (Mona), Cornell University, and the private sector organization, the Jamaica Agricultural Development (JADF), was initiated.

The major goal of the program was to develop new papaya varieties with durable resistance to PRSV using CP genes of the virus. Somatic papaya embryos (derived from "Sunrise" solo hybrids) were transformed with the CP gene of the virus collected from one of the traditional papaya growing regions of the island Caymans.

Two versions of the viral CP gene were cloned and bombarded into papaya tissues at the New York State Agricultural Experiment Station (NYSAES), Geneva, NY; sion (CPT) and an untranslatable version (CP_{NT}), based on the status of literature on transgenic virus resistance at the time. The literature then differentiated between CP-mediated and RNA-mediated protection elicited by CP_T and *CPNT* viral genes, respectively.

It was speculated that the mechanism of resistance in the former case was conferred by the CP, is broad spectrum against distantly related viruses, and is effective at both low and high levels of virus inoculum. Different lines of evidence also suggested that resistance exhibited by transgenic plants not carrying *a* CP_T version of the viral *CP* gene, is RNA mediated and results from the induction of a specific cellular RNA degradation mechanism that may be induced by prior infection.

Because the diversity of PRSV in the country was not known, both CP versions were utilized in the transformation experiments to assure field resistance. Within 1 yr after bombardment, some 39 R0 transformants were acclimatized under greenhouse conditions at Cornell University.

CP transcript was detected in Northern blot analyses with total RNA from established plants; the signal intensities were higher for some transgenic line designations than others. High levels of resistance (78%) to manual inoculations with the homologous virus were observed under greenhouse conditions with plants carrying the CP_T gene. Lower levels of resistance (10%) were obtained with transgenic plants carrying the CP_{NT} and a recovery phenotype (15%) that was characterized by the development of chlorotic spots over the leaf lamina 20-58 d following inoculations with all subsequent new leaves free of symptom expression.

Those R0 clones that were not used in the infectivity assays were subsequently shipped to Jamaica for field testing in the same region of the island from which the homo-logous isolate was collected. It should be noted that Jamaica did not have the regulatory framework in place for overseeing the importation and testing of genetically modified organisms when the project started in 1994. In 1997, the National Biosafety Committee (NBC) was set up as a subcommittee of the National Commission of Science, and Technology and the Plants (Importation) Control Regulations was passed in Parliament under Section 38 of the Plants (Quarantine) Act to allow for the importation and controlled field testing of the transgenic papaya.

Therefore, the transgenic plants remained under greenhouse conditions at Cornell University for 2 yr while the appropriate biosafety regulations were put in place and a permit issued. On transfer to Jamaica (1998), the primary transformants (R_0) were acclimatized for 2–3 wk under shadehouse conditions and subsequently transferred to the field.

A plot of about 0.2 hectares was enclosed by fencing on the commercial Brampton Farm in St. Catherine. Thirteen rows of transgenic (184 plants) and nontransformed "Sunrise" solo papaya (33 plants) were randomly established in the plot. The test plot was surrounded by a border row consisting of nontransformed "Sunrise" solo plants and some 60 hectares of mangoes.

No mechanical inoculations were done and papaya fields in the nearby regions served as the primary source of PRSV inoculum. Within 5 mo after establishment, PRSV was detected in the plot. Even though the permit issued allowed for field evaluation of the materials, the researchers were instructed by the NBC to remove the infected trees.

The major concern was whether infected plants in the plot would act as a reservoir for the virus and contribute to the spread of the virus in the area. It was only after a series of discussions with members of the NBC, papaya growers, and a review of the literature on the field-testing of transgenic papaya in Hawaii that the NBC revoked the injunction and assessment of resistance was allowed. By 17 mo, all nontransformed trees showed symptoms typical of PRSV infection.

As for the transgenic trees, a similar trend in the levels of resistance conferred by the CP genes under greenhouse conditions was observed in the field. Transgenic lines with the CP_T gene showed strong resistance (80%) against field infections while those with the CP_{NT} gene were not as resistant (44%), showing tolerance, that is, attenuated symptoms, or severe symptoms after a delay (18-24 mo) in the onset of disease.

In a second trial (1999), R_1 offspring exhibited similar levels of resistance as the parental lines (58%), improved resistance (26%), or lowered resistance (16%). Together, these observations suggested that even though the protein expression product is not involved in the resistance mechanism, resistance was only achieved by transgene messenger RNA (mRNA) from a CP_T gene.

Other studies have similarly shown that the CP_{NT} gene is not as successful as the CP_T gene in conferring transgenic resistance. For instance, all 13 transgenic oilseed rape lines carrying CP_{NT} were as susceptible to *Turnip mosaic virus* as control plants; only transgenic lines carrying the CP_T gene effectively conferred resistance against the virus.

Likewise, *Potato mop-top virus* CP_{NT} transgenic potato exhibited mild symptom expression whereas CP_T lines were completely resistant and did not develop symptoms. The horticultural characteristics of the Jamaican transgenic papaya were also assessed relative to the nontransformed trees. Numerous lines were found to exhibit agronomic traits suitable for local and export markets. There were gynodioecious trees that produced fruit of red flesh (average weights between 260 and 535 g) and also gynodioecious trees with larger fruits (>535 g) with red or yellow flesh. Nutritional components (vitamin C, brixes) and the antinutritional component, benzyl isothiocyanate (BITC) were also examined.

Statistical analyses conducted at the interval indicate that these components of the transgenic lines were not significantly different from the nontransformed controls. Although total sugars were lower in fruits from line 52.3 ($p < 0.05$) and fats slightly higher in fruits from line 52.24 ($p = 0.037$), the values were within the range documented for papaya. Another examined characteristic of the transgenic lines of practical benefit related to biosafety.

One of the concerns with foods derived through genetic engineering is whether transgenic products are safe for human consumption and there is no potential of introducing allergens into the food supply. Although the safety of NPTII, GUS, and viral CP gene products is documented, it is reassuring that neither CP nor NPTII protein was detected in the edible pulp of fruits from the Jamaican transgenic papaya trees.

Surprisingly, however, GUS activity was consistently detected in only one transgenic line. Encouraging data were also obtained in a 12-wk dietary study with adult Wistar rats. The rats were divided into three groups; (a) rats fed normal diet (a marketed laboratory rodent diet recommended for rats, mice, and hamsters), (b) normal diet supplemented with commercial papaya pul diet and transgenic papaya pulp. At the end of the study, blood, liver, and kidney were collected and total plasma protein determined as well as the activities of transaminases and phosphatases.

Plasma proteins are mad are important in maintaining the pressure between fluids of the vascular system and surrounding cells. Although an insensitive indicator, measurement of the total plasma protein can point to liver dysfunction. On the other hand, levels of the enzyme transaminases (aspartate and alanine transaminases) and phosphatases (acid and alkaline phosphatases) in the plasma are a measure of liver damage; high levels are suggestive of disease or injury of the organ.

In the analyses, there were no significant differences ($p < 0.05$) in the levels of acid and alkaline phosphatases in the liver of rats fed normal rat diet compared to those fed the different preparations of the transgenic and commercial papaya. The same trend was observed for the phosphatases in the kidney and plasma.

Moreover, the protein content of the plasma was not statistically different among the groups. Together, data suggesting no negative effects of the papaya supplements (commercial or transgenic) on tissue and organ integrity has been obtained. The study shown above represents the results of 6 yr of research focused on the development and transfer to the field in Jamaica of PRSV-resistant transgenic papaya lines. The resistance exhibited by the Jamaican transgenic papaya lines should be useful for commercial production and efforts are presently focused on stabilizing interesting lines by continued self-pollination.

Moreover, Southern blot data recently obtained in the lab (Tennant et al., unpublished data) suggest a single insertion of the CP gene in some of the resistant transgenic lines (e.g., line 52.3). Invariably, this will be of benefit in transferring the resistant trait to local papaya varieties, namely the large fruited "Santa Cruz giant" and "Cedro" varieties.

These varieties are as susceptible to PRSV as the commercial "Sunrise" solo variety and are grown for the local market, particularly for the hotel and baking industries. Given these results, the collaborators put together a timeline for moving the transgenic papaya toward deregulation and commercial release.

In 2000, a field trial aimed at further field evaluation and building the seed supply would be set up at Bramptom Farm, in 2001 data would be submitted to the NBC for a decision on deregulation and setting up trials on farmers' orchards, and in 2002, pending a positive ruling from the NBC and the establishment of the necessary regulatory framework, large-scale field release of transgenic seeds to papaya growers.

However, the project is 2 yr behind the proposed schedule. The third field trial has been set up as planned but the question at hand is whether researchers will be allowed to conduct field-tests on growers' orchards and whether the transgenic papaya will be deregulated and released commercially in Jamaica. Although Jamaica was ahead of its CARICOM counterparts and Small Island Developing States (SID) in having established a National Biosafety Committee in 1997 and initiated field-testing genetically modified organisms in 1998, the country now appears to be lagging behind.

The required regulatory guidelines for the release of genetically modified organisms have not progressed despite formulation in 2000. Discussions on how to facilitate the extension of the field trials to growers' orchards (provided satisfactory review of the data generated by the previous field tests) were initiated last year.

Meetings with Parliamentary Counsel representatives advised the NBC on examining the Act to accommodate this Post Quarantine stage or confined testing of the transgenic papaya on growers' orchards. No final decisions have been communicated to date. Moreover, one of the collaborators financing the project (JADF) has requested that the next phase of the project involving field-testing on growers' orchards across the island, be postponed until the NBC has deregulated the transgenic papaya.

The organization foresees legal repercussions with some growers should the transgenic materials be released under temporary provisions by the NBC. However, there is interest among papaya farmers for the research and release of materials. Growers have visited the field trials and identified transgenic trees exhibiting acceptable commercial traits.

Moreover, a recent survey of papaya growers producing on farms for the local (26%) or export market (74%), report that 80% are anxious to receive the transgenic papaya seeds and to participate in setting up experimental plots with the transgenic papaya on their orchards. The other 20% are hesitant of the introduction the material into commerce.

This is because their major markets are in Europe and they fear genetic contamination of their nontransformed materials. Some of these growers have said that their European buyers explicitly stated that fruits from farms with transgenic crops (papaya or otherwise) would not be taken. Another reason given by these growers against adopting the transgenic papaya, is that they may not be safe for consumption and that the transgenic trees actually carry the virus (i.e., they are "immunized") and will spread the virus to other trees.

Interestingly, these farmers are situated in the Western end of the island, where the virus only recently moved in and the disease pressure is very low. They are the ones presently exporting papaya fruit and maintaining the industry. In a wider survey of the Jamaican public in the corporate area and outskirts, 40% are willing to try genetically modified products, 49% will not try these products, and 11% were not able to say whether they would purchase these products or not (Pinnock et al., unpublished data).

The respondents felt that genetic engineering could probably have a positive effect on the quality of life (73%) and were in favour of the application of the technology to medicine (73%), the improvement of ornamentals (52%), and plant defenses (69%), rather than the improvement in the taste of foods (39%) and improvement of livestock (32%). Thus, there is interest in the project from some papaya farmers and consumers, so the researchers will continue developing a transgenic product with acceptable commercial performance while the regulatory processes "catch up."

The research will focus into developing homozygous transgenic lines and in transferring the transgenes to other local varieties, as well as developing new varieties using portions of viral genes to elicit resistance by chimeric transgenes (Gonsalves et al., unpublished results) and safer selection markers.

Most importantly, the NBC has mounted an education program to inform the public on genetic engineering; activities at schools, as well as radio and television broadcasts, have been initiated to clear up some of the misconceptions on the technology. But as the old adage goes, only time will tell whether transgenic papaya will eventually be deregulated in Jamaica.

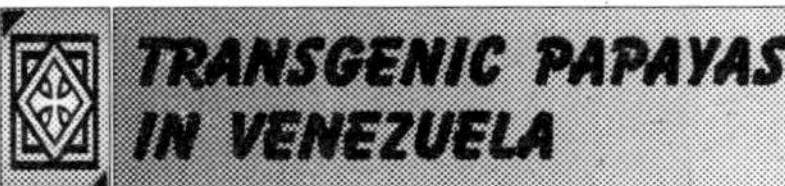

TRANSGENIC PAPAYAS IN VENEZUELA

Among the three countries discussed in this chapter, Venezuela is the one producing the most but exporting the least of papaya. In 2002 Venezuela produced 120,000 Mt of papayas, and the United States (mainly Hawaii) and Jamaica produced 20,684 and 8,637 Mt, respectively.

However, yields were higher for the United States and Jamaica (300,000 and 230,000 kg/ha) as compared to Venezuela (176,000 kg/ha), which exported papayas for an added value of only $472,000 (1000 Mt), whereas Jamaica and the United States obtained gains as high as $3.3 million (2200 Mt) and $17.2 million (8300 Mt), respectively.

The United States also imported papayas valued $62 million in the year 2001. Venezuela is a country with an economy based on the exploitation of oil and where agriculture represents just 5% of the GDP. Papaya is a very popular fruit and its production high but, as shown before, almost entirely devoted to local consumption.

Commercial and domestic orchards are severely attacked by PRSV, and the most efficient control measure is eradication. Big farmers use to switch to other crops when roguing does not suffice as a control measure of PRSV, but small orchards are kept until the virus decimate the trees. In Merida, state papayas are mainly produced in the southwest, close to Lake Maracaibo.

Although reports on early infection by PRSV in the area are lacking, in the beginning of the 1990s the main papaya orchards were severe virus, and industrial papaya production declined dramatically. In an effort aimed at developing the infrastructure required to help advance projects in Bio technology in Venezuela the Inter American Development Bank issued Universidad de Los Andes (ULA) with a grant for the creation of transgenic papayas resistant to local varieties of PRSV.

Transgenic papayas were obtained in Venezuela (Universidad de Los Andes, ULA) thanks in part to a collaborative effort with Cornell University that started in 1992. In 1993, the CP gene of two different Venezuelan geographical isolates from Merida were cloned at the New York State Agricultural Experiment Station (NYSAES) in Geneva, NY.

The two isolates were collected in low land areas where papaya is grown domestically (PRSV LA in Lagunillas, Merida) or commercially (PRSV EV in El Vigia, Merida). Both isolates were the donors of CP genes that were cloned in their translatable, antisense or sense/untranslatable forms. Although the actual role of the protein was not clear when the transformations were done (1994-1995), only the translatable constructs were used for further experiments.

Contrary to the Jamaica's case, plant material was transformed in Venezuela instead of NYSAES, after a transfer agreement of biological material was signed between ULA and Cornell University. The embryo donors for transformation via *Agrobacteriurn* tumefaciens were open-pollinated papayas grown commercially in El Vigia and known locally as "Tailandia Roja" ("Thailand Red").

Plant material was donated by an industrial grower willing to assay the transge they were developed. Plant material was trcnsformed foll designed in ULA for this purpose, and few transformants were recovered (1995–1996) and subsequently crossed (1997–1998).

The progenies from several of these crosses were analyzed for three generations. Venezuela's

R0 transgenic papayas consisted of few independent lines with single-copy insertions of either the PRSV LA *CP* or PRSV EV *CP* gene. Female, male and hermaphrodite R0 plants were intercrossed or selfed in Merida to obtain the first seed-derived generation; R_1 and R_2 generations were obtained and analyzed by one of us at NYSAES .

Resistance seemed to be RNA mediated, and R_1 and R_2 plants showed promising levels of resistance not only to the homologous strain of the virus but also to distant geographical isolates of PRSV, such as one from Thailand with only 87–89% similarty to the Venezuelan transgenes. Among the R_2 plants, two hermaphrodite plants showing high levels and ranges of resistance were identified and kept for future multiplication and breeding programs. A replicate set of experiments was performed under greenhouse conditions in Merida, and resistance to the homologous strains of the virus was high.

Similar to the Jamaica case, but not the Hawaii case, neither the *NPTII* nor the viral *CP* are detectable in the transformed, PRSV-LA and -EV resistant transgenic papaya plants. Because the trans donated by an open pollinated plant, segregation of important horticultural traits was observed. For instance, differences in colour of the flesh, and size and shape of the fruits are common.

However, all fruits tested were sweet and pleasant (G. Fermin, personal observations). Further characterization of the transgenic fruits is stalled, for reasons that will be explained below. However, at this point it is worth recalling that Venezuelan papayas were transformed with PRSV CP genes isolated from viruses in areas where papayas are grown for domestic or industrial purposes (i.e., the transgenes are homologous to the virus present in the area of the potential release of the engineered plants), that they harbor a single copy of the transgenes nptII and *CP*, that these two proteins are not detectable in the transformed plants, that all of the transgenic plants lack reporter marker genes and that the safety concerns of their release and future consumption were diminished to its minimal potential.

Although transgene insertions by *Agrobacterium* mediated transformation are "cleaner" than the ones produced by biolistics, we still lack data supporting the idea than the single insertions harbored by the Venezuelan transgenic papayas are composed only of transgenes and T-DNA sequences with or without rearrangements.

The purpose of creating PRSV-resistant papayas was to offer a clean and reliable solution to the problem associated with the presence of a virulent strain of the virus in the papaya growing area of Merida. The next natural step was to release some experimental plants in the field and assess their degree of resistance under the local pressure of the virus.

It is worth recalling that in the year 2004, Venezuela still does not have a set of rules and regulations regarding the manipulation, release, import, and commercialization of GMOs. There is a Biosafety National Committee in charge of presenting some rules for the discussion of future national laws, but the work is being drastically slowed by some anti-GMO activists.

Curiously, the same kind of fierce opposition to transgenic papayas has not been heard of in Hawaii and Jamaica to the level we have suffered in Venezuelan. Yet, even in the absence of a national mandatory rule, the Venezuela transgenic papayas were the subject of a study aimed at analyzing all potential sources of risk related to their release, and a methodology to liberate them in the field at a small scale was proposed, years before the transgenic plants were developed.

A small plot of transgenic papayas was set at Lagunillas, Merida. The area of Lagunillas was chosen for this set of tests for various reasons: one of the virus samples was isolated in the area, papaya is being grown only domestically far from any commercial orchard (and from the expermental field), and the plot is close to the researchers in charge of the field tests.

The lack of a national legislation pertaining to the management of GMOs obliged some researchers to look for creative solutions. Laboratory work and field experiments with GMOs were not expressly prohibited in the country, but a special permit from the Ministry of Health of Venezuela (MHV) was granted to perform the field testing (1999–2000) in the small plot described above.

The plot was planted with R_1 individuals previously selected in the green house as PRSV-resistant. Later on, the plants continued to show very good performance in the field under the local pressure of the transgenic, PRSV-resistant papayas were set to flower unexpected problems started to emerge. Parallel to the development and characterization of the transgenic papayas for Venezuela, open opposition to GMOs from nongovermental organizations (NGOs) flourished in Merida by the end of the 1990s.

The appearance of misinformed but well-organized opposition groups, the poor preparedness of the public opinion, and the lack of supporting legislation on GMOs all conspired against the acceptance of the transgenic papayas, engineered in the cleanest way possible to solve a practical problem.

The main arguments against the transgenic papayas were that the experiments were being performed in secrecy, the risks associated with the creation and field release of the transgenic papayas were unknown, and the country lacked the appropriate regulations to deal with GMOs in a responsible and organized manner. Only the third claim had certain truth.

The news on the grant obtained to develop transgenic papayas in Merida was highly publicized nationally. Besides, all the data was publicly available to anyone interested, mainly as monographs from one of the authors of this chapter (G. Fermin) but also through articles in local journals and seminars across the country. The claim of secrecy is probably the most unfair of all.

Later, we understood that the accusation of secrecy was just another element of a very well-orchestrated campaign to discredit the scientists involved in the experiments and to predispose the public against the transgenic technology as such. The few transgenic papayas tested in the field were held responsible for the increase in mutatio rates, abortions "contamination" of food and water, and so on; scientists were held as criminals and their names painted on the city walls with degrading epithets.

The people of Lagunillas suffered the most, being constantly bombarded with a collection of "information" aimed at creating panic. Opponents did not accept the legality of the permit issued by MHV and the matter was discussed and solved by the State Legislature.

The scientists were allowe experiments until the plants set fruits, which could be collected and transferred to the laboratory. However, the plants should be incinerated after fruits were collected. A serious discussion about GMOs was lacking, the fierce battle ended up exhausting all reasoning, and legal decisions were not accepted nor respected by the infuriated people who were convinced by few NGO representatives that transgenic papayas would trigger serious health problems.

The most fanatical opponents of transgenic plants violently attacked the plot's guardian and set the small plot on fire in December 2001. At the time this chapter was written, transgenic papayas were still considered a "bad thing" and the legislation on GMOs was moving forward slower than necessity dictates.

Some NGOs call for a total ban on GMOs, and opinions in the government are divided. When the political uncertainty of the country dissipates, Venezuelans will probably recover the time to think about more pressing and fundamental issues, like how to speed the process of developing a sound Plant Biotechnology program that might help modernize Venezuelan agriculture.

In the meantime, a well-thought plan is under way to mitigate the damage done by few anti-GMO activists. The media, mainly the local media, is surprisingly better informed than the anti-GMO activist and crave for more reliable information on transgenic food.

Several interviews in newspapers and radio and local TV stations are helping dissipate the mystified negative aura that surrounds the issue. A more proactive approach to interact with State legislators is currently underway, and workshops to educate are being organized in the University on the potential risks associated with the release of transgenic plants.

Once the papaya growers feel enamored of the many benefits this technology might bring, we will probably reach a stage of more conscious and informed acceptance of this precious biotechnological good.

CONCLUDING REMARKS

We have presented three examples in the adoption of a transgenic crop: the engineered papayas resistant to PRSV. They differ in the degree of acceptance, despite the unequivocal success the transgenic papayas from Hawaii, Jamaica, and Venezuela have shown in resisting PRSV attack.

The Hawaiian transgenic papaya story often is viewed as a rather unique Hawaiian papaya industry had no alternative solution. However, what is often missed is that the papaya case represents efforts by researchers who were simply trying to do their job of addressing and developing solutions to a defined problem.

The fact that the problem became of crisis proportion and that the solution was rather dramatic should not be interpreted to mean that the transgenic approach should only be used as a last resort. Instead, it should be viewed as a component of a system for developing effective, safe, and sustainable efforts to address agricultural problems.

The Jamaican case is potentially a good example of a successful story on technology transfer, if it can overcome the serious challenges on the deregulatory stage to bring the papayas to the market.

However, Venezuela faces more fundamental problems: a society still in search of a political identity spends little time solving problems of longlasting effects (e.g., food production and safety).

From the three countries discussed here, only Venezuela has a very involved and aggressive group of anti-GMO activists.

If the success of the transgenic papayas in other latitudes has not convinced them on the many virtues of the biotechnological approach to solve agriculture-related problems, then the potential success of the Venezuelan transgenic papayas will probably do so.

PRODUCTION OF TRANSGENIC CROPS BY THE FLORAL-DIP METHOD

The production of transgenic plants in tissue culture requires careful preparation of plant tissues, a procedure of transforming individua screening system for selecting transformed plants.

In some crops, such as a lettuce, a well-established tissue culture system is available for the efficient production of transgenic plants at high transformation efficiencies (number of explants producing transformed shoots/total number of explants).

However, in crops such as radish, it has been difficult to regenerate shoots from transformed cells of seedling explants. At present, there are very few species that can be routinely transformed in the absence of a tissue culture based regeneration system.

One such plant, *Arabidopsis thaliana*, can be transformed by several in *plants* methods including vacuum infiltration, transformation of germinating seeds, and floral dipping. These systems have contributed greatly toward the isolation and understanding the functions of many plant genes.

In terms of *in plants* methods being applied to crop plants, there have been very few successful reports. The first successful report on the application of in *plants* transformation toward a crop was achieved in pakchoi (*Brassica rapa L. ssp. chinensis*) by the vacuum infiltration of an *Agrobacterium* suspension into flowering plants.

Despite confirming the transmission of the *bar* gene from the *Agrobacterium* into the progeny infiltrated plants, only two plants from 20,000 seeds sown yielded transformed plants (transformation efficiency of 0.0001%).

However, in a separate study, the application of infiltration of seedlings or flowering plants could be used successfully in the production of transformed plants of the model legume, *Medicago truncatula*.

Here, the frequencies of the progenies producing transformed plants were variable but significantly higher compared to the pakchoi system (flower infiltration method, 4.7–76%; seedling infiltration, 2.9–27.6%) and also compared to *Arabidopsis* (0.1–3%).

However, Southern blot analyses of these legume transformants revealed that these plants were a mixture of independent and sibling transformants (flower infiltration, 13–23% independent; seedling infiltration, 67–86%) and so the mechanism of transformation appears different in the two methods.

In addition, as the flower infiltration method in *Medicago truncatula* produces a large number of sibling transformants and in *Arabidopsis* the plants are usually independent hemizygotes for the transgene, it appears that the two species have different mechanisms of transgene integration.Here, a more recent *in plants* transformation method, floral-dipping, is described in the production of stably transformed plants of radish.

This procedure is simple, reproducible, and has been used successfully in the production of agronomically useful germplasms of this crop.

MATERIALS

Plant Material

1. Seeds of Korean variety "Jin Ju Dae Pyong" (Kyoungshin Seeds Co., Ltd., Seoul, South Korea).

Agrobacterium tumefaciens Strain

1. *Agrobacterium tumefaciens* strain AGL1 with pCAMBIA3301 is used in all plant transformation work. The vector has both the selectable marker *bar* and reporter *gusA* genes located between T-DNA borders under the control of the CamV 35S promoter. The binary vector is transferred into AGL1 by electroporation.

Agrobacterium *Culture Media*

1. Yeast-Extract Peptone (YEP) medium: 10 g/L of tryptone (Difco, Detroit, MI), 10 g/L of yeast extract (Difco), 5 g/L of NaCl.
2. Kanamycin sulfate (Sigma, St. Louis, MO): 50 mg/mL of stock in water. Sterilize by filtration through a 0.2-μm membrane. Store at –20°C for 3-4 mo.
3. Rifampicin (Sigma): 4 mg/mL of stock dissolved in methanol. Sterilize by filtration; store at –20°C for 3–4 mo.
4. Liquid culture medium: YEP medium supplemented with 50 mg/L of kanamycin and 50 mg/L of rifampicin.
5. Agar-solidified medium: YEP medium with 14 g/L of agar (Bacto-agar, Difco), 50 mg/L of kanamycin, and 100 mg/L of rifampicin.

Other Solutions

1. GUS assay buffer: 10 mM $Na_2EDTA \cdot H_2O$, 0.1% Triton X-100, 0.1% 5-bromo-4 chloro-3-indolyl-(-D-glucuronide (X-Gluc), 0.1 *M* NaH_2PO_4, 0.5 *M* $K_3Fe(CN)_6$.
2. Floral dip inoculation medium (IM): 5% (w/v) sucrose, 0.05% (v/v) Silwet L-77 (Osi Specialities, Danbury, CT) dissolved in water, pH 5.2.

METHODS

Preparation of Agrobacterium for Floral Dipping

1. From glycerol stocks, A. *tumefaciens* strain AGL1 carrying pCAMBIA3301 is streaked onto YEP agar-solidified medium containing antibiotics and cultured at 28°C for 2–3 d in the dark.
2. Using a platinum loop, a single bacterial colony is chosen to inoculate 10 mL of YEP liquid medium with selection contained in a 50 mL-capacity Falcon tube and incubated overnight on an orbital shaker at 180 rpm.
3. The culture is transferred to a conical flask containing 1 L of YEP medium with 50 mg/L of kanamycin and grown overnight as described.
4. Measure the optical density (OD) of the bacterial culture by spectrophotometry. An OD of

0.7–0.8 at a wavelength of 600 nm is sufficient for the transformation of radish.

5. The liquid culture is decanted into 300-mL-capacity sterile centrifuge tubes and centrifuged in a Beckman Centrifuge (20 min, 4800g, room temperature). Discard the supernatant by decanting into a container containing 1% domestic bleach (v/v) to kill residual bacteria.
6. Resuspend the pellet in the centrifuge tubes using 1 L of inoculation medium by gentle agitation.

Production of Radish Plants for Transformation

1. Seeds are sown in a soil mix of Vermiculite, Perlite, TKS2 soil-based compost, and peat (12:4:3:1; v/v) in a glasshouse under natural daylight supplemented with 61 µE/m²s daylight fluorescent illumination (26°C day/18°C night).
2. Plants at the four-leaf stage of development are transferred to (20-cm diameter, 30-cm high) pots containing Vermiculite, Perlite, TKS2 soil-based compost, and peat (2:1:1:2) to allow the plant to grow to maturity (one plant per pot).
3. After 7–14 d (six-leaf stage of development), plants are vernalized in a cold chamber (4 ± 2°C, 16-h photoperiod, 45 µE/m²s daylight fluorescent tubes) for 10 d to initiate bolting. Cold-treated plants are transferred to the glasshouse and allowed to bolt.
4. Plants should be watered with a general-purpose liquid feed at fortnight intervals to ensure the plants grow healthy.

Floral Dipping Procedure

1. Prior to dipping, all siliques, flowers, and floral-buds showing petal colour are removed from bolted plants using fine-pointed scissors.
2. Combine the suspended *Agrobacterium* cultures into either a 1- or 2-L measuring cylinder or beaker and carefully invert the potted plant into the infection medium so that the inflorescence is completely submerged.
3. Gently mix the inflorescence of the plant in IM for 5 s and then remove the plant from the bacterial suspension and return to the upright position. Insert a clear polyethylene bag immediately over the dipped inflorescence, removing as much of the air inside the bag, and then tie just below the inflorescence.
4. Place the plants in a shaded area away from direct sunlight and allow them to remain overnight.
5. Remove the bag, and then allow the plants to flower. For successful seed production, each flower needs to be hand pollinated with a fine paint brush for 2-3 consecutive days at a time when the pollen is free flowing (around midday).
6. Seeds should be harvested from siliques when they start to dry. Seeds should be incubated at 30°C for at least 10 d to aid ripening before they can be sown.

Selection of Transformed Plants

1. Seeds from floral-dipped plants are sown to soil and the origin of where the seeds were collected is noted (which plant dipped, location of bolt).

2. Fifteen days after sowing, the plants are sprayed with 0.03% (v/v) Basta® at weekly intervals for 3 wk to select those plants that are resistant to the herbicide.
3. Plants that show no necrosis on the leaves after the third spraying have a functional bar gene. To confirm further the transgenicity of these herbicide-resistant plants, leaf pieces are excised and analyzed by GUS histochemical staining. GUS assays are performed by immersing tissues in assay buffer. After overnight incubation at 37°C, the tissues are immersed in 95% ethanol (1–2 h) and then stored in 70% ethanol for 2–5 *d*.
4. From thousands of plants analyzed for GUS expression, plants that exhibit leaf necrosis (including wild-type) after herbicide spraying fail to show GUS expression and plants that are resistant to Basta® are all positive by GUS assays.
5. Herbicide-resistant plants are transferred to pots and allowed to grow to the six leaf stage of development, at which time the plants are vernalized to promote bolting as described.
6. At the time of flowering, the plants must be hand pollinated to promote successful seed set. The seeds are harvested, ripened, and then sown to soil and the resulting plants are screened for transgene activity (T2 generation).
7. Seeds from these transformed T2 lines are collected and sown to determine which lines are homozygous (all plants are herbicide resistant) and heterozygous (usually 75% of plants are Basta® resistant) for the transgenes.

Notes

1. Experiments conducted in culture have shown that A. tumefaciens strain AGL1 is virulent toward explants of Korean cultivar "Jin Ju Dae Pyong" and so this bacterium was used in such floral-dip studies.
2. Suspending the bacterial pellet by vigorous shaking should be avoided, as it may harm the efficiency of the bacteria to transform the plant.
3. In earlier studies, radish plants of various stages of bolt development (primary bolt, single stem, 3–9 cm height; secondary bolt, 10–15 cm; tertiary bolt, 16-24 cm) were dipped into a suspension of Agrobacterium to determine whether plant development stage is a critical factor in the optimization of transformation efficiency. Plants that exhibit a single stem with many immature floral buds at the time of dipping are the most responsive in terms of the production of transformed seeds. Optimal transformation efficiency (1.4%) is achieved by dipping such plants into a bacterial suspension containing 5% sucrose and 0.05% Silwet L-77. Secondary bolted plants are also amenable to transformation but the yield of transformed seeds is considerably lower (0.2% optimum). Later developing bolted plants usually fail to produce transformed seeds.
4. A liquid fertilizer applied to cold-treated plants at 2-wk intervals helps to prevent leaf chlorosis, which is often seen approx 7 d after the plants are transferred back to the glasshouse. The feed also allows the plant to produce a strong thick stem from which many floral buds develop.
5. The removal of siliques, flowers, and floral buds with petal colour at this stage helps to

reduce the number of harvested nontransformed seeds. In addition, such wounding within the inflorescence aids the bacteria to infect the plant.

6. The presence of a surfactant in the IM is critical for the successful transformation of immature flowers of radish. Silwet L-77, a trisiloxane, in the presence of 5% sucrose is the most effective surfactant treatment in the production of transformed seeds compared to nonionic detergents such as Pluronic F-68 and Tween-20. The use of Silwet L-77 at 0.05% is a significantly improved treatment compared to concentrations used at 0.01 and 0.1%. In the absence of a surfactant, no transformed seeds are produced when inoculated with Agrobacterium in the presence of sucrose.

7. Keeping the infected plant in the shade helps prevent the bag from overheating, which would have a strong negative effect on plant transformation. Removal of air in the bag allows the bacteria to move around the infloresce the area capable of being infected by the bacterial agent.

8. The radish variety "Jin Ju dae Pyong" when grown in a glasshouse free of pollinating insects will not produce seeds owing to a sexual incompatibility problem. The use of a fine paintbrush to hand-pollinate the flowers is essential for the production of seeds and subsequently the production of transformed plants.

9. To be able to understand the mechanism of how Agrobacterium infects the inflo-rescence of radish, seeds from floral-dipped plants must be categorized in terms of where the seeds originated from the infected plants. Southern blot analyses have revealed that between 50 and 60% of all transformed plants were siblings for the transgene (had the same T-DNA insertion pattern) and that such populations of siblings originated from siliques from the same bolt or inflorescence stem. The case of *Arabidopsis*, which usually produces independently transformed seeds, suggests that radish and *Arabidopsis* follow different mechanisms of how the developing seeds are genetically transformed.

10. A few days prior to cold treatment, leaves from putative transformed plants are harvested for molecular analysis. Southern analysis is used to understand both the insertion pattern of the transgenes (independent or sibling transformant) and also to determine copy number (usually one or two copies).

TRANSFORMED TOMATO PLANT

Organogenesis from transformed explants is a very common method for generating transgenic plants. This technique is often favoured because, unlike embryogenesis, morphogenesis from organ explants has been developed for many different plant species. Thus, established organogenesis systems are often easily adapted into genetic transformation protocols.

In addition, regeneration from pieces of nonmeristematic transformed tissue may present a lower risk of chimerism than regeneration from plant meristems. The technique was first popularized by Horsch et al., who described a simplified method for *Agro-bacterium-mediated* transformation of plants.

Since then, organogenesis has become the preferred method for the regeneration of transgenic individuals in many species including tobacco, petunia, tomato, potato, cauliflower, squash, cotton, chrysanthemum, sunflower, and apple.

This chapter uses *Agrobacterium tumefaciens-mediated* transformation of tomato, *Lycopersicon esculentum*, to illustrate the methods for organogenesis from transformed explants. The earliest reports of *Agrobacterium-mediated* transformation of tomato were by Horsch et al. and McCormick et al. nearly 20 yr ago.

Other methods for introducing foreign DNA into the crop, including microinje bombardment, electroporation, and polyethylene glycol (PEG)-mediated transformation of protoplasts, have also been described.

However, none of these techniques has proven to be as popular as the *Agrobacterium-mediated* method, which is favoured for its practicality, effectiveness, and efficiency. The ability to genetically engineer tomato

has been of great value because of its agronomic and economic importance and its usefulness as a model system.

Tomato was one of the first crop species with a molecular genetic map and, more recently, is the subject of new areas of study including functional genomics, proteomics, and metabolomics. Genetic transformation of tomato played an integral role in the map-based cloning of the first disease resistance and quantitative trait loci isolated in plants and in the identification of genes for many other important agronomic, morphological, and developmental traits.

Agrobacterium-mediated transformation of tomato with an antisense polygalacturonase construct was also used to develop the first commercial transgenic plant product, the Flavr Savr tomato. In addition, the technique has been used to introduce very large fragments of DNA (up to 150 kb) into the tomato genome. Since the early reports of transformation, several groups have described various factors that affect the efficiency of tomato transformation.

Conditions that influence tomato transformation include the choice and age of explants, the length of preculture, the strain and concentration of the *Agrobacterium tumefaciens* culture used for cocultivation, the length of cocultivation and medium used, the use of a petunia or tobacco suspension culture feeder layer, the orientation (adaxial side up vs abaxial side up) of cotyledon explants, the gelling agent, the plate se frequency of transfer to fresh selective medium.

However, it is important to note that conditions that result in an efficient transformation system for one genotype, do not always translate into an efficient system for other genotypes. Our work required the development of an efficient transformation system for several different tomato lines for complementation analyses; testing of new vector systems; and studying value-added traits, promoter efficacy, and functional genomics.

Based on information from reports in the literature, we developed a standard tomato transformation protocol that has been routinely used by several research groups for transformation of various *L. esculentum* freshmarket cultivars including Moneymaker, Yellow Pear, Rio Grande, Momor, the processing line E6203, and Micro-Tom which was developed for the ornamental market.

Average transformation efficiencies (the percent of explants that give rise to transformed plantlets) using this protocol ranged from 10 to 15%.

MATERIALS

Tissue Culture Media

Stocks

1. 1X Murashige and Skoog (MS) basal salts mixture powder. If MS salts in powdered form are not available, MS major salt, minor salt and iron stocks can be prepared and stored at 4°C for several months. MS major salt stock (10X): 19 g/L of KNO_3, 16.5 g/L of NH_4NO_3, 4.4 g/L $CaCl_2 \cdot 2H_2O$, 3.7 g/L $MgS0_4.7H_20$, 1.7 g/L of KH_2PO_4. MS minor salt stock (100X): 0.62 g/L of H_3BO_3, 2.23 g/L$MnSO_4{\cdot}4H_2O$, 0.86 g/L of ZnSO4$\cdot 7H_2O$, 0.083 g/L of KI, 0.025g/L of H2O,1 mL of $CuSO_4{\cdot}5H_2O$ stock (2.5 mg/mL), 1 mL $CoC1_2.6H_2O$ stock (2.5mg/mL). MS iron stock (200X): 8.6 g/L of ethylenediaminetetraacetic acid (EDTA) ferric-sodium salt, light sensitive; store in brown bottle.

2. 0.4 mg/mL of thiamine-HCl, dissolve in H_2O; store at 4°C for up to 1 mo.
3. 0.5 mg/mL of pyridoxine-HCl, dissolve in H_2O; store at –20°C.
4. 0.5 mg/mL of nicotinic acid, dissolve in H_2O; store at –20°C.
5. 1 mg/mL of 2,4-dichlorophenoxyacetic acid (2,4-D), dissolve in H_2O, store at 4°C for up to 1 mo.
6. 1 mg/mL of kinetin, dissolve in a few drops of 1 M HCl; store at -20°C.
7. 2 mg/mL of glycine, dissolve in H_2O; store at –20°C.
8. 0.25 mg/mL of folic acid, dissolve in H_2O; store at –20°C.
9. 0.5 mg/mL of D-biotin, dissolve in H_2O, store at –20°C.
10. 1000X Nitsch vitamin stock: 2 g/L of glycine, 10 g/L of nicotinic acid, 0.5 g/L of pyridoxine HCl, 0.5 g/L of thiamine-HCl, 0.5 g/L of folic acid, 40 mg/L of D-biotin; adjust pH to 7.0 to clear solution, and store at –20°C.
11. 1 mg/mL of zeatin, dissolve in a few drops of 1 M HCl, filter –20°C.
12. Appropriate selective agent stock for vector, dissolve as necessary, filter sterilize, store at –20°C. (This protocol has been successfully used with kanamycin, hygromycin, and bialaphos as selective agents. All of these compounds are toxic and should be handled with appropriate care.)
13. 300 mg/L of timentin (a mixture of ticarcillin disodium and potassium clavulanate) or 500 mg/L of carbenicillin, dissolve in H_2O, filter –20°C. These compounds are toxic and should be handled with appropriate care.

Media

1. 1/2 MSO medium: 1/2X MS salts; 100 mg/L of myoinositol, 2 mg/L of thiamineHCl, 0.5 mg/L of pyridoxine-HCl, 0.5 mg/L of nicotinic acid, 1 % and sucrose, 0.8% agar, pH 5.8. Autoclave and store at room temperature for up to 1 mo.
2. KC Biological MS (KCMS) medium: 1X MS salts, 100 mg/L of myoinositol, 1.3 mg/L of thiamine-HCl, 0.2 mg/L of 2,4-D, 200 mg/L of KH_2PO_4, 0.1 mg/L of kinetin, and 3% sucrose. For solid medium, add 0.52% Agargel, pH 5.5. Autoclave and store at room temperature for up to 1 mo.
3. Luria Bertani (LB) medium: 10 g/L of Bacto-tryptone, 5 g/L of yeast extract, 10 g/L of NaCl, 1.5% Bacto-agar Difco (BD, Franklin Lakes, NJ). Autoclave, cool to 55°C, add appropriate filter-sterilized selection agent, and store at 4°C (length of time depends on selection agent).
4. Yeast extract medium (YE): 400 mg/L of yeast extract, 10 g/L of mannitol, 100 mg/L of NaCl, 200 mg/L of $MgSO_4 \bullet 7H_20$, 500 mg/L of KH_2PO_4. Autoclave and store at room temperature for up to 3 mo.
5. MS liquid medium: 1X MS salts, 100 mg/l myoinositol, 2 mg/L glycine, 0.5 mg/L of nicotinic acid, 0.5 mg/L of pyridoxine-HCl, 0.4 mg/L of thiamine-HCl, 0.25 mg/L of folic acid, 0.05 mg/L of D-biotin, 3% sucrose, pH 5.6, autoclave, store at room temperature for up to 3 mo.

6. 2Z medium: 1X MS salts, 100 mg/L of myoinositol, 1X Nitsch vitamins, 2% sucrose, 0.52% Agargel, pH 6.0. Autoclave; cool to 55°C; and add filter-sterilized stocks of vector-selective agent, timentin, or carbenicillin and 2 mg/L of zeatin. Store at room temperature for up to 1 wk.
7. 1Z medium: same composition as 2Z except zeatin is reduced to 1 mg/L. Store at room temperature for up to 1 wk.
8. Selective rooting medium: 1X MS salts, 1X Nitsch vitamins, 3% sucrose, 0.8% Bacto-agar, pH 6.0. Autoclave, cool to 55°C, and add filter-sterilized stocks of vector selective agent and timentin or carbenicillin. Store at room temperature for up to 1 mo.

Preparation of Plant Material

1. Seeds from *L. esculentum* line(s) of choice.
2. 20% Household bleach (1.05% sodium hypochlorite) plus 0.1% Tween-20.
3. Sterile distilled water.
4. Solid 1/2 MSO medium in Magenta (GA7) boxes.

Pre culture

1. Tobacco suspension culture (approx 7 d after subculture) grown in liquid KCMS medium.
2. Solid KCMS medium in 100 x 15 mm petri dishes.
3. Sterile Whatman filter paper (7-cm circles).
4. Tomato seedlings (6-8 d old).
5. Sterile Petri dishes or paper towels.
6. Sterile distilled water.

Preparation of Agrobacterium tumefaciens

1. *A. tumefaciens* strain containing tumor-inducing (Ti) plasmid harbouring the gene of interest.
2. LB medium supplemented with appropriate selection agent in 100 × 15 mm Petri dishes.
3. YM liquid medium.
4. MS liquid medium.

Infection and Cocultivation

1. Precultured tomato explants prepared as described elsewhere in this chapter.
2. *A. tumefaciens* culture prepared as described elsewhere in this chapter.
3. Sterile Magenta boxes or other wide-mouthed containers.
4. Sterile paper towels.

Selective Plant Regeneration

1. Selective 2Z medium in 100 × 15 mm Petri dishes.

2. Selective 1Z medium in 100 × 20 mm Petri dishes and Magenta boxes.
3. Micropore filter tape (3M Corporation, St. Paul, MN).
4. Selective rooting medium in Magenta boxes.

Transfer to Soil

1. Four-inch pots.
2. Sterile potting mix.
3. Clear plastic bags or containers.

METHODS

Preparation of Plant Material

1. Surface sterilize seeds in 20% household bleach for 10 min and rinse three times with sterile distilled water.
2. Culture on 1/2 MSO medium in Magenta boxes (25 seeds/box). Maintain at 24 ± 2°C, under a 16-h photoperiod (cool white fluorescent lights, 60-100 E/m^2/s for 6-8 d depending on the tomato line being used. Cotyledons should be expanded and seedlings should be used before first true leaves emerge.

Pre culture

Preparation of Feeder Plates

1. One day prior to explant preparation, pipet approx 2 mL of a 7-d-old tobacco suspension culture onto solid KCMS medium in Petri dishes.
2. Seal plates with Parafilm and incubate overnight in the dark at 24°C ± 2°C.
3. Cover the suspension culture with a 7-cm circle of sterile Whatman filter paper.

Preparation of Explants

1. Remove seedlings to damp sterile paper towel or Petri dish containing sterile water and excise cotyledons.
2. Cut both ends of cotyledon to remove the tip and petiole. If cotyledon sections are longer than 1 cm, cut them in half.
3. Place cotyledon explants on feeder layer plates prepared the previous day as described elsewhere in this chapter. At this stage as many as 80 explants can be cultured on a single plate; however, fewer should be used if contamination may be a problem.
4. Seal plates with Parafilm and culture for 1 d at 24 ± 2°C, under a 16-h photoperiod.

Preparation of A. tumefaciens

1. Streak *A. tumefaciens* strain onto fresh plate of LB medium containing the appropriate selection agent and incubate for 48 h at 28°C.
2. Transfer four well-formed colonies to a flask containing 50 mL of YM liquid medium

supplemented with the appropriate selection agent and maintain in a shaking incubator at 28°C until an OD600 of 0.4–0.6 is reached (usually over night).

3. Centrifuge the cells at 8000g for 10 min at 20°C.
4. Resuspend pellet in 50 mL of MS liquid medium.

Infection and Cocultivation

1. Transfer cotyledon explants prepared as described elsewhere in this chapter to 25 mL of prepared *Agrobacterium* suspension in a sterile Magenta box or similar widemouthed container and incubate for 5 min.
2. Remove bacterial suspension with a sterile pipet.
3. Blot explants on sterile paper towels and place with the adaxial sides down (that is, upside down) on the original feeder plates.
4. Maintain at 19–25°C in the dark for 48 h of cocultivation.

Selective Plant Regeneration

1. Transfer cotyledon explants to plates containing selective 2Z medium with the adaxial sides facing up. A total of 25 explants are cultured on each plate at this stage.
2. Seal plates with micropore tape and maintain at 24 ± 2°C under a 16-h photoperiod of cool white fluorescent lights.
3. After 3 wk, transfer the cultures to plates containing 1Z medium.
4. Transfer explants to fresh medium at 3-wk intervals. Discard explants that are completely brown or bleached and trim off dead tissue from regenerating explants. Approximately 10 explants are cultured on each plate at this stage—the exact number depends on the size of the callus and regenerating tissue.
5. When shoots begin to regenerate from the callus, transfer cultures to 1Z medium in Magenta boxes.
6. Excise shoots from callus when they are approx 2 cm tall and transfer to selective rooting medium in Magenta boxes.
7. Maintain plants at 24 ± 2°C under a 16-h photoperiod of cool white fluorescent lights.

Transfer to Soil

1. When the selected transgenic lines have well-formed root systems, they can be transferred to soil.
2. Remove a plant from its culture vessel and gently wash the medium from the roots. Use tepid water.
3. Transfer each plant to a 4-in. pot containing a sterile potting mix.
4. Cover each plant with a clear plastic container or a plastic bag secured to the pot to provide a humid environment. Transfer to a growth chamber or to a shaded area of a greenhouse. Do not place in direct sunlight.

5. After 1 wk, gradually lift the plastic container each day during the next week. If a plastic bag is used, cut a small hole each day during the course of the next week, then remove the bag.

Timetable for Tomato Transformation

This timetable is based on a 6-d germination period. Adjust the times to the germination period of your tomato seeds.

D 1. Sterilize seeds and transfer to 1/2 MSO medium.

D 5. Streak *Agrobacterium* onto LB selective medium.

D 6. Prepare feeder layer plates.

D 7. Prepare cotyledon explants.Inoculate liquid overnight culture.

D 8. Prepare *Agrobacterium* suspension for infection.
Infect and cocultivate cotyledon explants.

D 10. Transfer cotyledon explants to selective 2Z medium.

D 31. Transfer cultures to 1Z medium. Continue to transfer to fresh 1Z medium at 3-wk intervals.

Notes

1. If contamination is a problem, the seeds can be soaked for 2 min in 70% ethanol before the bleach treatment.
2. The tobacco suspension culture (NT I) is maintained in liquid KCMS medium on a rotary shaker and is subcultured weekly by adding 2 mL of the old culture to 48 mL of fresh medium.
3. Remove only as many seedlings as can be prepared in a few minutes, as they wilt quickly.
4. Hypocotyls may also be used but they regenerate more slowly and tend to produce more nontransformed shoots.
5. Explant orientation at this step is only for convenience.
6. Culture at lower temperatures has been associated with higher transformation efficiencies.
7. Explant orientation at this step has a significant effect on transformation efficiency. Explants placed adaxial side up curl into the medium, make better contact with it, and produce more transformed shoots.
8. Micropore tape gives higher transformation efficiency than Parafilm.
9. Zeatin level is reduced at this step to conserve resources. The reduction does not have a negative effect on transformation efficiency.
10. Transfer is necessary every 3 wk as plates sealed with micropore tape dry out quickly. More frequent transfers do not increase transformation rate.
11. Do not transfer shoots without meristems because they will not develop meristems in rooting medium.

12. To ensure that the plants are free of *Agrobacterium* before transfer to soil, shoots can be rooted twice on medium containing timentin or carbenicillin and then once on medium lacking this antibiotic. Plantlets that show *Agrobacterium* contamination at this stage should be destroyed by autoclaving. Alternatively, a PCR assay with *Agrobacterium-specific* primers can be done to ensure that plants are not contaminated.
13. Cover each plant with either the plastic container or a bag immediately after transfer to soil. The plants wilt quickly after removal from the culture containers. For plastic containers, Magenta boxes can be used or even clear plastic bottles that have the top of the bottle removed.
14. If placed in direct sunlight, heat will build up under the plastic containers or bags and kill the plants. If a growth chamber or shaded area of a greenhouse is not available, then maintenance in a lab setting under lights will be sufficient for the first 2 wk. After that period, they can be moved to a greenhouse.

12 Chapter

LOCALIZATION OF TRNSGENIC IN PLANTS CHROMOSOMES

Methods for the production of transgenic plants are continually being improved, increasing the chances of obtaining plants with desirable low numbers of transgene copies and the required transgene expression profiles. One aspect of the genetic transformation process that is at present very difficult or impossible to control, is the exact insertion site of the transgene in the host genome.

As the transgene insertion site may play an important role in determining the stability and expression characteristics of the introduced gene, an understanding of the location of transgenes within the host genome is often desirable. Fluorescence *in situ* hybridization (FISH) is a powerful technique that enables transgenes to be located to specific chromosomes and to specific chromosome regions.

The technique involves the preparation of a labeled probe, homologous to the target sequence, hybridization of the probe to a suitable chromosome spread, and then detection of the hybridized probe. The method provides a good starting point for more detailed analysis of the exact transgene insertion site using either genetic mapping techniques or a range of molecular techniques.

The first *in situ* hybridizations of nucleic acid probes to chromosomes preparations used radioactive probes. These methods have now been superseded by nonradioactive techniques including the use of fluorescence. The FISH technique offers numerous advantages including more rapid and sensitive detection that has in turn allowed low- and single-copy transgene sequences to be detected in plant chromosome preparations.

Other advantages include improved safety and the ability to combine several probes, labeled with differen a single experiment. There have now been many reports of the use of FISH to detect the location of transgenes in a range of plants, for example, petunia, tobacco, rice, *Vicia faba*, barley, wheat, triticale, and oat.

In oat and barley, the FISH technique was able to reveal complex transgene integration sites where linked transgene insertions were interspersed with genomic DNA. In addition, the technique may be used to identify plants homozygous for the transgene at an early stage.

Recent reviews give excellent summaries of the use of the technique to localize transgenes. Schwarzacher and Heslop-Harrison provide a comprehensive practical guide to all aspects of *in situ* hybridization and this should be referred to for further details.

This chapter concentrates on providing details of a protocol that has been successfully used to detect transgenes in wheat and barley and that is sensitive enough to detect single-transgene copies.

MATERIALS

Mitotic Metaphase Chromosome Preparation

1. Fixative solution: three parts 100% ethanol to 1 part glacial acetic acid.
2. 1X Enzyme buffer: 4 mM citric acid, 6 mM sodium citrate.
3. Cellulase (Sigma, St. Louis, MO).
4. Pectinase (Sigma).
5. 45% acetic acid.

Probe Preparation

1. 1X TAE buffer: 0.04 M Tris-acetate, 1 mM EDTA. A 1X working solution is made by dilution of a 50x concentrated stock solution in water (50X TAE: 242 g of Tris, 57.1 mL of glacial acetic acid, 100 mL of 0.5 MEDTA at pH 8.0).
2. QIAquick polymerase chain reaction (PCR) Purification Kit (Qiagen,Valencia, CA).
3. 10X Nick translation buffer: 0.5 *M* Tris-HCl pH 7.8, 0.05 *M* $MgCl_2$, 5 mg/mL bovine serum albumin (BSA).
4. Unlabeled nucleotide mix: 1:1:1 ratio of 0.5 mM dATP, dCTP, dGTP in 100 mM Tris-HCl, pH 7.5.
5. 0.1 M dithiothreitol (DTT).
6. DNA polymerase I/ DNase I (Gibco BRL, Invitrogen, Carlsbad, CA).
7. 1 mM Biotin-11-dUTP (Sigma).
8. Digoxigenin-11-dUTP 1 mM (Roche Diagnostics, Nutley, NY).
9. dTTP 1 mM (in 100 mM Tris-HCl, pH 7.5).
10. QIAquick Nucleotide Removal Kit (Qiagen).
11. 10X PCR buffer (Amersham, Piscataway, NJ).
12. dNTP stock containing 2 mM each of dATP, dTTP, dCTP, dGTP.

13. 0.2 mM M13 17-bp reverse primer (Amersham).
14. 0.2 mM M13 single-stranded 17-bp primer (Amersham).
15. 1X TE: 10 mM Tris-HCl, 1 mM EDTA, pH 8.
16. *Taq* DNA polymerase (5 U/μL) (Pharmacia, Pfizer, New York, NY).
17. Buffer 1: 0.1 *M* Tris-HCl, pH 7.5, 0.15 *M* NaCl.
18. Buffer 2: 0.5% (w/v) blocking reagent (Roche) in buffer 1.
19. Buffer 3: 0.1 *M* Tris-HCl, pH 9.5, 0.1 *M NaCl*, 0.05 *M* $MgCl_2$.
20. Anti-biotin-AP Fab fragment (Roche).
21. Anti-digoxigenin-AP Fab fragment (Roche).
22. Detection solution: 22.5 μL of nitroblue tetrazolium (NBT) 50 mg/mL of (Promega, Madison, WI), 17.5 μL of 5-bromo-4-chloro-3-(BCIP) 50 mg/mL of (Promega), 4.96 mL of buffer 3. (Prepare immediately prior to use).

Hybridization

1. 100 tg/mL of RNase A (Sigma).
2. 2X Saline sodium citrate (SSC): 0.3 M NaCl, 30 mM sodium citrate at pH 7.0).
3. Pepsin (25 tg/μL) (Sigma) in 0.01 M HCl.
4. Depolymerized paraformaldehyde (Sigma) 4% in water. To prepare a 50-mL solution: add 2 g of paraformaldehyde to 30 mL of sterile water and heat to 50°C, add 10 mL of 0.1 M NaOH to dissolve. Make up to 50 mL with sterile water.
5. 100% Formamide (Sigma).
6. 50% Dextran sulfate (Sigma).
7. 20X SSC: 3 M NaCl, 0.3 M sodium citrate at pH 7.0.
8. 10% Sodium dodecyl sulfate (SDS).
9. Sheared salmon sperm DNA (5 tg/μL). 10. 20% (v/v) Formamide in 0.1X SSC.

Probe Detection

1. 4X SSC, 0.2% Tween-20 (Sigma).
2. 5% BSA (Sigma) in 4X SSC, 0.2% Tween-20.
3. Extra-avidin conjugated to Cy3 (Sigma).
4. Anti-digoxigenin conjugated to fluorescein (Roche).
5. 4-6-Diamidino-2-phenylindole (DAPI) (2 *μg/mL*) (Sigma). A stock solution was prepared at 100 *μg/mL* in sterile water and stored at –20°C. The working solution was prepared at 2 *μg/mL* by diluting the stock solution in McIlvaine's buffer, pH 7.0 (82 mL of 200 mM Na2HPO4 and 18 mL of 100 mM citric acid).
6. Antifade solution (Citifluor).

METHODS

The methods outlined below describe the four main steps of the FISH technique.

Mitotic Metaphase Chromosome Preparation

Plant Material

Chromosome preparations were made from the root tips of germinating seed lings as these give a good accumulation of metaphase chromosomes. The method of chromosome preparation follows the squashing protocol of Schwarzacher and Leitch with some modifications.

1. Germinate the seeds on moist filter paper in Petri dishes at 25°C for 24 h, then transfer to 4°C for 24 h and transfer back to 25°C again for 24-30 h according to species. At this stage the root tips should be approx 1 cm long.
2. Cut the root tips from each seed. Using forceps, place root tips in 10-mL vials of aerated, icy, distilled water. Pack the vials in ice and store at 4°C for 24 h.
3. Remove the root tips from the ice water and immediately place in fixative solution overnight at room temperature. Thereafter, store the fixed material at 4°C.

Chromosome Preparation

High-quality chromosome preparations are required for *in situ* hybridization. The presence of cytoplasm and other cellular debris will reduce the quality of results.

1. Transfer root tips from the fixative solution into clean vials using forceps.
2. Remove the fixative by washing the root tips in 1X enzyme buffer (4 *MM* citric acid and 6 mM sodium citrate) for 3 × 5 min.
3. Digest the root tips in 1% (w/v) cellulase (Sigma) and 20% (v/v) pectinase (Sigma) in 1X enzyme buffer for 90–100 min at 37°C.
4. Following digestion, remove the enzyme solution from the vial with a pipette and replace it with 1X enzyme buffer. Incubate the digested material in 1X enzyme buffer at room temperature for at least 15 min before preparing the chromosome spread.
5. Prepare microscope slides by clearing with 100% ethanol. Chromosome spreads are made under a dissecting microscope.
6. Take enough root tip material for one preparation (usually a single root tip).
7. To the root tip, add a one or two drops of 45% acetic acid to disperse cell cytoplasm and after 2-3 min, using a fine syringe needle and working under a dissecting microscope, remove the root cap and as much outer material as possible to leave the central meristematic tissue on the slide.
8. Gently spread the meristematic tissue around a small area on the slide and cover the material with a clean cover slip. Firmly press down on the coverslip applying even pressure.
9. Place the slides with the chromosome spreads onto a metal tray placed onto a bed of

dry ice for at least 10 min or until the slides turn opaque, then flick the coverslip off using a razor blade.

10. Allow the slides to air-dry and observe the chromosome spreads using phasecontrast microscopy. Good preparations should appear with high contrast. Any residual cytoplasm around chromosomes can be seen as a gray shadow. Only slides with at least 10 good chromosome spreads should be used for *in situ* hybridization. Whenever possible cytological preparations should be used immediately. However, slides can be stored at 4°C in a dry environment for several weeks. If stored at -20°C they may be kept for several years.

Probe Preparation

The method used for probe preparation is described by Salvo-Garrido et al.. Essentially the method involves using, as a probe, digested fragments from the plasmid used for transformation. This approach led to significant improvements in the detection of low copy number transgenes over methods using intact plasmids as probes. The method described yields 2 × 50 μL probes.

Plasmid Digestion

1. Prior to plasmid digestion, check the plasmid to be used as a probe by running it on a 1% agarose gel in 1X TAE buffer. The gel should show clean plasmid bands of the expected size with no contamination. Determine the amount of plasmid needed to result in 1 tg per probe. Note that the probe sample goes through two purification steps that each reduce yield. Starting with approx 4 tg of plasmid gives good results.
2. Digest the plasmid with appropriate enzymes that will release fragments of 1-3 kb. Use either one or two different enzymes to give the required fragments and carry out the digestion at an appropriate temperature for the enzymes chosen in a total volume f 50 μL for approx 3 h. Use an enzyme buffer appropriate to the enzymes chosen.
3. Following digestion, run 2 μL of the digestion products on a 1% agarose gel in 1X TAE. Check that the expected digestion products are obtained.

Purification of the Digested Plasmid

Purify the digested plasmid using a QIAquick PCR Purification Kit. Follow the manufacturer's instructions.

Labeling of the Probe

Plasmid fragments are labeled by nick translation using DNase 1-DNA poly

merase activity to incorporate labeled dUTP together with unlabeled dATP, dCTP, and dGTP.

1. Mix the following components in an Eppendorf tube:
   a. 10 μL of 10X nick translation buffer.
   b. 10 μL of unlabeled nucleotide mix.
   c. 2μL of 0.1 M DTTc. 2 μL of 0.1M DTT.
   d. 50 μL of digested plasmid.

e. 10 μL DNA polymerase–DNase I.

f. Either 5 μL of 1 mM biotin-11-dUTP or 2 μL of 1 mM dioxigenin-11-dUTP mixed with dTTP.(1 mM in 100 mM Tris-HCl, pH 7.5) to a concentration of 0.35 mM dioxigenin-11-dUTP, 0.65 mM dTTP. Sterile water to give a total volume of 100 μL.

2. Mix the contents of the tube gently and centrifuge briefly in an ultracentrifuge to bring the contents to the bottom of the tube.
3. Incubate at 15°C for 90 min.

Probe Purification

It is necessary to purify the probe to remove any unincorporated nucleotides.

1. Purify the probe using a QIAquick Nucleotide Removal Kit. As the total volume of the nick translation mix is 100 μL it is necessary to split this into 2 × 50 μL so as not to overload the QIAquick columns. Follow the manufacturer's instructions. After purification, 50 μL is eluted from each column. Each 50-μL aliquot should contain at least the minimum of 1 μg required for each probe.

Preparation of Marker Probes

A range of marker probes may be used to aid identification of the individual chromosomes so that the transgenes can be allocated to specific chromosomes and chromosome arms. For example, in barley, the probes pTa7l (18S-5.8S 26S rDNA) and pTa794 (5S rDNA), which hybridize to ribosomal DNA, give distinct and specific chromosomal patterns that unambiguously identify all seven individual chromosomes . These barley markers will be used as examples to describe the preparation of labeled marker probes.

1. pTa 71 is linearizsed with *EcoRV* and can be labelled by nick translation using the same protocol described above for the plasmid. Alternatively, pTa 794 can be labeled by PCR, and the protocol for PCR labeling of this marker is given below as an example.
2. Add the following to an Eppendorf tube keeping all components on ice:
   a. 5 μL 10X PCR buffer.
   b. 1.5 μL dNTP stockb. 1.5 μL dNTP stock.
   c. 1.5 μL 1 mM biotin- 11 -dUTP **or** digoxigenin-11-dUTP.
   d. 1.5 μL 0.2 mM M13 reverse primer.
   e. 1.5 μL 0.2 mM M13 single stranded primer.
   f. 1 μL miniprep DNA of pTA794 diluted 1: 100 in 1X TE.
   g. 0.5 μL *Taq* DNA polymerase.
   h. 37.5 μL water = 50 μL total volume.
3. Carry out the PCR reaction using the following conditions:
   a. 94°C 5 min.
   b. 94°C, 30 s. c. 56°C, 30 s d. 72°C, 90 s.

e. To step b for 30 cycles. f. 72°C, 5 min.

Thereafter, store at -20°C or keep on ice when in use.

Checking the Incorporation of the Label in the Probe

1. Soak a small square of Hybond N+ membrane in buffer 1 in a sterile Petri dish for 5 min then blot dry between filter paper.
2. In pencil, draw one circle on the membrane for each probe you wish to check. Load 1 µL of probe inside the circle and allow 10 min to air-dry.
3. Incubate the membrane in bufferl for 1 min, then buffer 2 for 30 min at room temperature.
4. Incubate membrane for 30 min at 37°C in a 1:500 dilution of anti-biotin-AP Fab fragment in buffer 1 for the detection of biotin or a 1:1000 dilution of anti digoxigenin-AP Fab fragment in buffer 1 for the detection of digoxigenin.
5. Wash membrane in buffer 1, 3 x 5 min, then transfer the membrane to buffer 3 for 2 min.
6. Incubate membrane in the detection solution for 10 min in the dark.
7. Wash the membrane in water and air-dry.
8. Dark spots inside the circles on the membrane indicate that the labeled nucleotide has been incorporated into the probe.

Hybridization

Pretreatment

Pretreatment is required to reduce nonspecific hybridization of probe to nontarget nucleic acid and to reduce nonspecific interactions with proteins that may bind to the probe.

1. Place slides in a humid chamber at 37°C.
2. To each slide add 200 µL of 100 tg/mL of RNase A in 2X SSC to the area containing the chromosome squash and cover with a plastic cover slip.
3. Place the chromosomes preparations back in the humid chamber in a 37°C oven for 1 h.
4. After RNase treatment, wash for 5 min with 2X SSC at 37°C and at 120 rpm in ashaking 37°C water bath to remove the plastic coverslip.
5. Add 200 µL of 25 µg/µL pepsin in 0.01 M HCl to the squash area on each slide, cover with a plastic cover slip, and incubate for 10 min at 37°C.
6. Wash the slides in water for 2 min to stop the reaction and to float off the coverslips, then wash slides in 2X SSC for 3× 5 min.

Prehybridization Fixation

Incubate slides in 4% freshly depolymerized paraformaldehyde in water for 10 min at room temperature, then wash for 3 × 5 min in 2X SSC. This step is to stabilise the chromosomes prior to dehydration

Dehydration

Dehydrate the slides successively in 70, 90, then 100% ethanol for 3 min each and allow to air-dry. Do not leave them for too long to dry or they will rehydrate. Dehydration prevents dilution of the probe.

Hybridization

1. Prepare the probe mix for hybridization by adding the following to an Eppendorf tube:
a. 20 µL 100% formamide.
b. 8 µL 50% dextran sulfate.
c. 4 µL 20X SSC.
d. 0.5 µL 10% SDS.
e. 2 µL sheared salmon sperm DNA (5 µg/µL).
f. 1 µL probe (1 µg/µL).
g. 4.5 µL water = 40 µL total volume (this is the quantity needed for one slide). When making up the hybridization mix, first add the water to the dextran sulfate to dilute it so that it is easier to handle.
2. Denature the hybridization mix at 70°C for 15 min, pulse spin to bring the contents to the bottom of the tube, then transfer tc ice immediately.
3. Add 40 µL of the hybridization mix to the squash area on each slide as quickly as possible, cover with a plastic coverslip, and place in a thermocycler (Omnislide). Use the following conditions in the thermal cycler: a. 78°C, 10 min.
b. 50°C, 1 min.
c. 45°C, 90 s.
d. 40°C, 2 min.
e. 38°C, 5 min.
f. 37°C, for 16 h overnight.

Posthybridization Washes

Posthybridization washes are carried out to remove any nonspecific or weakly bound probeweakly bound probe.

1. Float cover slips away by washing in 2X SSC, in a 37°C water bath, shaking at 120 rpm.
2. Remove the SSC and wash in a stringent wash solution (20% [v/v] formamidein 0.1X SSC) for 2 × 5 min at 37-42°C and 120 rpm.
3. Wash slides with O.1X SSC for 3 x 5 min in a 37°C waterbath with shaking at 120 rpm and then allow to cool to room temperature.

Probe Detection

1. Places slides in 4X SSC, 0.2% Tween-20 for 5 min. Add 200 µL of 5% BSA in 4X SSC,

0.2% Tween-20 to each slide and incubate the slides for 5 min at room temperature covered with a plastic cover slip.

2. Biotin-labeled probes can be detected with extra-avidin conjugated to Cy3 and digoxigenin labeled probes can be detected with anti-digoxigenin conjugated to fluorescein. Using the same cover slip, incubate the slides for 1 h at 37°C with the respective antibody and/or avidin at a concentration of 2 tg/mL in a total volume of 50 µL 5% BSA in 4X SSC, 0.2% Tween-20. As the antibody is light and temperature sensitive, prepare the mix in a foil-covered Eppendorf and keep on ice. To add the antibody, leave the plastic cover slip on the slide and pour away any excess 5% BSA in 4X SSC 0.2% Tween-20. Then using forceps, carefully lift the corner of the cover slip and add the 50 µL of antibody and 5% BSA in 4X SSC, 0.2% Tween-20.
3. After 1 h, float away the coverslips by washing the slides in 4X SSC 0.2% Tween-20 for 5 min at 120 rpm, then wash for a further 2 x 5 min in 4X SSC 0.2% Tween-20 at 120 rpm and allow to cool.

Counterstaining and Mountants

1. Add 100 µL DAPI (2 tg/mL) to the squash area on each slide and incubate for 10 min at room temperature in the dark to visualize the chromosomes.
2. Rinse the slides in 4X SSC, 0.2% Tween-20, add a drop of antifade solution, and apply a cover slip. To seal the cover slip, paint the edges of the cover slip with nail varnish. Store slides at 4°C in a dry environment. Slides can be viewed immediately but can also be kept at 4°C for viewing later. Slides will keep for several months.

Microscopic Visualisation

1. Examine slides using a suitable microscope (e.g., Nikon Microphot-SA) as per the manufacturer's instructions. Choose a suitable filter set for the fluorochrome being visualized. An example of using *in situ* hybridization to identify transgene insertion sites in barley is shown elsewhere in this Chapter. Biotin-labeled pTa794 was used as a marker probe, which identified a subtelomeric region on the long arm of chromosome 4H. Digoxigenin-labeled fragments of the inserted plasmid were used as a transgene probe. In this way, the transgene insertion site was localized to a subtelomeric region of the short arm of chromosome 4H.
2. Photographic images can be captured on Fuji 400 ISO (ASA) 35-mm film or similar film using an automatic camera on a photomicrographic attachment.

Notes

1. Germination should be carried out using sterile distilled water and in sterile Petridishes. If necessary, the seed can be surface sterilized by rinsing with 70% ethanol, washing two times in sterile distilled water, rinsing for 1 min in sodium hypochlorite solution (6% [w/v] available chlorine) followed by at least three washes in sterile distilled water to prevent contamination of seeds.
2. Take care not to touch the end of the root tips, as this may cause cell damage.

TR1
5' g t a a t g c a t g a c g t t a t t t a t g a g a t g g g t t t t t a t g a t t a g a g t c c

TR2
c g c a a t t a t a c a t t t a a t a c g c g a t a g a a a a c a a a a t a t a c g c g c a a a

TR3 TR4
c t a g g a t a a a t t a t c g c g c g c g g t g t c a t c t a t g t t a c t a g a t c g g g a a

Right border
t t a a a c t a t c a g t g t t t g a c a g g a t a t t g g c g g g t a a a c c t a a g –3'

TL1 TL2
5' a t t a g g g t t c c t a t a g g g t t t c g c t c a t g t g t t g a g c a t a t a a g a a a

c c c t t a g t a t g t a t t g t t t g t a a a a t a c t t c t a t c a a t a a a a t t t c

TL3 Tl4
t a a t t c c t a a a a c c a a a a t c c a g t a c t a a a a t c c a g a t c c c c c g a a t t a

a t t c g g c g t t a a t t c a g t a c a t t a a a a a c g t c c g c a a t g t g t t a t t a a g

Left border
t t g t c t a a g c g t c a a t t t g t t t a c a c c a c a a t a t a t c c t g c c a c c a – 3'

Figure 12.1: Specific primers used for TAIL-PCR of insertion flanking sequences. The TDNA is from the pCAMBIA binary vectors. Primers TR1/TL1, TR2/TL2, TR3/TL3, and TR4/TL4 are corresponding to the P1, P2, P3, and P4 shown in Figure elsewhere in this chapter.

3. It is really important that the water is cold enough to contain ice crystals. Shake the water vigorously to aerate it before putting in the roottips. A colchicine treatment is an alternative to the ice water treatment .
4. Make up the fixative solution fresh each time. It is also recommended to use fresh ethanol as an opened solution may absorb moisture from the air, which dilutes the ethanol.
5. Root tips can easily be transferred from vial to vial, without damaging the root cap, by using a Pasteur pipet.
6. Adjust the digestion time and concentration of the enzyme to suit the material. It is often necessary to increase the concentration of cellulase to 2%. The longer the material has been fixed, the longer digestion is required.
7. If the root cap is difficult to remove. this indicates that digestion should have been carried out for a little longer.
8. It is important to make sure that squashes are flat. Good indications of an unflat squash are air bubbles under the cover slip. Try to tap out any air bubbles using a syringe needle before pressing down on the cover slip.
9. Before placing the slide on dry ice. etch the surface of the slide, around the edge of the cover slip using a diamond pencil. This will ensure that, once the cover slip has been removed, one will still know the location of the squash on the slide. The metal tray should be placed on a bed of level dry ice, so that the surface freezes evenly. To help prevent cytological preparation loss, is it very important to flick off the cover slip while the slide if still frozen, that is, opaque. Despite this measure, material can be lost from the slide when removing the cover slip. However, if root tips have been incubated in icy water to

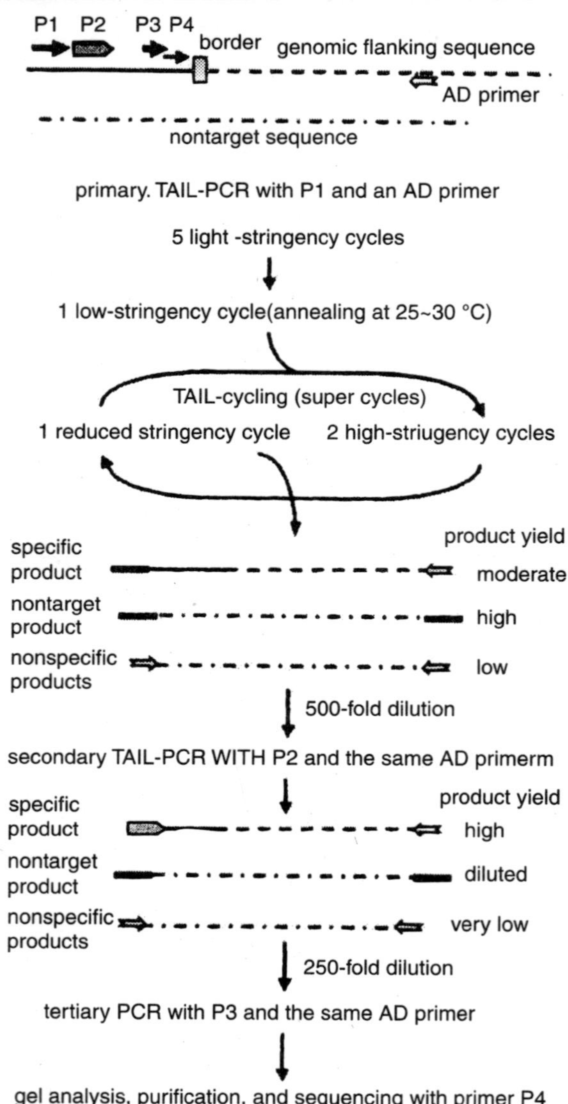

Figure 12.2: Schematic diagram of TAIL-PCR procedure. In high-stringency cycles, the high-temperature annealing favors the specific primer having higher T_m for priming, resulting in linear amplification for target molecules and no amplification for nonspecific ones. By interspersing reduced-stringency cycles to allow AD primer for priming, double-stranded molecules can be formed, and the preferential amplification of target molecules becomes logarithmic. In the secondary and tertiary PCR, the nontarget product primed by P1 at both ends fails to be reamplified owing to the lack of the P1 primer.

allow accumulation of metaphase chromo somes then there should be plenty of cells left on the slide with chromosomes at the correct stage of division so if a few are lost it should not be a problem.

10. Use sterile water in the preparation of buffers and solutions to ensure that they are free

from contamination.

11. Good laboratory practice is important, enzymes should be kept on ice and frozen buffers and sample stocks should be thawed at room temperature and thereafter be kept on ice.
12. It is very important that the temperature does not go above 16°C as this reduces the efficiency of nick translation. Therefore, it is better to incubate the mix at 15°C in a PCR machine rather than in a water bath to achieve better temperature control.
13. Prior to hybridization pretreatments, place slides in a humid chamber at 37°C. The humid chamber can be a metal tray lined with filter paper that is soaked in sterile distilled water.
14. Plastic cover slips can be cut from plastic autoclave bags (e.g., Guest Medical).
15. Be gentle during the washing steps, as violent washing will remove cells from the slide.
16. Take care to dispose of the waste paraformaldehyde solution correctly.
17. The denaturation step is the critical step and the other steps in the thermocycler just allow a slow cool down to 37°C. Therefore, the use of a thermocycler is not essential for denaturation. If a thermocycler is used add only sterile water to the thermocycler humid chamber. The presence of dextran sulphate in the hybridization solution makes the solution viscose and, therefore, difficult to spread over the squash area without disturbing the material on the slide. To overcome this, pipet the hybridization mixture onto the slide in 8 × 5 μL aliquots and when applying a plastic cover slip, ensure the aliquots of hybridization mixture join so that no air bubbles remain over the squash area.
18. Preincubate all solutions at 37°C before use. Note that waste formamide solution must be disposed of correctly.
19. Caution: Care is needed in handling DAPI, as it is carcinogenic.
20. These steps should be carried out as quickly as possible as the fluorescence of the slide will be lost by exposure to light.
21. FISH signals can fade quickly so that it is often only possible to take one or two pictures of any particular cell.

AMPLIFICATION OF GENOMIC SEQUENCES FLANKING T-DNA INSERTIONS BY THERMAL ASYMMETRIC INTERLACED POLYMERASE CHAIN REACTION

DNA tagging by T-DNA and transposon insertions has become an important approach for study of functional genomics in plants. With this approach large numbers of DNA-insertion lines and important mutations have been created in *Arabidopsis* and rice.

To identify the genes tagged by DNA insertions, it is necessary to recover genomic sequences flanking the insertion tags. However, the tagged gene sequences cannot be obtained simply by regular specific polymerase chain reaction (PCR) procedures because the genomic flanking sequences are unknown.

So far several PCR-based methods such as inverse PCR and thermal asymmetric interlaced

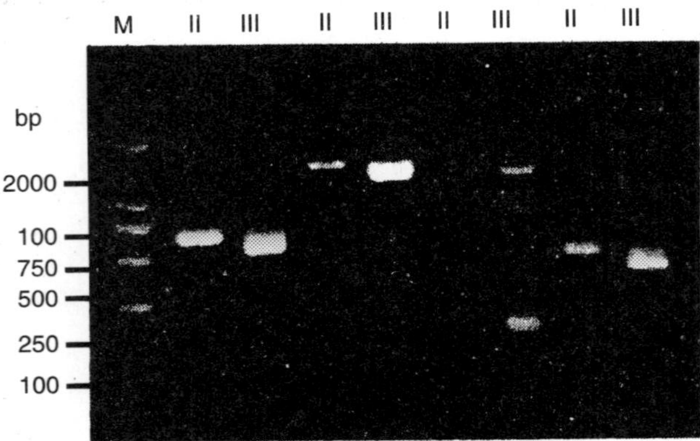

Figure 12.3: Agarose gel analysis of TAIL-PCR products from T-DNA tagging rice lines. Ex-Taq was used for the PCR reactions. Each set of two lanes contains products from secondary (II) and tertiary (III) reactions. The product specificity is confirmed by the size shift between lanes II and III. Multiple product bands observed in some samples may be nested fragments derived from single insertion by annealing of the AD primer at more than one site along the target sequence molecules, or different sequences if there are multiple insertions. Lane M, Molecular weight marker.

(TAIL) PCR have been developed for amplification of unknown DNA fragments fanked by known sequences. With the advantages of simplicity and high efficiency, TAIL-PCR has been widely used for molecular biology studies.

This chapter presents a detailed protocol for the TAIL-PCR with the examples of amplification of rice genomic sequences tagged by T-DNA using the pCAMBIA binary vectors. Specific products isolated by TAIL-PCR can be used as templates for direct sequencing or for cloning.

MATERIALS

1. Genomic DNAs prepared by conventional methods (working DNA samples are diluted with 5 mM Tris-HCl, pH 8.0, or 1/4X TE buffer in concentrations of 3050 ng/mL).
2. Specific primers TR1 and TL1 (1.5 μM); TR2 and TL2 (2.0 μM); TR3 and TL3 (2.5 μM), and AD primers (20 μM). The primers are diluted with 5 mM Tris-HCl, pH 85 0mM Tris-HCl, pH 8.0
3. *Ex-Taq* DNA polymerase (Takara, Japan).
4. 10X *Ex-Taq* buffer (MgCl2 free).
5. 25 mM MgCl2.
6. dNTPs mixture (2.5 mM each of dATP, dCTP, dGTP and dTTP).
7. Low melting agarose.
8. β-Agarase.
9. Gel purification ki9. Gel purification kit.
10. Sequencing kit.

METHODS

TAIL-PCR uses three nested specific primers in cons together with an arbitrary degenerate (AD) primer having a lower T,,, (meltingtemperature), so that the relative amplification efficiencies of specific products can be thermally controlled.

In the primary reaction, one lowstringency PCR cycle is conducted to create one or more annealing sites for the AD primer in the targeted sequence. Specific product is then preferentially amplified over nonspecific ones by swapping of two high-stringency PCR cycles with one reduced-stringency PCR cycle.

This is based on theprinciple that in the high-stringency PCR cycles with high annealing temperature only the specific primer having a higher T_m can efficient molecules, whereas the AD primer is much less efficient for annealing because of its lower T_m.

The seminested PCR amplifications help to achieve higher specificity. By two rounds of TAIL-PCR reactions, specific products that are primed at one end by specific primers and the other end by AD primer are amplified to levels visible on agarose gel.

Primer Design

1. Specific primers: Three specific primers are set on the region near the T-DNA right or left border. A fourth specific primer is used for the sequencing. The first and second specific primers are designed to have 22-26 nt with T,„ of 62–65°C as calculated by the formula: T_m = 69.3 + 0.41(G/C)% - 650/L (L = primer length). The third specific primer can be shorter (18-20 nt) because normal PCR cycling is used for the tertiary reaction. Ilsewhere in this Chapter shows the locations of the specific primers for T-DNA of the pCAMBIA binary vectors.

   A distance of 50–70 bp is set between the first and second specific primers to facilitate confirmation of the product specificity by the differential shift on agarose gel.

2. AD primers. AD primers are designed to have 15–16 nt with average T_m, of approx 45°C. Degenerate bases are introduced in the primers of degeneracy.

AD1:	5'-NTCGA(G/C)T(A/T)T(G/C)G(A/T)GTT-3'
AD2-1:	5'-NGACGA(G/C)(A/T)GANA(A/T)GAA-3'
AD2-2:	5'-NGACGA(G/C)(A/T)GANA(A/T)GTT-3'
AD2-3:	5'-NGACGA(G/C)(A/T)GANA(A/T)GAC-3'
AD2-4:	5'-NGACGA(G/C)(A/T)GANA(A/T)CAA-3'
AD2-5:	5'-NGACGA(G/C)(A/T)GANA(A/T)CTT-3'
AD3:	5'-NGTA(A/T)AA(G/C)GTNT(G/C)CAA-3'

PCR Reactions

Primary Reaction

1. Prepare primary reaction mixture, each reaction (25 μL) consisting of:

Table 12.1: Cycling Conditions Used for TAIL-PCR

Reaction	File no.	Cycle no.	Thermal condition
Primary	1	1	92°C (3 min), 95°C (1 min)
	2	5	94°C (30 s), 65°C (1 min), 72°C (2.5 min)
	3	1	94°C (30 s), 25–30°C (3 min), ramping to 72°C over 2 min, 72°C (2.5 min)
	4	14	94°C (15 s), 65°C (1 min), 72°C (2.5 min), 94°C (15 s), 65°C (1 min), 72°C (2.5 min), 94°C (15 s), 44°C (1 min), 72°C (2.5 min)
	5	1	72°C (5 min)
Secondary	6	11–12	94°C (15 s), 65°C (1 min), 72°C (2.5 min), 94°C (15 s), 65°C (1 min), 72°C (2.5min), 94°C (15 s), 45°C (1 min), 72°C (2.5 min)
	5	1	72°C (5 min)
Tertiary	7	12–14	94°C (40 s), 45°C (1 min), 72°C (2.5 min)
	5	1	72°C (5 min)

a. 2.5 µL of 10X PCR buffer.

b. 2.0 µL of 25 mM MgCl2.

c. 2.0 µL of dNTPs mixture.

d. 2.5 µL of specific primer TR1 or TL1.

e. 2.5 µL of any of the AD primers (50 pmol, final concentration 2 *µM*)

f. 0.75–0.8 U of *Ex-Taq* DNA polymerase.

g. dH_2O to 25 µL.

2. To each reaction add 1 µL (30–50 ng) of genomic DNA.
3. Perform primary amplification with thermal conditions as summarized in Table elsewhere in this chapter.
4. Run 10 mL of the product on 1.0% agarose gel.

Secondary Reaction

1. Prepare secondary reaction mixture, each reaction (25 µL) consisting of:

a. 2.5 µL of 10X PCR buffer.

b. 2.0 µL of 25 mM MgCl2.

   c. 2.0 μL of dNTPs mixture.
   d. 2.5 μL of specific primer TR2 or TL2 (5 pmol).
   e. 2 μL of the same AD primer (40 pmol).
   f. 0.7 U of *Ex-Taq* DNA polymerase.
   g. dH_2O to 25 mL.
2. Dilute 1 μL of the primary PCR product with 20 μL of H_2O and add 1 μL of the diluted DNA to each reaction.
3. Perform secondary amplification with thermal conditions as summarized elsewhere in this Chapter.

Tertiary Reaction

1. Prepare tertiary PCR mixtures, each reaction (25 μL) consisting of:
   a. 2.5 μL of 10X PCR buffer.
   b. 2.0 μL of 25 mM MgCl2.
   c. 2.0 μL of dNTPs mixture.
   d. 2.5 **μL** of specific primer TR3 or TL3 (6.25 pmol).
   e. 1.5 μL of the same AD primer (30 pmol).
   f. 0.6 U of *Taq* DNA polymerase.
   g. dH_2O to 25 mL.
2. Dilute 1 μL of the secondary PCR product with 10 μL of H_2O and add 1 μL of the diluted DNA to each reaction.
3. Perform tertiary amplification using thermal conditions as summarized elsewhere in this Chapter.

Agarose Gel Analysis

The secondary and tertiary products (10 gL) were run on 1.0 % agarose gel. The specificity of the products is confirmed by the expected size differences between the secondary and tertiary products.

Scaled-Up Tertiary Amplification

If specific products are detected, the tertiary amplification is repeated with larger scale of 50 gL per reaction.

Agarose Gel Purification

Purification by Low-Melting Agarose Gel

1. Run the tertiary PCR products on 0.8% low-melting agarose 9g/mL of ethidium bromide.
2. Recover the gel bands containing target DNA fragments and wash twice for 20 min with 5 vol of 1X (-agarase buffer (10 mM Tris-HCl, pH 6.5, 1 MM EDTA).

3. Melt the gel at 70°C for 5 min, and digest the gel at 40°C for [6]0.2 U of.
4. Extract the digested gel with an equal volume of equilibrat chloroform).
5. Extract the DNA with equal or 2/3 volumes of chloroform (optional).
6. Add 0.1 volume of 3 M sodium acetate and 2 vol of ethanol to precipitate the DNA.
7. Resuspend the DNA in 10–15 μL of 5 mM Tris-HCl, pH8.0 (without EDTA), and run 1–2 μL of the DNA on agarose gel with DNA of known amounts (e.g., 10, 20, 40 ng) to check the concentrations of the recovered DNA.

Purification Using a Gel Purification Kit

Run the tertiary PCR products on 0.8% agarose gel containing 0.3 gg/mL ethidium bromide, recover the gel bands containing target DNA fragments, and purify the DNA using a gel purification kit.

Purification by Direct Ethanol-Precipitation

If the tertiary PCR products show single bands, the products can be purifiedby phenol–chloroform extraction and ethanol precipitation.

Sequencing

The purified TAIL-PCR products can be used for direct sequencing using the fourth specific primer, or cloned into a TA-cloning vector.

Notes

1. *Ex-Taq* has higher performance and can amplify larger fragments. However, other *Taq* DNA polymerases such as Ampli-Taq also can be used for TAIL-PCR. The amount of the *Taq* enzyme for the primary reaction is about 10-20% more than normal PCR.
2. If the successful rate of TAIL-PCR is low in combination with any AD primers, try to design another specific primer for the primary reaction.
3. Because the AD2-1 primer usually gives higher successful rate, AD2-2, AD2-3, AD2-4, and AD2-5 are derived from AD2-1 by modifying the bases at the 3'-end.
4. When other *Taq* polymerases are used, the final concentration of $MgCl_2$ in the primary reaction is 2 mM, and those in the secondary and tertiary reactions are 1.5 mM.
5. The concentration of primer TR1 or TL1 should not be higher than 0.2 μM.
6. The visible products of the primary PCR are nontarget products primed at both ends by TR1 or TL1, which are created from nonspecific priming and amplified with highest efficiency. The target ones primed at one end at another end by AD are still at low levels of yield and nonvisible on agarose gel. Agarose gel analysis of primary products is usually unnecessary for routine amplification.
7. Control reactions with AD primer or specific primers only also can be set, but they are usually unnecessary.
8. If nonspecific products primed at both ends by AD primer are observed, which show the same fragment sizes between the secondary and tertiary reactions, try to decrease the concentration of the AD primer.

9. To reduce the cost, the amount of β-agarase for the gel digestion is decreased and the gel may not be digested completely. However, the partially digested gel can be completely removed by phenol extraction. Although low-melting agarose without digestion with β-agarase also can be removed by repeated phenol extraction, the recovery rate of DNA is relatively low.

10. To reduce the cost for sequencing, about one fourth of a standard reaction of the kit is recommended. For example, use 2 μL of ABI-Big-Dye Sequence Mix and about 30–50 ng of template DNA in 8-μL reaction volume for the sequencing.

Chapter 13

ANALYSIS OF GENE EXPRESSION

Advances in sequencing technologies resulted in extensive collections of gene sequences from a plethora of eukaryotic plant pathogens, including oomycete and fungal species. One major thrust in this post genomics era is to identify genes that are important for pathogenesis and virulence.

One class of such genes encodes so-called effectors that manipulate host cell structure and function either by facilitating infection (virulence factors) or by triggering defense responses (avirulence factors or elicitors).

Typically, ectopic expression of single effector genes in plant cells leads to phenotypic effects. For example, expression of avirulence (Avr) genes in plant cells that contain the matching resistance (R) gene usually results in the hypersensitive response (HR [I]).

Also, expression of effector genes in susceptible hosts can lead to phenotypic responses that may reflect virulence function. The use of plants for heterologous gene expression traditionally involved integration of a transgene into the plant genome.

The main setback of this approach is the considerable time required for generating stable transgenic plants. The time interval may vary from weeks to months depending on the plant species. Alternatively, ectopic gene expression also can be accomplished using such transient expression systems as *Agrobacterium tumefaciens-based* transient transformation (agroinfiltration), viral expression systems (agroinfection), and particle bombardment.

Transient expression systems have a number of advantages over stable transformation. These assays are rapid and simple to perform. They can be applied to fully differentiated plant tissues, thus allowing the analysis of cell death-inducing genes without inducible promoters. Additionally, these assays are not influenced by chromosomal positional effects (8). Therefore transient expression assays have become popular in the study of plant–microbe interactions and also have been applied to high-throughput analyses. This chapter describes two methods, agroinfiltration and binary *Potato virus* X (PVX) expression (PVX agroinfection), that have proved successful in our studies on effectors of *Phytophthora.*

Despite some limitations, these assays are crucial to modern molecular plant pathology research. In addition, they meet the demand for efficient and robust high throughput functional analysis in plants.

Agroinfiltration

Agrobacterium tumefaciens is the most commonly used agent in plant transformation experiments. This bacterium is a ubiquitous pathogen of plants. It enters through natural wounds and causes tumors (crown galls) at infection sites.

Translocation of transfer DNA (T-DNA) from a Ti plasmid (i.e., tumorinducing plasmid) occurs after the virulence machinery of the bacterium is activated by low-molecular-weight phenolic compounds and monosaccharides that are released from wounded plant cells, combined with a slightly acidic environment.

The agroinfiltration assay involves incubations of *A. tumefaciens* cell suspensions with 3'-5'-dimethoxy-4'-hydroxy acetophenone (*acetosyringone*). This phenolic compound mimics plant wounding, thereby inducing *vir* gene expression.

This treatment is followed by the infiltration of cell suspensions into leaf panels, allowing transformation of accessible plant cells and leading to expression of the transgene(s) contained in the T-DNA region. Although chromosomal integration of T-DNA elements takes place during transformation, it is not known whether this is required for expression to occur.

Nevertheless, the majority of plant cells in the infiltrated region express the transgene. Ectopic expression of single pathogen genes in plant cells often leads to phenotypic effects. For instance, expression of bacterial, fungal, or oomycete Avr genes in plant cells that contain the matching R gene results in the HR. In situations in which an expressed effector gene is not recognized, other phenotypic changes, such as chlorosis, cell enlargement, cell division, or necrosis, can be observed.

In either case, phenotypic assessments of infiltrated leaf areas can help identify effector genes and aid in subsequent functional characterizations.

PVX Agroinfection

A number of plant viruses can be used as vehicles for transient gene expression in plants. RNA viruses can multiply to very high levels in infected plants, which makes them ideal vectors for gene expression. To engineer viral vectors, viral RNA genomes are reverse transcribed in vitro and cloned as full-length complementary DNAs in transcription vectors.Insertion of foreign genes into plant viral genomes can be achieved using the following methods:

1. Gene replacement, in which nonessential viral genes such as the coat protein gene are replaced by the gene of interest.

2. Gene insertion, in which the gene of interest is placed under the control of an additional strong subgenomic promoter.
3. Gene fusion, in which the gene of interest is translationally fused with a viral gene.

Among plant RNA viruses, PVX is widely used for expressing virulence and avirulence genes from viruses, bacteria, fungi, and oomycetes. The PVX genome was modified by incorporating a duplicated coat protein promoter sequence followed by a multiple cloning site for insertion of the gene of interest.

Original constructs required in vitro transcription of PVX RNA followed by rubbing inoculation onto plant leaves. However, more recently, David Baulcombe and collaborators (Sainsbury Lab, Norwich, UK) developed binary PVX vectors in which the full-length PVX genome, flanked by the *Cauliflower mosaic virus (i.e.,* CaMV) 35S promoter and the nopaline synthase terminator, was cloned in the T-DNA of an *A. tumefaciens* binary vector.

Viral infection is initiated by wound inoculation of the recombinant *A. tumefaciens* strain onto leaves of host plants resulting in transfer of the T-DNA containing the PVX genome into plant cells. The PVX genome is then transcribed from the 35S promoter, resulting in virus particles that can move from one plant cell to another and spread systemically in the inoculated plants.

Expression of the inserted gene is achieved during viral replication. The PVX agroinfection assay has emerged as a robust and reliable system to identify virulence and *Avr* genes from microbial and viral pathogens.

Expression screens using the PVX vector facilitated the isolation and study of *Avr* and effector genes from fungal, oomycete, bacterial, and viral plant pathogens.

MATERIALS

Agroinfiltration

1. *Nicotiana benthamiana* seeds.
2. LB solid agar media plates supplemented with 50 μg of kanamycin and 25 mg of rifampicin/mL.
3. *A. tumefaciens* strain GV3 101.
4. *A. tumefaciens* strain GV3 101 containing binary vector constructs.
5. YEB medium: 5 g of beef extract, 1 g of yeast extract, 5 g of bacteriological peptone, 5 g of sucrose, and 2 mL of 1 M $MgSO_2$/L.
6. 3'-5' Dimethoxy-4'-hydroxy acetophenone (acetosyringone): 100 mM stock in dimethyl formamide or 70% ethanol.
7. 2-[N-Morpholino] ethane sulfonic acid (MES).
8. MMA infiltration medium: 5 g of MS salts, 1.95 g of MES, 20 g of sucrose, pH adjusted to 5.6 with 1 M NaOH, and 200 μM acetosyringone/L. 9. 1-mL Syringe.

PVX Agroinfection

1. *N. benthamiana* seeds.

2. LB solid agar media plates supplemented with 50 µg kanamycin/mL.
3. *A. tumefaciens* strain GV3 101.
4. GV3 101 harboring pGR106 or pGR106 carrying a reporter gene as a negative control.
5. GV3 101 harboring pGR106-INF1 as a positive control. 6. Sterile toothpicks.

METHODS

Agroinfiltration

Growing N. benthamiana Plants for Agroinfiltration

1. Germinate *N. benthamiana* seeds in soil in a pot at 22 to 25°C with high light intensity. Cover the pots with cheesecloth to prevent drying and to provide adequate moisture.
2. After germination, remove cheesecloth and allow plants to grow for approx 1 to 2 wk.
3. Transplant 2-wk-old seedlings individually into separate Styrofoam cups containing soil and allow them to grow until they reach eight-leaf stage.

Agroinfiltration Assay Procedure

1. Streak recombinant *A. tumefaciens* strains onto LB solid agar media plates supplemented with 50 µg kanamycin and 25 µg rifampicin/mL and incubate at 28°C for 2 to 3 d.
2. Inoculate 3 mL of YEB cultures containing 50 µg of kanamycin and 25 µg of rifampicin/mL, with the recombinant *A. tumefaciens* strains and grow overnight (28°C, approx 225 rmp).
3. Inoculate large YEB media suspensions, containing 50 µg of kanamycin, 25 µg rifampicin/mL, and 2 µM acetosyringone with the overnight culture. Grow cultures overnight at 28°C to an OD600 of approx 1.
4. Harvest the cells by centrifugation (4000g for 10 min), pour off the supernatant and resuspend the pellet in MMA medium to an OD of 2.
5. Incubate and shake cells at room temperature for 1 to 3 h.
6. Place *A. tumefaciens* suspensions into a syringe. Carefully invert the leaf and hold the lower side up. Support the infiltration site with your index finger and place the syringe against the leaf and index finger. While applying gentle pressure to the leaf, inject the suspension slowly from the syringe. Successful infiltration can be seen as the *Agrobacterium* suspension spreads from the infiltration site into the leaf. A movie on "how to agroinfiltrate" is availableat.
7. Incubate the plants in a growth chamber or confined space at 22°C.
8. Response should be visible in 2 to 3 d after infiltration.

PX Agroinfection

Groing N. benthamiana Plants forgroinfection

1. Germinate *N. benthamiana* seeds in soil in a pot at 22 to 25°C with high light intensity. Cover the pots with cheesecloth to prevent drying and to provide adequate moisture.

2. After germination, remove cheesecloth and allow plants to grow for approx 1 to 2 wk.
3. Transplant 2-wk-old seedlings individually into separate Styrofoam cups containing soil.
4. Allow plants to grow until they reach four-leaf stage.

Agroinfection Assay Procedure

1. Streak recombinant *A. tumefaciens* strains onto LB solid agar media plates supplemented with 50 mg of kanamycin/mL and incubate at 28°C for 2 to 3 d.
2. Toothpick-inoculate individual clones on the lower leaves of *N. benthamiana* plants by dipping a wooden sterile toothpick in a culture of the recombinant *A. tumefaciens* strain and piercing the leaves on both sides of the mid vein.
3. Incubate the plants in a growth chamber or confined space at 22°C.
4. Response should be visible starting from 7 d after inoculation.
5. Strains carrying the recombinant constructs should be examined for altered viral symptoms and compared with the control strains. Strains carrying the vector pGR106 induce systemic mosaic symptoms, and strains carrying pGR106-INF1 induce local HR lesions (5).

NOTES

Agroinfiltration

1. We prefer to use GV3 101 because it electroporates at high frequency (>10^8 cfu/µg of DNA). This streamlines the cloning procedure, as ligation mixtures can be directly electroporated into *Agrobacterium*.
2. Several binary vectors can be used. Vectors based on the pCB300 series (19) or the pAvr9 vector (1) allow high expression of the candidate gene and have worked well in our hands.
3. Agroinfiltration can be used with large (>2 kb) genes. PVX does not permit expression of genes exceeding 2 kb.
4. It is critical to include proper controls for each experiment. An *A. tumefaciens* strain containing a vector without gene insert is recommended as a negative control. Binary vectors containing genes expressing marker proteins, such as (3-glucuronidase (i.e., GUS) or green fluorescent protein, can be used to verify the level of transformation by agroinfiltration.
5. Several transgenes can be delivered into the same cell with agroinfiltration system facilitating simultaneous expression of interacting proteins (i.e., Avr and R proteins) or assembly of multimeric proteins.
6. It is recommended to make fresh MMA media by adding acetosyringone just before washing and incubation of the cell suspensions.
7. *N. benthamiana* plants that have healthy and fully developed leaves are desired in this assay. Infiltration of senescing leaves can lead to necrosis and reduced transformation rates.
8. The amount of infiltration media to be used depends on the size of the experiment and infiltration efficiency.

9. Infiltration with dense *A. tumefaciens* suspensions can lead to background necrosis. These problems can be avoided by using suspensions with lower OD600 values.
10. It is strongly recommended to practice and refine one's infiltration technique with water. Some users prefer to cause a slight wound on the leaf using a needle or a razor blade at the site of injection, which will facilitate infiltration of the bacterial solution.
11. Incubation temperatures of infiltrated plants should not exceed 28°C. *A. tumefaciens* transformation efficiency and transgene expression peaks at 22°C (20).
12. Detectable transgene expression should occur 2 to 3 d after infiltration. However, the timing of phenotypic changes varies depending on the effector tested.
13. The agroinfiltration system, unlike viral vectors, does not permit systemic expression of the foreign gene.

PVX Agroinfection

14. PVX agroinfection is limited to host plants, such as *N. benthamiana, Nicotiana tabacum, Lycopersicon esculentum,* and *Solanum tuberosum.*
15. We use GV3101 because it electroporates at high frequency (>10^8 cfu/µg of DNA). This streamlines the cloning procedure since ligation mixtures can be directly electroporated into *Agrobacterium.*
16. Younger plants at three to four leaf stages are preferable for inoculation if systemic symptoms are sought. For local responses, multiple clones can be inoculated on a single leaf. *See also* the method described by Takken et al. (9) for inoculation of 96 clones in tobacco leaves.
17. Always use pGR106 empty vector or a pGR106 carrying a reporter gene, such as gfp, as a negative control. The presence of an insert slows down virus infection so the use of a control vector carrying an insert of the same size as the candidate gene might be more appropriate than the empty vector.
18. It is advisable to use fresh cultures that are not older than 4 d.
19. Excess amount of *A. tumefaciens* can be used for toothpick inoculation. 20. Three leaves per plant can be used to serve as triplicates. 21. Inoculate a minimum of four plants for each construct.
22. Incubation temperatures of infected plants should not exceed 28°C. *A. tumefaciens* transformation efficiency peaks at 22°C. This is also the optimal temperature forvirus replication.
23. Symptoms should be scored every day from 7 d postinoculation and until 15 to 21 d after inoculation.

USE OF ROBUST-LONG SERIAL ANALYSIS OF GENE EXPRESSION TO IDENTIFY NOVEL FUNGAL AND PLANT GENES INVOLVED IN HOST-PATHOGEN INTERACTIONS

Many fungal genes play an important role in their pathogenic effects on host plants. Similarly, many host genes participate in the defense response to fungal infection. Identification and characterization of these fungal and host genes will lead to a better understanding of the molecular events during fungal-plant interactions.

In the recent years, several large-scale genomics approaches have been developed for transcriptome analysis of plant and fungal genomes.

These methods include expressed sequenced tag (EST) sequencing, micorarray, serial analysis of gene expression (SAGE), and massively parallel signature sequencing (MPSS). Compared with SAGE and MPSS methods, EST and microarray procedures are relatively simple techniques for expression analysis.

EST sequencing was the first method for gene discovery and expression analysis, and it has been recently used in studies on fungal– plant interactions. However, high levels of gene redundancy in the EST sequences and the inability to detect low-abundance transcripts has limited that method for deep transcriptome analysis.

Microarrays are an efficient tool for profiling and they have been used in many plant–microbe interaction studies. One of the disadvantages of microarrays is that a large set of EST or genomic sequences must be available for the microarray design.

In addition, RNA variants, such as alternative splicing transcripts, cannot be detected using microarrays (10). Recently, MPSS has been used to characterize the *Arabidopsis* genome, and many other genes uncharacterized by ESTs sequencing were identified.

However, MPSS libraries can only be constructed by Solexa, and the MPSS technique is exclusively licensed to a company for several important crop plants.

Compared with the aforementioned methods, SAGE is a unique method that is not only ideal for large-scale expression profiling but is also easy to use in any molecular laboratory.

SAGE is based on two basic principles: isolation of a short sequence tag (14 or 21 bp) from the 3' region of a transcript and the concatenation of multiple tags in a serial fashion for sequencing. It is a high-throughput method for evaluating the different levels of transcripts without previous sequence information.

Of importance, the transcript variations from alternative initiation and termination, alternative splicing, trans-splicing, and antisense transcription can be revealed by using SAGE. In the last decade, SAGE has been applied mainly in mammalian systems.

There have only been several plant SAGE libraries reported because of some difficulties in the library construction. Low cloning efficiency and the small inserts of SAGE clones are two major problems in SAGE library construction. Recently, we developed an efficient and rapid method called robustlong serial analysis of gene expression (RL-SAGE) that solved these two problems.

Using our RL-SAGE protocol, 10 libraries of rice blast fungus *Maganporthe grisea,* rice, and maize plants were made. Here, we report the major steps for RNA preparation from fungal and plant tissues, RL-SAGE library construction, di-tag and individual tag extraction, annotation and matching of RL-SAGE tags, and digital display of RL-SAGE tags in the *Magnaporthe grisea Oryza sativa* (MGOS) database.

The methods optimized for fungal and plant RL-SAGE library construction and analysis in our laboratory should be easily applied to other organisms.

MATERIALS

Fungus Infection Assay

M. grisea Infection

1. An avirulent isolate C9240 (from H. Leung, International Rice Research Institute, Philippines).
2. A virulent strain Che86061 (from G. Lu, Fujian Province, China). 3. Nipponbare seeds.
4. Conviron growth chamber.
5. Oatmeal agar (50 g of oatmeal +15 g of agar).
6. Conidiospores.
7. 0.01% Tween-20.
8. Plastic container.
9. Parafilm.
10. Distilled water.
11. Microscope.
12. Microscopic slides and cover slips.
13. Hemocytometer.

Rice Sheath Blight Infection Assay

1. Detached leaves.
2. Filter papers.
3. Petri plates.
4. *Rhizoctonia solani* strain RRO 102. 5. Jasmine 85 seeds.
6. Water agar.

M. grisea Mycelia Liquid Culture

1. Strain 70-25 (from Ralph Dean, North Carolina State University).
2. Liquid medium: 0.2% (w/v) yeast extract and 1% (w/v) sucrose for 3 d (28°C at 200 rpm.

Isolation of Total RNA and Messenger RNA From *M. grisea* Mycelia and Infected Rice Leaf Tissue

1. Liquid nitrogen.
2. Trizol solution (Invitrogen, Carlsbad, CA).
3. Chloroform.
4. Iso-propanol.

5. 95% and 75% ethanol.
6. Diethyl pyrocarbonate (DEPC)-treated H2⁰
7. mRNA isolation kit (Qiagen Inc., Valencia, CA).

RL-SGE Library Construction

1. I-SAGE/I-LongSAGE kit with magnetic stand (Invitrogen, Carlsbad, CA).
2. Siliconized (non-sticky) 1.5-mL tubes (Ambion, Inc, Austin, TX).
3. Streptavidin beads (Dynal Biotech Inc., Lake Success, NY).
4. *NlaIII* and *MmeI* (New England Biolabs, Inc., Beverly, MA).
5. PAGE-purifiedd oligos (Integrated DNA Technologies Inc, Coralville, IA).

   Linker 1A:

   5'-TTTGGATTTGCTGGTGCAGTACAACTAGGCTTAATATCCGACATG-3'

   Linker 1B:

   5'-TCGGATATTAAGCCTAGTTGTACTGCACCAGCAAATCC-C7 aminomodified-3'

   Linker 2A:

   5'-TTTCTGCTCGAATTCAAGCTTCTAACGATGTACGTCCGACATG-3'

   Linker 2B:

   5'-TCGGACGTACATCGTTAGAAGCTTGAATTCGAGCAG-C7 aminomodified-3'
6. PCR primers:

   Primer 1: 5'-biotin GTGCTCGTGGGATTTGCTGGTGCAGTACA-3'

   Primer 2: 5'-biotin GAGCTCGTGCTGCTCGAATTCAAGCTTCT-3'
7. Magnetic stand (Invitrogen, Carlsbad, CA).
8. Dynal oligo(dT) magnetic beads (5 mg/mL in phosphate-buffered saline containing 0.02% sodium azide).
9. Wash buffer A: 10 mM Tris-HCl, pH 7.5 0.15 M LiCl, 1 mM ethylene diamine tetraacetic acid (EDTA), 0.1% lithium dodecyl sulfate, 10 μg/mL glycogen.
10. Wash buffer B: 10 mM Tris-HCl, pH 7.5, 150 mM LiCl, 1 mM EDTA, 10 μg/mL glycogen.
11. Wash buffer C: 5 mM Tris-HCl, pH 7.5, 0.5 mM EDTA, 1 M NaCl, 1% sodium dodecyl sulfate, 10 μg/mL mussel glycogen.
12. Wash buffer D: 5 mM Tris-HCl, pH 7.5, 0.5 mM EDTA, 1 M NaCl, and 200 μg/mL bovine serum albumin.
13. Lysis/binding buffer: 100 mM Tris-HCl, pH 7.5, 500 mM LiCl, 10 mM EDTA, 1% lithium dodecyl sulfate, and 5 mM dithiothreitol.
14. SOC medium: 0.5% yeast extract, 10 mM NaCl, 2.5 mM KCl, 10 mM $MgCl_2$, 10 mM $MgSO_4$, 20 mM glucose.

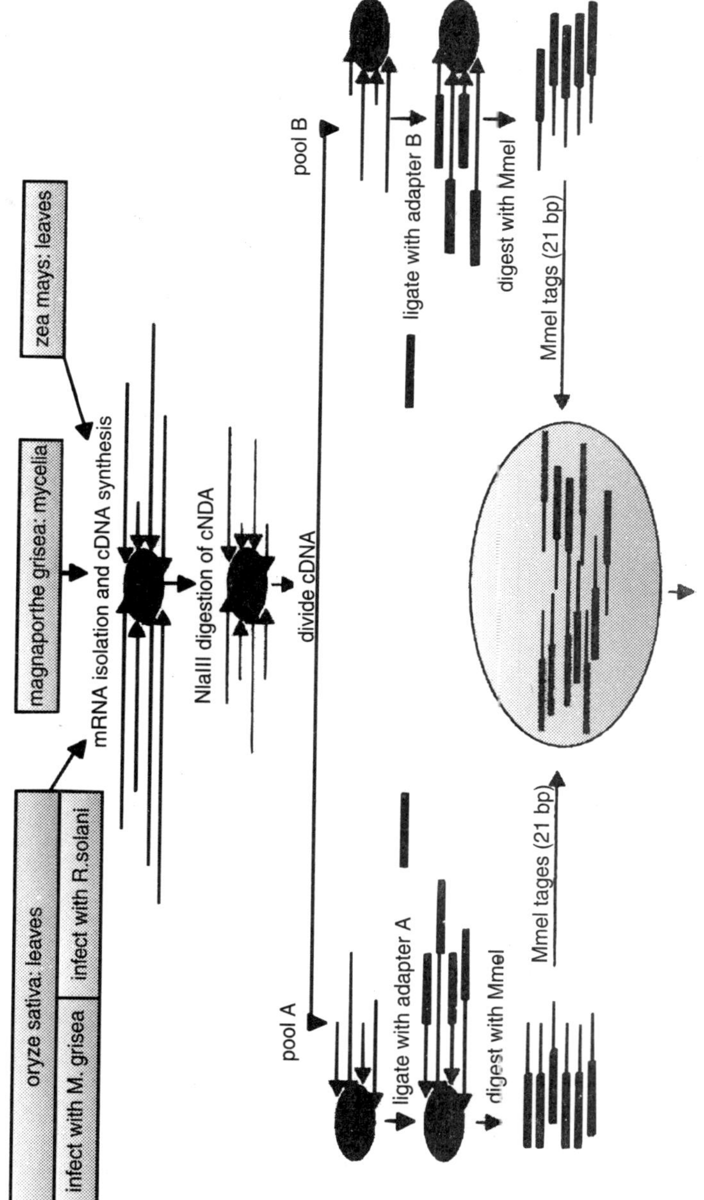

Figure 13.1: Diagrammatic representation of robust-long serial analysis of gene expression (RL-SAGE) methodology and tag annotation. Total RNA from rice leaves, mycelium tissue of M. grisea and maize leaves (see Sections 2 and 3) is isolated. Messenger RNA (mRNA) is captured on magnetic beads containing oligo(dT) 18 (shown as oval) Single- and double-stranded complementary DNA (cDNA) are synthesized according to the I-SAGE kit (Invitrogen) instructions. The cDNA with NlaIII (recognition site CATG) is digested and only 3' fragments on the magnetic beads are retained. Then the cDNA is divided into two parts, pool A and pool B, which are ligated with adapters A and B, respectively. The ligated mix is digested with MmeI (type IIS restriction enzyme), which recognizes TCCGAC sequences in the adapter and cleaves 20 bases away in the cDNA sequence. The purified tags are self-ligated to form di-tags. The linker sequences were removed from the polymerase chain reaction (PCR) amplified di-tags forgeneration of concatemers. The size-selected concatemers are cloned into the SphI site of the pZERO-1 vector (Invitrogen) RL-SAGE clones are sequenced using M13 reverse and forward primers. SAGESpy software, Fig. 2; Stahlbergh et al., unpublished) is then used to isolate di-tags and tags from the sequences. RL-SAGE tags are matched to genomic and EST sequences and the digital expression level of each tag is displayed on the Magnaporthe grisea Oryza sativa *database. (Figure contd.)*

15. Zeocin antibiotic (100 mg/mL in water).
16. TE buffer: 10 mM Tris-HCl, pH 7.5, 1 mM EDTA, pH 8.0.
17. pZero-1 (1 µg/µL; *see* I-SAGE kit instructions).
18. Platinum *Taq* DNA polymerase (Invitrogen).
19. Phenol:chloroform:isoamyl alcohol (25:24:1,v/v).

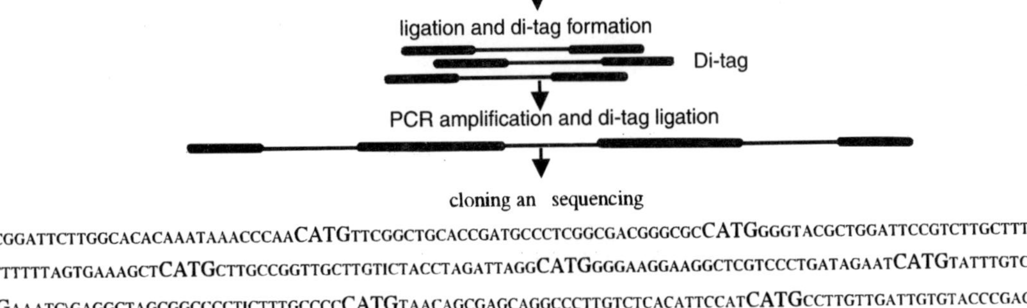

Figure 13.1: Contd.

20. Vertical gel electrophoresis apparatus for large format gels (15 × 17 cm).
21. 40% (w/v) acrylamide:bisacrylamide solution (29:1).
22 10% (w/v) ammonium persulfate.
23. TEMED.
24. 5X TBE running buffer: 89 mM Tris, 89 mM boric acid, 2 mM EDTA (15).
25. 5X TBE sample buffer: 18 mM Tris base, 18 mM boric acid, 0.4 mM EDTA, 3% Ficoll Type 400, 0.02% bromophenol blue, 0.02% xylene cyanol.
26. 0.5-mm Spacer and comb.
27. One-shot TOP 10 electrocompetent cells (Invitrogen).
28. Low-salt luria broth agar plates with Zeocin composition: 1% Tryptone, 0.5% yeast extract, 0.5% NaCl, pH 7.5.
29. 7.5 M Ammonium acetate.
30. Freeze medium: 2.5% w/v luria broth, 13 mM KH_2PO_4, 36 mM K_2HPO_4, 1.7 mM sodium citrate, 6.8 mM $(NH_4)_2SO_4$, and 4.4% v/v glycerol.

METHODS

Leaf Tissue From M. grisea-Infected Rice Plants (Nipponbare)

1. Grow Nipponbare seedling for 21 d in a Conviron growth chamber at 26°C durng the day, 80% relative humidity, 12 h light at 200 gmol photons $m^{-2} s^{-1}$ and 20°C at night, 60% relative humidity.
2. Collect ascospores from *M. grisea* isolate Che86061 (virulent) or C9240-2 (avirulent) mycelium, which has grown on oatmeal agar plates for 2 wk.
3. Inoculate the experimental plants with 2×10^5 spores/mL in 0.01% Tween-20 solution.
4. Inoculate the control plants with only the 0.01% Tween-20 solution.
5. Keep the inoculated plants in a sealed plastic container in the dark for 24 h with 100% humidity and then move the plants to normal growth chamber conditions for 5 to 6 d for disease development.
6. Harvest-infected leaves at 24 and 96 h after inoculation for RNA isolation.

Leaf Tissue for R. solani infection (Jasmine 85)

1. Grow Jasmine 85 plants to V 11 stage in a greenhouse.
2. Detach approx 14 cm of the second youngest leaves of the rice plants and immediately place in a plastic container (24 × 24 × 1.8 cm) with wet filter papers (23 × 23 cm).
3. Use a 1-mL Eppendorf tip to excise a 0.8-cm-diameter potato dextrose agar plug containing mycelium of *R. solani* isolate PRO102 and remove the plug with a sterile toothpick.
4. Place the 0.8-cm-diameter agar plug on the abaxial leaf surface of the collected leaves

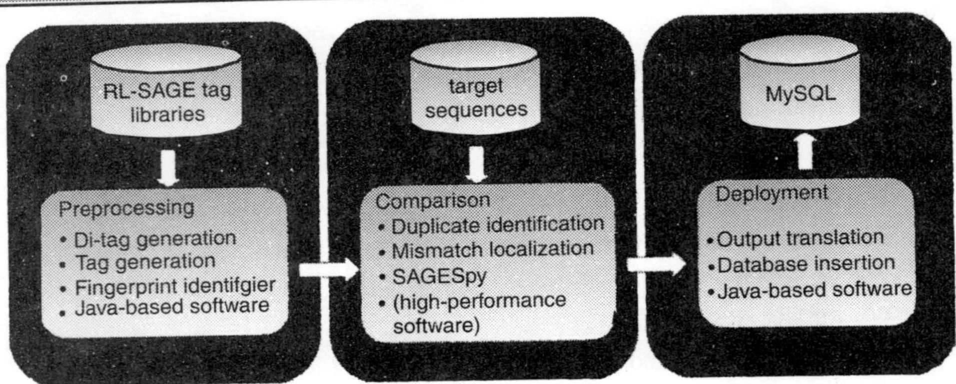

Figure 13.2: Robust–long serial analysis of gene expression (RL-SAGE) tag libraries are preprocessed using Java-based programs for generating di-tags, and tags, and for assigning sequence fingerprint identifiers. The output is compared with target sequences (e.g., TIGR expressed sequence tags, KOME FL-cDNA, genomic DNA) using high-performance SAGESpy software, and an XML-based output file. The XML output file is translated and imported into the MySQL database for subsequent use for search and query support.

and seal the container with parafilm to maintain high humidity. For control leaves, use a 0.8-cm-diameter potato dextrose agar plug without any pathogens for inoculation.

5. Keep the container under cool white fluorescence light at 21 to 24°C. Harvest leaf tissue near the infection area for RNA isolation after 16 h of inoculation.

M. grisea Mycelia Tissue From In Vitro Culture

1. Inoculate fungus (70–25 mycelium) in a liquid medium (0.2% [w/v] yeast extract and 1% [w/v] sucrose).
2. Shake for 3 d at 28°C at 200 rpm.
3. Centrifuge the medium to collect the mycelium tissue, and freeze it in liquid nitrogen for RNA isolation.

Isolation of Total RNA and Purification of mRNA

1. Grind approx 2 g of leaf or mycelium tissue into a fine powder using liquid nitrogen and immediately transfer into 15 mL of Trizol solution.
2. Mix well and incubate at room temperature for 10 min.
3. Add 4 mL of chloroform, incubate at room temperature for 5 min, and then centrifuge for 20 min (9000g) at 4°C.
4. Transfer supernatant into another 25-mL tube containing 10 mL of ice-cold isopropanol, mix well, and then incubate on ice for 10 min.
5. Centrifuge for 15 min (9000g) at 4°C.
6. Wash the RNA pellet with 15 mL of 75% ethanol, centrifuge for 10 min (9000g), and then discard the alcohol.
7. Dry the RNA pellet at room temperature for 10 to 15 min.

8. Dissolve the RNA pellet in 700 µL of DEPC-treated H_2O at 65°C for 10 min.
9. Quantify the amount of total RNa using a spectrophotometer by taking OD at 260/280 nm, or estimate the amount on agarose gels.
10. Isolate the polyA+ mRNA using Qiagen mRNA purification kit according to manufacturer's instructions.

RL-SGE Library Construction

The detailed RL-SAGE protocol is available at the Plant-Microbe Interactome database. The major modified steps are described in the RL-SAGE method. The diagrammatic representation of the RL-SAGE procedure is shown in Figure elsewhere in this chapter The major steps in RL-SAGE library construction are described in this section.

1. Use approx 50 ng of mRNA for each RL-SAGE library procedure.
2. Equilibrate the oligo(dT) beads in lysis/binding buffer.
3. Synthesize complementary DNA (cDNA) using the reagents provided in the I-longSAGE kit.
4. Digest the cDNA with NlaIII and divide into two parts. Ligate pool A with adapter A and pool B with adapter B.
5. Release 21-bp tags from cDNA by digestion with *MmeI* (Type IIS restriction enzyme), and purify the tags from a 16% polycrylamide gel electrophoresis (PAGE; w/v).
6. Generate di-tags by overnight ligation of the tags from pool A and B.
7. Perform 20 to 50 di-tag PCRs (50-µL reactions).
8. Isolate the di-tag band (138 bp) on a 12% PAGE gel.
9. Remove linker sequences from di-tags by digesting with *NlaIII.*
10. Purify the di-tag band (38 to 40 bp] on a 16% PAGE gel.
11. Purify di-tags again using Streptavidin magnetic beads.
12. Self-ligate di-tags to generate concatemers.
13. Perform a partial digestion of concatemers with *NlaIII.*
14. Purify more than 500 by concatemer bands on a 6% PAGE gel.
15. Prepare pZero-1 vector by *SphI* digestion, and ligate concatemers.
16. Transform the ligation mix using TOP 10 electrocompetent cells (Invitrogen).
17. Randomly pick 20 clones to check the size of inserts.
18. Pick 3000 to 7000 clones for sequencing.

SAGESpy Software for RL-SAGE Tag Annotation

SAGEspy is a set of high-performance applications employing the Cray Bioinformatics Library (CBL) and their Portable Cray Bioinformatics Library (PCBL) counterparts for conducting large-volume serial analysis of gene expression comparisons. The program was optimized for use on the Cray

SV 1 and Cray X1 systems employing OpenMP parallelism to speed the solution. Figure elsewhere in this chapter shows the major steps in RL-SAGE data processing using SAGESpy. The specific steps in the SAGE data analysis are described:

1. Get the sequences of RL-SAGE clones.
2. Isolate di-tags that are 40- to 42-bp long from the sequences of the RL-SAGE clones.
3. Isolate individual RL-SAGE tags from the di-tags.
4. Convert the RL-SAGE tags into FASTA format.
5. Isolate virtual sense (from sense strand or Watson stand of the DNA) and antisense (or Crick strand of DNA) tags from a known nucleotide database (EST or genome sequence).
6. Convert all the virtual SAGE tags into FASTA format.
7. Use the SAGESpy software to match the experimental RL-SAGE tags (derived from the libraries of rice, *M. grisea,* or maize) with the virtual SAGE tags to calculate the matching rate of the experimental tags.
8. Extract output files in FASTA format containing the list of tags, which hit target sequences and also list of tags with no hit in the database.
9. Identify novel transcripts, antisense transcripts, and alternatively spliced transcripts.
10. Repeat the matching analysis allowing one or two mismatches are allowed.

Digital Northern Analsis of RL-SGE Tags at MGOS Database

The MGOS database is a unique database that hosts EST and SAGE data of both rice and *M. grisea.* MGOS also contains the genome sequences of both rice and *M. grisea* (Soderlund et al., unpublished), which are displayed in the GMOD genome browser.

The RL-SAGE tags from the four rice libraries (OSJNGb, d, f, and g) and one *M. grisea* library (MG_SGa) are shown in figure elsewhere in this chapter. Tags from each library are processed and displayed on the SAGE page of MGOS database. The browser contains tracks for the annotated genes and the RL-SAGE tags, along with other evidence, such as EST alignments. Annotation of RL-SAGE tags is as follows:

1. The tag hits an annotated gene, and hence inherits its UniProt annotation.
2. It does not hit a gene, but it does hit the genome.

The flanking region is searched against UniProt, and if it hits a UniProt gene it uses that annotation. All unique tags are grouped together from all five libraries based on exact hits. These tags can be queried from the query page, which allows a versatile set of queries, such as the following:

1. Show all tags that have at least one or more occurrences from a selected subset of the libraries. This allows the user to view the shared tags across a set of libraries.
2. Show all tags that have at least N or more occurrences in a selected subset of the libraries and do not have any tags from a different subset of the libraries. For example, the user may want to see the tags that are in both of the susceptible libraries but are not in the normal library.

3. Show the tags that have p-values less than N, where the p value is calculated based on one of the methods provided by IDEG6.
4. Show the tags that have one or more hits to the genome (or ESTs).
5. Show all tags that have a specific annotation. Any combination of the above.

A table of the tags that satisfy the query is displayed, where the columns of information shown are based on those selected by the user. For example, they may select the percentage of tags per library, UniProt matching ID, and overall quality of the tag. From the table, the user may select a specific tag, which then takes them to the tag's page, containing all of the information about it.

Notes

1. The RL-SAGE protocol was successfully used for library construction in rice, maize, and *M. grisea* in our laboratory. Two to three libraries could easily be constructed per month.
2. This method was optimized for a small amount of mRNA sample (approx 50 ng) for when only a limited amount of mRNA available for library construction is available.
3. Like in any other SAGE protocol, RL-SAGE also uses a single anchoring enzyme, *NlaIII* (CATG), which may occur in every 256 bp. Therefore, a small portion (approx 3%) of transcripts in the genome might be uncovered during transcriptome analysis. To overcome this limitation, one could make another library using a different anchoring enzyme such as *DpnII.*
4. Using this procedure, just 20 di-tag PCRs are sufficient to generate enough di-tags for concatenation
5. Partial digestion of concatemers with *NlaIII* greatly improved the ligation efficiency of the concatemers (>150,000 clones from 10 μL of ligation mix) and increased the insert size (average of 1.0 kb). If all of the 150,000 clones are sequenced, 4.5 million tags can be recovered. Because of the high cost of *sequencing,* sequencing of 3000 to 5000 clones per library should be sufficient (100,000 to 200,000 individual tags).
6. Because of increased ligation efficiency, very few empty clones are found in the libraries. Therefore, clones can be picked randomly without using a PCR-screening strategy to remove empty *clones.* This improvement saves a lot of time when generating a RL-SAGE library. Following our method, two to three RL-SAGE libraries can be easily generated within a month.

Chapter 14

RESISTANCE TO PLANT VIRUSES

Plant viruses are intracellular, molecular obligate parasites and cause significant economic losses worldwide. Traditional approaches for managing plant virus diseases include avoiding virus-infected material, chemical control of arthropod vectors and, when available, use of virus-resistance in cultivated crops.

However, all of these are labor intensive, and chemical control of insect vectors is becoming more expensive with potential undesirable side effects, including environmental hazards and the generation of insecticide resistance in vector populations and those of other insect pests.

The observation of crossprotection (1), wherein the inoculation of mild virus strains on plants provided protection from more severe strains, suggested that alternative approaches were possible. Several hypotheses were proposed to explain the resistance obtained by crossprotection, but methods to test them were not possible for many years.

However, the development of techniques for genetic manipulation of higher plants by using binary vectors derived from *Agrobacterium tumefaciens* allowed the examination of several hypotheses associated with crossprotection.

In one such study, expressing the *Tobacco mosaic virus* (TMV) coat protein in *Nicotiana tabacum* plants delayed the onset of symptoms on the transgenic plants when subsequently challenged by TMV and produced many studies designed to understand crossprotection as well as to develop virusresistant crop plants of value for agriculture.

Subsequently, several different strategies have been attempted for developing virus-resistant plants and this chapter attempts to describe these strategies and their molecular mechanisms.

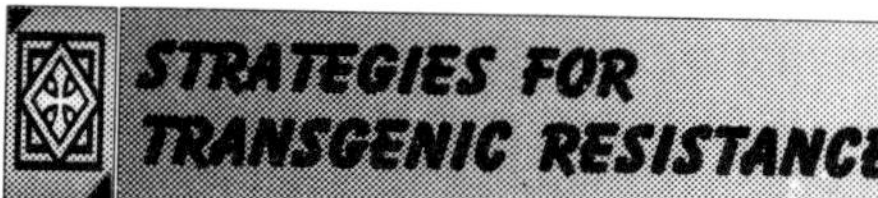

STRATEGIES FOR TRANSGENIC RESISTANCE

Coat Protein-Mediated Resistance

It has been two decades since the first report of coat protein (CP)-mediated resistance to TMV in N. *tabacum* was described, which was immediately followed by reports of CP-mediated resistance for many other plant viruses, most of which have been reviewed by Beachy et al. and Wilson. In most instances, obtaining resistance seemed to be the primary objective, and studies aimed at understanding the mechanism(s) of resistance are rather scant.

At least for TMV, a model virus for understanding plant-virus interactions for the past century, the mechanisms by which transgene expressed CP interferes have been described. The inability of antisense RNA or nontranslatable + sense RNA sequence of CP made a strong case for the expression of protein as the factor in imparting resistance.

It is widely believed that polysome-mediated virion disassembly is the first step in the establishment of the TMV infection. Studies using TMV RNA and virus particles on protoplasts derived from TMV CP-expressing plants indicated that resistance was primarily to virus particles, because of a likely interference in the initial events during the infection process.

Subsequent studies using in vitro encapsidated TMV genomic RNA containing reporter genes also suggested that virion disassembly was primarily reduced in transgenic plants. However, resistance was limited to the inoculated cell, with no apparent effects on cell-to-cell spread. Expression of CP appears to be a necessity for successful transgenic resistance to TMV, but reports with other viruses in which nontranslatable CP cistron RNA conferred resistance in transgenic plants suggested that protein expression was not always an absolute requirement for effective resistance.

Still, the initial successes with CP-mediated resistance led to the creation of plants expressing different virus CP sequences and resistance to a wide array of RNA viruses. However, CP-mediated resistance does not appear to work for plant-infecting DNA viruses.

Still, the success of CP-mediated protection has resulted in the commercialization and squash lines resistant to *Cucumber mosaic virus* (CMV), *Zucchini yellow mosaic virus* (ZYMV), and *Watermelon mosaic virus* (WMV), and transgenic papaya resistant to *Papaya ringspot virus.*

Antisense and Sense RNA

Antisense RNA constructs essentially are complementary (c)DNA sequences fused to a promoter so as to be expressed as RNAs complementary to the virus genomic RNA. In addition to various genes, plant virus genomes contain nontranslated region (NTR) sequences required for initiating translation and for genome amplification.

Although antisense approaches were not very effective when directed to the TMV CP region, inclusion of the TMV genomic RNA 3'-end NTR proved to be effective (6). It has to be noted that viral RNA-dependent RNA polymerase binds to the 3'-end NTR of the genomic RNA to initiate the minus-sense RNA synthesis, which is subsequently copied into the genomic RNA.

Any delay in the synthesis of RNA would eventually lead to lower virus titer in a given time and hinder the systemic spread of the virus. In addition to CP genes, several virus nonstructural genes, expressed either as translatable or nontranslatable transcripts, often conferred virus resistance to transgenic plants. One of the best examples is the expression of the TMV 54-kDa gene, corresponding to the carboxyl end of the 183-kDa replicase.

Transgenic plants expressing this gene, irrespective of the copy number, showed high levels of resistance to TMV. However, no protein product corresponding to this gene was detectable and the plants were not resistant to a distantly related TMV strain. It was found that in these plants TMV RNA was restricted to the inoculated cells.

However, these plants also were found to be exhibiting posttranscriptional gene silencing (PTGS) and the silencing was specific to the antisense strand but not to the + sense strand. Examples of the sense and antisense RNA-mediated resistance have been reviewed by Wilson and Baulcombe.

Interference From Heterologous Viral Genes

Several plant viruses encode specialized proteins known as cell-to-cell movement proteins (MPs). These proteins either interact with secondary plasmodesmata, the intercellular connections between adjacent plant cells, or form tubules to allow intercellular trafficking of virions and/or ribonucleoprotein complexes comprising viral RNA and one or more of virus-encoded proteins.

In addition, MPs also bind to RNA and/or DNA. The RNA/DNA binding domain and the plasmodesmata modifying domain being different, the mutants of MPs were thought to bind to plasmodesmata and interfere with wild-type MP-mediated plasmodesmatal trafficking of virus RNA/DNA.

A perceivable advantage of this strategy is a broad-spectrum resistance to diverse plant viruses that are dependent on the same type of plasmodesmata for the establishment of infection. Expression of defective TMV 30-kDa MPs, in addition to conferring resistance to TMV, also was able to confer resistance to *Tobacco rattle virus, Tobacco ringspot virus* (Family Comoviridae), *Alfalfa mosaic virus* (Family Bromoviridae), *Peanut chlorotic streak virus* (Family Caulimoviridae), and CMV.

Among the viruses tested, the members of Comoviridae and Caulimoviridae, instead of using a plasmodesmata-modifying protein, encode a specialized protein that forms tubules allowing trafficking of virions from the infected to adjacent cells.

Truncated MPs of plant-infecting DNA viruses of the genus *Begomovirus* (Family Geminiviridae) also have resulted in imparting resistance to homologous as well as heterologous viruses. However, it is not known how these plants perform under field conditions.

Also, because plants have evolved plasmodesmata as the intercellular communication conduit, interference by MPs may affect plant communication leading to undesirable transgene effects.

Posttranscriptional Gene Silencing

PTGS was first observed in transgenic *Petunia* plants as a coordinated and reciprocal inactivation of host genes and transgenes encoding homologous RNA. However, PTGS or RNA silencing is a recently recognized strategy for developing virus-resistant plants.

Double-stranded RNA generated from a replicating virus, a transgene, or an aberrant RNA can act as a key initiator molecule that is subsequently processed by an RNaseIII-like enzyme to produce 25 nt RNAs known as small antisense RNAs, which were subsequently recognized as small interfering RNAs (siRNAs).

The RNA-induced silencing complex, a key component of which is an endonuclease, is then guided by the siRNAs to specifically cleave homologous RNAs. Involvement of PTGS in virus protection was first evident in transgenic plants using potyviral CP cDNA sequence. Lindbo et al. first proposed PTGS as an antiviral state in plants.

This is best achieved when plants are transformed with constructs that express a self-complementary RNA, containing sequences homologous to the target plant virus. Transgene constructs encoding intron-spliced RNA with hairpin structure provided stable silencing to nearly 100% efficiency against homologous plant viruses.

Hairpin constructs can be made using generic vectors such as pHANNIBAL and pHELLSGATE, and 98 to 853 nt sense/antisense arms in hairpin constructs were efficient in silencing 90 to 100% of independent transgenic plants.

In addition to transgene expression, transient expression of doublestranded RNA corresponding to viral sequences, either by mechanical inoculation or by *Agrobacterium-mediated* leaf infiltration, can also impart resistance to plant viruses and has been reviewed recently.

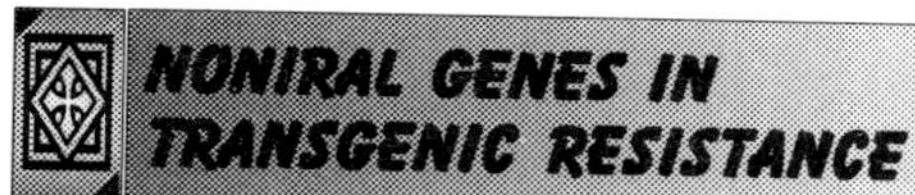

NONIRAL GENES IN TRANSGENIC RESISTANCE

Genes of Plant Origin That Confer Virus Resistance

The *R* (resistance) genes are dominant plant genes that encode proteins that participate in a general surveillance to identify a specific pathogen. The presence of an incompatible reaction in a specific plant *R* gene–virus combination results in a hypersensitive reaction (HR) where the cells surrounding the initially infected cell are programmed to die, resulting in the isolation of viral infections.

Often such responses in resistant plants trigger a systemic acquired resistance that signals a highly resistant state for further virus challenge. The discovery of the resistance gene *N* in *Nicotiana glutinosa* and its subsequent use in the breeding program of cultivated tobacco (*N. tabacum)* virtually eliminated the incidence of TMV.

The *N* gene when expressed as a transgene conferred resistance in susceptible genotypes of *N. tabacum,* as well as in tomato *(Lycopersicon esculentum)*. Similar to the *N* gene, the *Rx* gene encoded product confers resistance to *Potato virus X* (PVX) in potato and in transgenic *Nicotiana benthamiana* and *N. tabacum*. The *Rx* gene product, structurally similar to other *R* gene products, restricts virus in an HR independent fashion.

However the gene product has the ability to cause cell death. Despite the association of several *R* genes conferring virus resistance, apart from *N* and *Rx* genes, *Tm-2*^{2} is the only other plant *R* gene that has so far been cloned and shown to confer virus resistance. In addition to *R* genes, transgenic resistance also has been demonstrated by way of expressing genes that confer resistance to systemic movement of plant viruses, but not involving HR.

The gene products of *RTM1* and *RTM2* of *Arabidopsis thaliana,* a model plant for studies on plant development and genetics, resemble the a chain of jacalin, a plant lectin, and a small plant heat shock protein, respectively. These two genes together restrict the long distance movement of *Tobacco etch virus* in A. *thaliana.*

However, utility of these genes in transgenic resistance is yet to be shown. The requirement for *cis* and trans-acting virus-encoded proteases in processing viral proteins suggests that protease inhibitors could be used to control virus interactions. Polyprotein processing is a common feature among many viruses but especially for those of families Potyviridae and Comoviridae.

A rice cystatin, which inhibits cysteine proteinases, when expressed in tobacco was able to impart resistance to *Potato virus Y* (PVY) and *Tobacco etch virus.* Transgenic plants were not resistant to TMV indicating that the resistance was specific to the two tested viruses of the family Potyviridae.

Ribosome-Inactivating Proteins

Ribosome-inactivating proteins (RIPs) are RNA N-glycosidases that specifically cause removal of the purine base at $_{A4324}$ of 28S rRNA, which results in the separation of the 3'-end of the substrate RNA, thus rendering the rRNA incapable of participating in the translation of mRNA in eukaryotes. Many cultivated crop plants accumulate RIPs in their seeds and various other parts.

Pokeweed *(Phytolacca americana) leaf* extract described as antiviral in nature when co-inoculated with TMV, contained a RIP, pokeweed antiviral protein (PAP). The cDNA sequence for PAP was subsequently isolated and cloned and transgenic tobacco and potato plants expressing PAP under the influence of a constitutive promoter showed remarkable resistance to infection by PVX and PVY upon mechanical inoculation.

Although these plants also showed resistance to mechanically inoculated CMV, resistance also was noticeable to a varying degree for infection by aphid-transmitted *Potato leaf roll virus* and PVY. In pokeweed, PAP is exported to the cellular matrix and, thus, the pokeweed ribosomes are protected.

In transgenic tobacco, upon mechanical injury, transgene-expressed PAP can gain entry into cytoplasm and neutralize the ribosomes before viral RNA can be translated and, thus, resistance is achieved. It is interesting to note that several RIPs of plant origin actually are polynucleotide N-glycosylases that can remove adenine bases from DNA and RNA substrates, including those of genomic RNAs of *Artichoke mottled crinkle virus* and TMV and, thus, might affect viral RNAs in the cytoplasm of infected plant cells.

A C-terminal deletion mutant of PAP was found to be unable to depurinate host rRNA; however, it was able to inhibit viral infection. It was recently found that PAP also can depurinate capped mRNA in plants and thus possess distinct antiviral properties. It is also likely that all RIPs are not targeted to the intercellular matrix, or the signal sequence for required for such targeting might be removed to render them cytotoxic.

Then virus-regulated expression of a transgenic RIP could provide virus protection by targeting the infected cells for ribosome inactivation. The AC2 gene product of bipartite geminiviruses (Genus Begomovirus, Family Geminiviridae) is a transactivator of rightward transcription and is required for the efficient expression of the CP, from DNA-A, and BV1, from DNA-B, respectively.

The cDNA for a RIP, dianthin, cloned from *Dianthus caryophyllus,* when fused to the promoter of AV1, conferred resistance to *African cassava mosaic virus* in N. *benthamiana.* It is not known whether transactivated RIP-mediated resistance is effective against insect-transmitted virus, the predominant way of geminivirus transmission in nature, nor is it known as to effects on vectors feeding on such plants.

Even if some plants were affected, because of the damage caused by the cytotoxicity, such damage could likely be tolerated because of the plasticity of growing plants, which might be a good way to control secondary virus spread in the field.

Apart from dianthin and PAP, trichosanthin is another RIP that has been shown to offer protection against CMV and TMV in transgenic tobacco. Although consumer concerns about the presence of RIPs in food and feed is a strong factor in the development of transgenic crops, it should be noted that RIPs are well distributed among plant kingdom and most grain crops including wheat and barley accumulate RIPs in their seeds.

Miscellaneous Genes

Apart from R genes and RIPs, several other ways have been shown to have potential for transgenic resistance. This list includes satellites, defective interfering RNAs, ScFv antibodies, and mouse 2–5A synthetase.

POINTERS FOR STRATEGIES FOR ENGINEERING VIRUS RESISTANCE

General Methods for Construction of Plasmid Vectors

Initial Cloning of Genes

It is now certain that it is very much possible to obtain transgene-derived virus resistance using genes and nucleic acid sequences derived from plant viruses, their resistant hosts, and even unrelated organisms.

The choice of transgene is dependent on several factors, including the plant host variety or inbred line to be transformed, the type of resistance required, production site of the crop (i.e., greenhouse or open field), the type of commercial commodity obtained from the transgenic crop, and consumer acceptance of the transgenic crop.

Resistance obtained by expression of the CP, RNA and self-complementary RNA, is largely dependent on homology of the virus isolates/strains, prevailing in the target area, with respect to the transgene. It is relatively straightforward to construct desired virus resistance genes if the sequence of the virus is known, and this has been described before in detail.

However, in brief, one can first design two primers for cloning viral genes/sequences of choice using any of the primer design programs. Incorporating restriction sites into the primers to facilitate subsequent cloning is important to consider at this point.

Some of the commercially available software that can be used are GCG (Accelrys Inc., San Diego, CA), OLIGO (Molecular Biology Insights Inc. Cascade, CO), Vector NT (Informax Inc., Frederick, MD), and Visual Cloning (Redasoft Corporation, Bradford, Ontario, Canada).

If the target virus has an RNA genome, the reverse primer has to be used for reverse

Table 14.1: Examples of Transgenic Resistance Using Gene Sequences Derived From Sources Other Than Plant Viruses

Gene	Source	Virus
A. R genes of plants		
N gene	*Nicotiana glutinosa*	TMV
TM-2^2 gene	*Lycopersicon esculentum* ToMV	
Rx gene	*Solanum tuberosum*	PVX
B. Enzymes and inhibitors of plant origin		
Pokeweed antiviral protein	*Phytolacca americana*	PVX, CMV, PVY
Dianthin	*Dianthus caryophyllus*	ACMV
Trichosanthin	*Trichosanthes kirilowii*	CMV, TMV
Cysteine proteinase inhibitor	*Oryza sativa*	TEV, PVY
C. Mammalian proteins		
2-5A synthetase	*Rattus rattus*	PVX, PVS, PVY
scFv	*Mus musculus*	ACMV TSWV

transcription followed by polymerase chain reaction (PCR) using a proof reading thermostable DNA polymerase, for e.g., *Pfu* (Stratagene Inc., La Jolla, CA), *Vent* (New England Biolabs Inc., Beverly, MA). Amplification of longer genes (>3 kb) would need a mixture of *Taq* and proofreading thermostable DNA polymerase, for e.g., ElonGase (Invitrogen Inc., Carlsbad, CA), TripleMaster PCR system (Brinkmann Inc., Westbury, NY).

The PCR product can be cloned into an intermediate vector for the verification of the sequence and restriction sites. There are many options for the selection of the binary vector that is going to be used for the transformation, as well as for transcription promoters and terminators, enhancers of transcription and translation, and markers for selection of transformed plants.

Many of the plant virus sequences, gene sequences used in binary vectors, and transformation and regeneration protocols have been patented. It is best to verify intellectual property rights before embarking on generation of transgenic plants. If the purpose is to express protein from the cloned gene, a proper translational context should be provided around the start codon.

The sequence of 5'-..AACA ATG..-3' is generally optimal; however, altering the context for some proteins has not necessarily enhanced the protein levels.

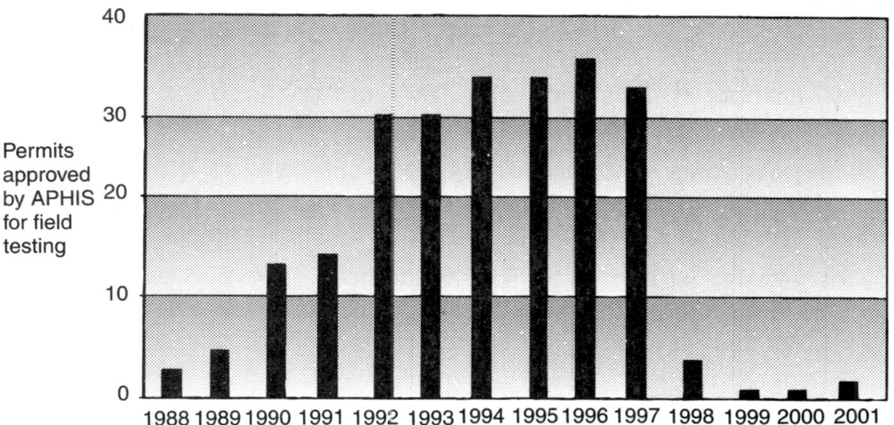

Figure 14.1: Bar diagram representing the trend in the number of applications approved by the Animal Plant Health Inspection Service of the United States Department of Agriculture for field-testing of transgenic plants for virus resistance over the years indicated.

Choice of Gene Deliery System

Currently, A. *tumefaciens-mediated* gene delivery and biolistic transformation are the most widely used systems for the delivery of foreign DNA and stable expression in plants. Although these methods are dependent on tissue culture, selection and regeneration of target plant species, electroporation of foreign DNA into meristematic tissue circumvents the need for tissue culture and regeneration.

However, only a fraction of the seeds obtained from the primary transformant are transgenic and thus requires careful screening of the progeny.

Virus Eolution and Engineered Resistance

Most plant virus infections generally occur as the result of entry of several thousand-virus particles into the host. Of these, it is certain that there is nucleotide sequence and biological variation among the infecting population.

The occurrence of a heterologous population of virus sequence variants between and within plants (quasi-species nature of some plant viruses) can be a potential problem for transgenic resistance. Of course, this can also be a problem for R gene mediated resistance, and therefore must be considered in any resistancebased control strategy.

Viruses have evolved to use this variation to take advantage of their environment, some of the population may be able to infect and replicate even under selection pressures imposed by natural and/or transgenic resistance.

Thus, understanding virus population dynamics and performing rigorous field trials in crop production sites can help to understand if resistance will be effective, and it must be accepted that even transgenic resistance is not going to be perfect or solve all problems related to controlling plant viruses or other pests/pathogens. Despite the rapid and significant advancements in developing engineered resistance to plant viruses, only a small number of transgenic virus resistant crop plants have been adopted in agriculture.

In the United States, only five transgenic crops, papaya resistant to *Papaya ringspot virus;* potato resistant to *Potato leaf roll virus;* potato resistant to PVY; squash resistant to WMV and ZYMV; and squash resistant to CMV, WMV, and ZYMV, are approved for complete deregulation for commercial cultivation.

However, during the years 1988 to 2001, 242 applications were granted approval by the Animal and Plant Health Inspection Service Unit of the United States Department of Agriculture for the field testing of transgenic crop plants potentially resistant to plant viruses.

Since 1997, the requests for permits for field testing have decreased considerably, which can be attributed in part to the proprietary rights on the promoters, marker genes and sequences of viruses, the lack of interest to identify the operating mechanism of resistance, and/or public perception.

However, transgenic resistance is perhaps the best thing to happen as a new approach for controlling plant virus diseases, and in many cases offers more environmentally sound approaches for disease control and in some instances the only possibility. It could go a long way to help keep crop losses stemming from plant viruses much lower than at the present.

VIRUS-INDUCED GENE SILENCING IN PLANT ROOTS

Virus-induced gene silencing (VIGS), also known as posttranscriptional gene silencing, is an epigenetic phenomenon that was first described in plants and was referred to as cosuppression. Later, it was also discovered to be the cause of cross protection from viral infections. Cross protection, also known as pathogen-derived resistance, has been observed in plants infected with a mild strain of a virus that were found to be immune to subsequent infection by severe strains of the virus.

Posttranscriptional gene silencing is a universal mechanism of sequence-specific degradation of endogenous RNA identified in several other organisms. In plants, VIGS of endogenous genes by recombinant viruses carrying portions of plant complementary DNA (cDNAs), quickly emerged as a tool to rapidly reduce message levels of endogenous genes to assess their functional role in a number of plant species.

The process begins with the introduction of infectious viral particles, carrying a portion of a gene of interest, into a young seedling. As the virus multiplies and the plant grows, the virus spreads to new growing parts and induces VIGS.

Several plant viruses have been engineered as vectors for use in VIGS, including tobacco mosaic virus, potato virus X, tomato golden mosaic virus, tobacco rattle virus (TRV) , barley stripe mosaic virus, cabbage leaf curl virus, and satellite virus-induced silencing system. Among these, TRV infects a large number of plant species and probably has the widest known host range of any plant virus.

TRV belongs to the tobravirus group of plant viruses that are characterized by a positive-sense single-stranded RNA and a bipartite genome, RNAl and RNA2. RNAl encodes genes required for both replication and movement, whereas RNA2 encodes the coat protein and two nonstructural proteins.

TRV can spread in both floral meristematic tissue and root tips, indicating systemic movement throughout plant tissues. This virus is also transmitted by nematodes of the genera *Trichodorus* and *Paratrichodorus,* indicating that TRV is most likely present more uniformly in roots, in particular near root tips, where these nematodes feed.

The nematode transmission of TRV and its efficient translocation in roots makes this virus an excellent candidate for use as a VIGS vector to silence genes in roots. Recently, TRV-VIGS vectors have been improved to efficiently silence endogenous genes. RNAI and RNA2 cDNAs have been inserted behind duplicated CaMV 35S promoter followed by a ribozyme and nopaline synthase terminator at the C-terminal end.

Regions of RNA2 coding for nonessential structural genes, including the region associated with nematode transmission, were replaced with a multiple cloning site. These constructs were inserted into binary vectors, pTRV 1 and pTRV2, and transformed into *Agrobacterium tumefaciens. Agrobacterium* clones containing pTRV 1 and pTRV2 are grown separately and mixed at equal concentration in infiltration buffer immediately before inoculation.

This and other virus vectors have been used to silence genes encoding metabolic enzymes,

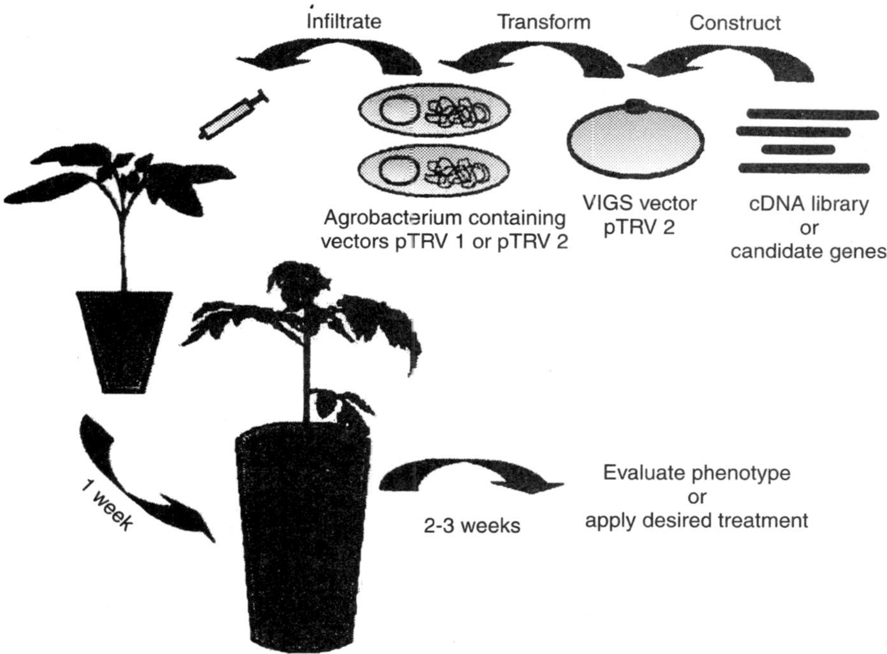

Figure 14.2: Schematic diagram of steps involved in tobacco rattle virus (TRV)-based virus-induced gene silencing assays in roots. TRV has a bipartite genome and both are modified and inserted into a binary vector resulting in pTRV1 and pTRV2. A cDNA fragment is cloned into the pTRV2 binary vector and transformed into Agrobacterium tumefaciens. pTRV1 is also transformed into A. tumefaciens. Bacterial clones containing each vector are grown separately and combined immediately before infiltration into leaflets of 2-wk-old tomato seedlings, grown in seedling flats. The virus moves to the roots and the root tips as it spreads systemically. One week after infiltration, seedlings are transplanted into larger containers. Two weeks after transplanting, plants are ready for root assays.

such as the phytoene desaturase *(PDS)* gene that results in photobleaching, and genes required for disease resistance. Most of the genes targeted by VIGS display phenotypes in aboveground parts of plants.

We have used TRV-based vector to silence the nematode resistance gene *Mi-1* in tomato roots (Kaloshian, unpublished results). *Mi-1* encodes a protein with coiled-coil, nucleotide-binding site, and leucine-rich repeat domains.

In addition to conferring resistance to three species of root-knot nematodes *(Meloidogyne* spp.) it also confers resistance to potato aphid, *Macrosiphum euphorbiae,* and whitefly, *Bemisia tabaci.* The Mi-1/tomato system provided a unique system that allowed us to assay for gene silencing in both leaves and roots of a plant. This chapter describes a method to silence genes in tomato roots using TRV-based vector pTRV 1 and pTRV2).

MATERIALS

Planting and Groth Material

1. Tomato seeds.
2. Organic soil mix.
3. Seedling flats, with 1-in.^{2} well size.
4. Slow-release fertilizer Osmocote® (17-6-10; Sierra Chemical Company, Milpitas, CA).
5. Tomato MiracleGro® (Stern's MiracleGro Products, Port Washington, NY).
6. Pots.
7. Growth chamber.

Cloning Into TRV Vector

1. TRV vector: pTRV 1 and pTRV2.
2. A 150- to 700-nucleotide DNA fragment of the target gene.
3. Restriction enzyme(s) and buffer.
4. Agarose.
5. 10X TBE (1.0 L): To make 1.0 L of 1OX TBE, mix 108 g of Tris base, 55 g boric acid, 40 mL of 0.5 M ethylenediaminetetraacetic acid (EDTA), pH 8.0, add double distilled H_2O to 1 L and adjust pH to 8.3 by adding boric acid.
6. QIAquick Gel Extraction Kit (QIAGEN).
7. T4 ligase and buffer.
8. *Escherichia* coli-competent cells.
9. Luria broth (LB) medium (1.0 L): 10 g bacto-tryptone, 5 g bacto-yeast extract, 10 g NaCl.
10. Kanamycin.

Transformation Into Agrobacterium, Growth, and Preparation for infiltration

1. *A. tumefaciens* strain GV3101 electrocompetent cells.

2. Electroporator.
3. Gentamycin.
4. *Agrobacterium* strain GV3101 containing pTRV1.
5. LB medium.
6. 200 mM Acytosyringone (3'-5' dimethoxy 4'-hydroxy acetophenone) in dimethyl formamide.
7. 1 M $MgCl_2$.
8. 1 M MES (2-[N-Morpholino] ethane sulfonic acid).
9. 3-mL Syringe.

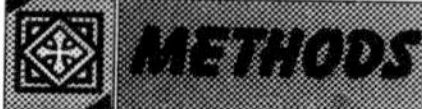

METHODS

Planting Tomato Seeds

1. Plant tomato seeds individually into seedling flats. Add Osmocote and cover with plastic wrap to avoid drying. Place seedling flat on a tray to catch the drainage water and help maintain adequate moisture. Seedlings with two fully developed leaves are used for VIGS assays.
2. Fertilize seedlings with Tomato MiracleGro weekly.

Construction of Virus Vector to Include Target Gene Fragment

1. Choose a 150- to 700-nucleotides region of the target gene.
2. Use appropriate restriction enzyme(s) to cut this fragment to generate either blunt ends or overhangs that could be cloned into the polylinker of pTRV2 vector.
3. Cut the virus vector, pTRV2, with restriction enzyme(s) to generate ends matching the insert.
4. Run both the insert and the virus vector restriction digests on an agarose gel and cut the fragments and elute DNA fragments using QIAquick Gel Extraction Kit (QIAGEN).
5. Ligate insert and vector for 4 to 12 h.
6. Transform the ligates into kanamycin-sensitive *E.* coli-competent cells, such as DH5a, , by either electroporation or heat shock method according to Sambrook et al..
7. Spread two different aliquots onto LB plates supplemented with antibiotics kanamycin (50 mg/L) and incubate overnight at 37°C.
8. Check the presence of insert in the TRV vector using polymerase chain reaction (PCR) or restriction digestion.
9. Inoculate a positive *E. coli* clone into 5 mL of LB supplemented with kanamycin (50 mg/L) and grow at 37°C for 12 to 16 h with 250 to 300 rpm vigorous shaking.
10. Isolate the recombinant vector as described in Sambrook et al.
11. Transform the recombinant vector into *A. tumefaciens* strain GV3101 competent cells using electroporation.

12. Add LB to 0.5 mL and shake at 250 rpm (28°C) for 2 h.
13. Collect the *Agrobacterium* cells by centrifugation at 11,750g in a microcentrifuge for 30 s and discard the supernatant.
14. Resuspend cells in the remaining supernatant and spread on LB plates supplemented with 50 mg/L kanamycin and 50 mg/L rifampicin and incubate at 28°C for 2 to 3 d.
15. Check the *Agrobacterium* transformants by PCR or restriction digestion for the presence of the TRV vector carrying the target gene.

Introducing Agrobacterium Into Tomato and Assay for VIGS

1. Inoculate individually a single colony of each *Agrobacterium* with pTRV1, pTRV2, and pTRV2 containing the target gene, into separate 10-mL test tubes containing 2 mL of LB medium supplemented with kanamycin (50 mg/L) and rifampicin (50 mg/L) and shake the tubes at 250 rpm (28°C) overnight.
2. Use the 2-mL overnight cultures to inoculate three 125-mL flasks containing 25 mL of LB medium with the same antibiotics as for the overnight cultures, supplemented with 10 mM MES and 20 μM acetosyringone, and grow overnight under the same conditions.
3. Harvest the bacterial cells in sterile disposable 50-mL conical tubes by centrifugation at 2800g for 10 min and resuspend in infiltration buffer containing: 10 mM $MgCl_2$, 10 mM MES, and 200 μM acetosyringone.
4. Adjust the concentration of each culture to OD600 of 0.8 to 1.0 and incubate at room temperature for 3 h.
5. Mix equal volumes (1:1 ratio) of pTRV1 and pTRV2 *Agrobacterium* solutions and use a 3-mL syringe without the needle to infiltrate the solution into the abaxial side of each expanded leaflet.
6. Move the seedlings onto a clean tray and make sure to place each batch of infiltrated seedlings on a separate tray. Maintain seedlings in a growth chamber at 21°C.
7. One week after infiltration, transplant seedlings into desired containers and planting medium and maintain at 21°C.
8. If further assay is needed to test for the VIGS phenotype, plants are ready for treatment 2 to 3 wk after transplanting. Apply the desired treatment and maintain plants under optimum conditions for the treatment.
9. At the time of phenotypic evaluation, sample root portion(s) with the expected symptoms for RNA analysis. Sample roots, freeze immediately in liquid nitrogen, and store at –80°C.

Notes

1. Use more than one *Solanum lycopersicum* cv. We have observed variation in VIGS efficiency among tomato cultivars (cvs). Best to germinate seeds in a mist chamber. If a mist chamber is not available, seedlings can be germinated in a greenhouse. You can plant a large number of seeds in a single seedling flat. Just before infiltration with *Agrobacterium,* cut sections of the flat with the desired number of seedling to assay with a single VIGS construct.

2. DNA fragments as small as 23 nucleotides have been used to silence the target gene. However, with smaller fragment sizes, silencing is ineffective and larger inserts are recommended for efficient silencing. If the target gene is a member of a gene family, then the 5' untranslated region should be used for silencing.
3. At the time of choosing the DNA region to target in VIGS, determine the region that will be used to assess the transcript level using reverse transcription PCR (RT-PCR). The region targeted for transcript analysis should be different than the one used for silencing, to avoid amplification of transcripts generated by the virus. If targeted sequences are limiting for lack of sequence information or unique sequences, in the case of gene families, overlapping regions could be used. In the latter case, one primer used in PCR should lie outside the targeted area for silencing. With large genes, when using oligo dT for RT, it is advisable to target the 3' region of a gene for PCR amplification to avoid problems with inefficient RT.
4. Currently, there are three TRV vectors available. The described approach for cloning could be used when targeting a limited number of genes using either one of the TRV vectors. For high-throughput VIGS using cDNA library screens, pTRV2-attR1-attR2 vector with Gateway (Invitrogen) cloning site allows a fast and efficient subcloning alternative.
5. It is highly useful to design primers flanking the pTRV2 polylinker to amplify the insert. If restriction digestion is used to check the presence of insert in the transformed colonies, the miniprep used for this purpose could also be used for the following step.
6. If the original TRV vector components, pTRV1 and pTRV2, are not in *A. tumefeciens* strain GV3101, then you need to isolate the pTRV1 vector and transform into strain GV3101.
7. A culture of *Agrobacterium* with empty pTRV2 vector is used as empty vector control. Although this TRV vector does not cause severe viral symptoms in the tomato cvs we have tested, we have seen differences in the viral symptoms among different cvs.
8. Adjust the concentration of pTRV1 and pTRV2 to the same OD600.
9. Based on establishment of *PDS* silencing in leaves, we identified 21°C to be the best temperature for VIGS in tomato. Although we have not tested the importance of temperature in TRV-VIGS in roots, we assume that efficiency in silencing in leaves also reflects silencing in roots.
10. Catch trays are recommended with TRV-VIGS to avoid contamination because the virus can be transmitted through direct root-to-root contact and root contact with drainage water.
11. For assays with root-knot nematodes, we use 32-oz plastic cups, with holes in the bottom, and sand.
12. There may be no need to maintain the temperature at 21°C after applying the additional treatment. For example, the outcome of most plant–pathogen or nematode assays is determined within the first few days after application of the inoculum, even if the assays require a period of time before they can be evaluated. Therefore, the result of the assay is based on the efficiency of silencing at the time of application of the inoculum.

13. In general, VIGS in tomato roots and leaves is not uniform. Therefore, it is important to sample roots with silenced phenotype to assess for transcript level.
14. The expression of VIGS in root tissue is maintained for months.

PROTEOMICS ANALYSES

Chloroplasts are essential organelles in vegetative tissues in plants. Although best known for their role in *photosynthesis, chloroplasts* synthesize many essential compounds, such as *plant hormones, fatty acids* and *lipids, amino acids, vitamins, purine* and *pyrimidine nucleotides, tetrapyrroles*, and *isoprenoids*.

Chloroplasts are required for nitrogen and sulfur assimilation and also contain numerous protein chaperones and targeting components, cofactor chaperones, assembly factors, peptidases, and proteases involved in protein (complex) biogenesis and homeostasis.

To facilitate chloroplast gene expression, chloroplasts contain proteins associated with plastid DNA and the plastid transcriptional and translation machinery, including many mRNA binding proteins involved in mRNA processing, stability, and translation.

It is predicted that all plastid types combined contain over 2000 to 4000 different proteins, with approx 550 a-helical proteins integral in thylakoid and inner envelope membranes and an unkown number peripherally associated with these membrane.

It is more than likely that the chloroplast stroma contains well over 1000 proteins. The thylakoid and envelope proteomes and associated proteins of chloroplasts *from Arabidopsis thaliana* have been analyzed in detail in a number of studies. The method described here is based on a protocol developed for *A. thaliana* by Dr. David Christopher and published protocols for isolation of chloroplasts from pea and spinach.

The method was further developed in our lab for large-scale chloroplast purification for proteome analysis and was published in Peltier et al.. Specific protocols have been developed by other laboratories (e.g.

by D. J. Schnell, K. Keegstra, and P. Jarvis) for purification of intact *Arabidopsis* chloroplasts for in vitro protein import essays.

The protocol in this chapter has not been tested for protein import activity but is optimized for chloroplast proteome yield and purity and minimal protein degradation.

MATERIALS

1. Medium A (grinding buffer; *see* **Note 1):** 50 mM HEPES-KOH (pH 8.0), 330 mM sorbitol, 2 mM EDTA-Na_2 (pH 8.0), 5 mM ascorbic acid, 5 mM cysteine; 0.05% bovine serum albumin (BSA).
2. Medium B (wash medium): 50 mM HEPES-KOH (pH 8.0), 330 mM sorbitol, 2 mM EDTA-Na_2 (pH 8.0)
3. Medium C (lysis medium): 10 mM HEPES-KOH (pH 8.0), 5 mM MgCl2, protease inhibitor cocktail.
4. 100% PF-Percoll: 0.8 g PEG 8000, 0.27 g Ficoll 400,000. Add Percoll to 27.5 mL end volume.
5. Percoll step gradients: PF Percoll (40% or 85%), 0.5 mM EDTA, 50 mM HEPESKOH (pH 8.0), 330 mM sorbitol.
6. Concentration spin device with 3-kDa cutoff filter, e.g., from Millipore-Amicon 3 kDa.

METHODS

For maximal yield of intact chloroplasts, grow the plants under short day length to stimulate vegetative growth and harvest the leaves during the first half of the light period. Prepare all media prior to harvesting of leaves and cool them down to 4°C in the cold room, preferentially on ice.

Avoid exposing the chloroplasts to light (work in dim green light) as much as possible, and keep buffers, rotors, and centrifuge tubes ice-cold. Work as fast as possible to avoid breakage of the chloroplasts. Chloroplasts can be purified in 30 to 40 min.

1. Grow *A. thaliana* plants under a 10-h light/14-h dark cycle at 23°C/17°C. For maximum leaf material, grow the plants for 5 to 6 wk and harvest prior to bolting.
2. Prepare all media, collect centrifuge tubes, and cool on ice. Cool down centrifuge and rotors.
3. Collect rosettes and cut leaves from rosettes. Make sure there is no soil attached. Wash in distilled water if needed.
4. Grind leaves 3X 10 s at half-speed in a Warren blender with sharp blades in ice cold grinding medium (medium A). Use approx 10 g of leaves (about equivalent to five full grown rosettes) per 100 mL grinding medium.
5. Filter the homogenate through a double layer of nylon cloth.
6. Collect crude chloroplasts by centrifugation for about 3 min at approx 1300g in a fixed angle rotor.
7. Resuspend the crude chloroplast pellet in medium B by "swirling" the suspension. Keep

Table15.1: Protease Inhibitor Cocktail

Inhibitor	Stock (mg/mL)	Medium	End-concentration (μg/μL)
Antipain	20	Water	50 *(74 tM)*
Bestatin	1	0.15 *M* NaCl	40 (130 *tM)*
Chymostatin	20	Dimethylsulfoxide	20
E64"	20	50% Ethanol	10
Leupeptin	1	Water	5
Phosphoramidon	20	Water	10
Pefabloc sc	100	Water	50
Aprotinin	10	Water	2

the volume as small as possible. Load the resuspended chloroplast on a Percoll step cushion, and spin for 10 min at 3750g in a swing-out rotor.

8. Collect the intact chloroplasts from the 40/85% Percoll interface using a pipet, and add medium B to dilute the Percoll. Keep the volume low, with chlorophyll concentrations between 1 and 3 mg/mL Spin the chloroplasts for 3 min at 1200g. Remove the supernatant. The pellet consists of intact chloroplasts.
9. For separation of thylakoid proteins and stromal proteins, add lysis buffer to about 1 to 3 mg chlorophyll/mL, and resuspend gently. Leave the suspension for 5 to 10 min on ice to swell the chloroplasts. These will burst and release stromal proteome. If needed, a Dounce homogenizer can be helpful to rupture the chloroplast. Be careful not to make too many strokes with the piston since this will partially rupture the thylakoid and release lumenal proteins into the stromal proteome.
10. Separate supernatant (containing mostly stroma) and pellet (containing the thylakoids) by a 20-min centrifugation at 10,000g. To remove envelopes and possible residual thylakoids from the stroma, carry out an additional spin of 25 min at 300,000g.
11. Concentrate stromal proteome in a concentration spin device with 3-kDa cutoff filter. The lysis buffer has a low salt and buffer concentration and is fully compatible with most protein separation steps used in proteome analysis [e.g., isoelectric focusing, sodium dodecyl sulfate polyacrylamide gel electrophoresis (SDSPAGE), blue native (BN)-PAGE or clear native (CN)-PAGE and in-solution digests.
12. For analysis of the thylakoid proteome, resuspend the pellet in the same low salt and weakly buffered medium, and, if needed, determine the chlorophyll concentration and/or protein concentration. Subsequent thylakoid proteome analysis will require removal of pigments (chlorophylls and carotenoids) and other lipophilic molecules (e.g., plastoquinone) and removal of lipids. This can be done by extraction with organic solvents (e.g., 70% acetone) or by incubation with ionic (e.g., SDS) or nonionic detergents (OGG, DM). Recent

examples of chloroplast membrane proteome analysis can be found elsewhere in this Chapter.

Notes

1. How to make 1 L of medium A: 60.1 g sorbitol; 50 mL 1 *M* stock of HEPESKOH (pH 8.0); 4 mL 0.5 *M* EDTA (autoclaved stock); fill up with double-distilled water to 1 L. Add fresh 0.88 g ascorbic acid; add 0.6 g cysteine; add 0.5 g BSA (can be omitted to minimize carryover to final intact chloroplast sample)
2. Use 40 to 50 full-grown rosettes per liter of grinding medium. For maximum yield, try to find an optimum grinding time and ratio between grinding medium/ grams of leaves.
3. It is best to prewet the cloth with grinding medium.
4. To make a Percoll step gradient, load 40% on top of PF Percoll 85% or load 85% with a syringe below 40% Percoll.
5. Chlorophyll determinations were done according to ref. 20. Resuspend 2 to 5 µL (v) of green chloroplast or thylakoid solution in 1 mL 80% ice-cold acetone. Vortex and keep the suspension on ice for 10 min in the dark. Spin for 5 min at 15,000g at 4°C. Collect the (green) supernatant and measure light absorption at 663.2 and 646.8 nm in a spectrometer. The chlorophyll (a + b) concentration (µg/mL) = $(1 + v)/v \times (7.15 \times A_{663.2} + 18.71 \times A_{646.8})$.
6. An alternative centrifugation regime is to collect thylakoids and envelopes together in one spin for 25 min at 300,000g. However, these g-values can lead to fairly hard thylakoid pellets and possible some extra release of lumenal and peripheral thylakoid proteins into the supernatant, the stroma.

ISOLATION AND SUBFRACTIONATION OF PLANT MITOCHONDRIA FOR PROTEOMIC ANALYSIS

Plant mitochondria provide energy in the form of ATP, provide a high flux of metabolic precursors in the form of tricarboxylic acids to the rest of the cell for N-assimilation and biosynthesis of amino acids, and have well-studied roles in plant development, productivity, fertility, susceptibility to disease, and programmed cell death.

Plant mitochondria house many hundreds of proteins identified by proteomics, and a total of 1000 to 1500 proteins are expected to be found in this cellular compartment. Techniques for the isolation of plant mitochondria without changing their morphological structure observed in vivo and maintaining their functional characteristics has required methods that avoid hyper- or hypoosmotic rupture of membranes and protection from harmful products released from other ruptured cellular compartments.

Several extensive methodology reviews and more specific methodology papers are already available on plant mitochondrial purification. Study of the composition, metabolism, transport processes and biogenesis of plant mitochondria requires an array of procedures for the subfractionation of mitochondria to provide information on localization and on association of proteins or enzymatic activities.

Here we present a basic appraisal of the classic methods for mitochondrial isolation and fractionation procedures to separate mitochondria into their four subcompartments.

MATERIALS

1. Plant tissue: Mitochondria can be isolated from virtually any plant tissue. However, depending on the requirements for purity and yield, particular tissues have their advantages:

   a. Nonphotosynthetic fleshy root and tuber tissues allow for large-scale, but lowpercentage-yield, plant mitochondrial isolation (approx 300 mg mitochondria) protein from 5-10 kg fresh weight [FW], 30-60 μg/g FW). Typically potato, sweet potato, turnip, and sugar beet are used for these purposes.

   b. Etiolated seedling tissues such as hypocotyls, cotyledons, roots, or coleoptiles are used, as they are free of dense chloroplast membranes, have lower phenolic content than green tissues, and provide high yields of functional mitochondria (approx 20 mg mitochondrial protein from 100 g FW, 200 μg/g FW).

   c. Interest in the function of mitochondria during photosynthesis and in plant species without abundant storage organs has led to methods for the isolation of chlorophyll -free mitochondria preparations from green tissues such as leaves and cotyledons (5-10 mg mitochondrial protein from 50-100 g FW, 100 μg/g FW). Once the plant material has been chosen, the tissue must be free from obvious fungal or bacterial contamination and should be washed in cold $_{H20}$ or in some cases surface-sterilized (for 5 min in a 1:20 dilution of a 14% w/v sodium hypochlorite stock solution). These steps ensure as little contamination as possible in further steps of the protocol, as bacteria will tend to copurify with mitochondria. All tissues should then be cooled to 2 to 4°C before proceeding.

2. Homogenization buffer solution: tissues should be processed as soon as possible following harvesting and cooling in order to maintain cell turgor and thus ensure maximal mitochondrial yields following homogenization. A basic homogenization medium consists of 0.3 to 0.4 M of osmoticum (sucrose or mannitol), 2 to 5 mM of divalent cation chelator (EDTA or EGTA), 25 to 50 mM of a basic pH buffer system (MOPS, TES or Na-pyrophosphate), and 5 to 20 mM of reductant (cysteine, ascorbate, dithiothreitol [DTT] or (β-mercaptoethanol), which is added just prior to homogenization.

   The osmoticum maintains the mitochondrial structure and prevents physical swelling and rupture of membranes, the buffer prevents acidification from the contents of ruptured vacuoles, the EDTA inhibits the function of phospholipases and various proteases, and the reductant prevents damage from oxidants present in the tissue or produced on homogenization. Although this basic medium is sufficient for tuber tissues, a variety of additions have been suggested to improve yields and protect mitochondria from damage during isolation. Etiolated seedling tissues and green tissues often require addition of 0.2 to 1% (w/v) bovine serum albumin (BSA) to remove free fatty acids, and 1 to 2% (w/v) polyvinylpyrolidone to remove phenolics that damage organelles in the initial homogenate.

As a result, we have routinely used a homogenization medium consisting of 0.4 M mannitol, 5 mM EGTA, 50 mM sodium pyrophosphate-KOH (pH), 10 mM cysteine, 0.5% BSA, and 1% PVP-40 with wide success in a variety of plant tissues.

This media can be freshly prepared, stored overnight at 4°C or frozen at-20°C, and stored for many weeks. Filtrates from homogenization can be filtered through a wide variety of muslin fabrics, Miracloth (Calbiochem, La Jolla, CA), or disposable clinical sheeting available from medical suppliers.

3. Wash buffer solution: a standard wash buffer solution is then used for resuspension of organelle pellets, as the base media for Percoll gradients and for washing purified organelle pellets. We have successfully used 0.3 M mannitol, 10 mM TES-KOH, pH 7.2, and 0.1% w/v BSA in a variety of plants.
4. Percoll gradient solutions: A gradient of Percoll is prepared in 0.3 M sucrose or 0.3 M mannitol supplemented with 10 mM TES, pH 7.2, 0.1% (w/v) BSA on the day of use. For a continuous gradient a gradient maker is required, for a discontinuous step gradient simple inverted syringe bodies will suffice.
5. Cytochrome oxidase assays:
    a. For the oxygen consumption-based assay the following is required: 0_2 electrode, 0.5 *M* Tris-HC1, pH 7.5, 5 mM Cyt c (oxidized), 0.5 *M* sodium ascorbate, 10% (w/v) Triton X-100, 0.1 M KCN
    b. For the spectrophotometric assay the following is required: Visible wavelength spectrophotometer, PD 10 desalting column, 0.5 M Tris-HCl, pH 7.5, 5 mM Cyt c, 10% (w/v) Triton X-100, 0.1 M KCN.
6. Subfractionation of mitochondrial compartments: further fractionation of mitochondria will require: wash medium without BSA (0.3 M mannitol, 10 MM TESKOH, pH 7.2), 2 M KC1 stock solution, low osmotic buffer (50 mM sucrose, 2 mM EDTA, 10 mM MOPS pH 6.5), 2 M sucrose, sodium carbonate soultion (100 mM Na_2CO_3), and a Dounce homogenizer.

METHODS

Following selection of the plant material and the homogenization medium (above), the most critical steps in yield of plant mitochondria are often found to be the method of homogenization, the pH of the medium, the ratio of homogenization medium to plant tissue, and a combination of the temperature of solutions and the time taken.

Homogenization

Depending on the tissue, homogenization can be accomplished by one or a combination of several methods:

1. Grinding in a precooled mortar and pestle.
2. Homogenization in a beaker with either a Moulinex mixer for 20 to 60 s or a Polytron blender at 50% full speed for 5× 2 s.
3. Homogenization in a Waring blender for 15 s at high speed, 2×15 s at low speed.

4. Juicing tuber material directly into 5X homogenization medium using a commercial vegetable/fruit juice extractor.
5. Grating of tissue by hand using a vegetable grater submerged under homogenization medium (usually required for mitochondrial extraction from soft fruits).

pH and Tissue Ratio

1. The pH value of the homogenization medium should be adjusted to approx 7.8 before homogenization to allow for acidification following cell breakage back to pH 7.0 to 7.6.
2. For some plant tissues, however, the amount of buffer may need to be increased to ensure that the pH of the final homogenate remains at or above pH 7.0, or the pH may need to be adjusted following homogenization with a stock of 5 M KOH or NaOH.
3. Homogenizing medium should be added in a ratio of 2 mL for each gram FW of non-green tissue or 4 mL per gram FW of green tissue. Decreases in these ratios can result in dramatic losses in yield.
4. Smaller ratios of medium to tissue (as low as 0.25 mL per gram FW) do work in the case of tuber tissues extracted by method 4 elsewhere in this Chapter probably because of the low level of phenolics and other toxins in tuber material.

Filtering and Differential Centrifugation to Obtain a Crude Organelle Pellet

1. The final homogenate should be filtered through four layers of muslin at 4°C. Direct filtrate via a funnel into a 4°C cooled beaker or conical flask; the remainder of the fluid in the cloth can be collected by wringing the cloth into the funnel. This is best performed in a 4 to 10°C cold room.
2. Transfer filtered homogenate into 50-, 250-, or 500-ML centrifuge tubes, depending on the volume of the preparation, and centrifuged in a precooled rotor for 5 min at approx 1500g in a fixed angle rotor in a preparative centrifuge at 4°C.
3. Decant supernatant gently into another set of centrifuge tubes, taking care not to transfer the pellet material, which contains starch, nuclei, and cell debris. Centrifuge supernatant for 15 min at approx 18,000g; discard the resulting high speed supernatant. The tan, yellow, or green pellet in each tube contains an unwashed crude organelle pellet.
4. Resuspend the pellet in 50 to 10 mL of a standard wash medium (e.g., 0.3 M mannitol, 10 mM TES-KOH, pH 7.2, and 0.1% w/v BSA) with the aid of a clean soft-bristle paint brush. If sucrose is used in the preparation, then sucrose can be used to replace mannitol in the standard wash medium (again at 0.3 M).
5. Transfer resuspended organelles to 50-mL centrifuge tubes, adjust volume to 40 mL with more wash medium, and centrifuge samples at 1000g for 5 min.
6. Transfer supernatants into another set of tubes, and sediment the organelles by centrifugation at approx 18,000g for 15 min. The high-speed supernatant is once again discarded and the washed crude organelles pellets can be uniformly resuspended in a small volume of wash medium as in step 5.

Density Gradient Purification of Mitochondria

The mitochondrial preparation just described is often adequate for a variety of respiratory measurements, but, depending on the tissue, it is frequently contaminated by thylakoid or amyloplast membranes, peroxisomes, glyoxysomes, endoplasmic reticulum, and occasionally bacteria. Further purification can be carried out using Percoll (GE Health) density gradients.

1. Layer-washed mitochondria, from up to 80 g of etiolated plant tissue or up to 40 g green plant tissue, over 35 mL of the chosen gradient solution in a 50-mL centrifuge tube.
2. Centrifuge at approx 35,000g for 45 min in an angle rotor of a preparative centrifuge without braking on the deceleration.
3. After centrifugation, mitochondria form a buff-coloured band below green chloroplast fragments or yellow-orange plastid membranes. Aspirate the mitochondria with a Pasteur pipet, avoiding collection of the yellow or green plastid fractions. Dilute suspension with at least 4 vol of standard wash medium, and centrifuge at approx 18,000g for 15 min in 50-mL tubes.
4. The resultant loose pellets should be resuspended in wash medium and centrifuged again at approx 18,000g for 15 min. The mitochondrial pellet resuspended in wash medium at a concentration of 5 to 20 mg mitochondrial protein/mL. This can be determined using a Bradford or Lowry assay.

Purity Determinations

Marker enzymes for contaminants commonly found in plant mitochondrial samples can be used to access the purity of preparations. Peroxisomes can be identified by catalase, hydroxy-pyruvate reductase, or glycolate oxidase activities. Chloroplasts can be identified by chlorophyll content and etioplasts by carotenoid content and/or alkaline pyrophosphatase activity, and glyoxysomes by antimycin-A-insensitive Cyt c reductase activity and plasma membranes by K^+/ATPase activity.

Cytosolic contamination is rare if density centrifugation is performed properly and care is taken in removal of mitochondrial fractions, but such contamination can easily be measured as alcohol dehydrogenase or lactate dehydrogenase activities. References to and methods for the assay of these enzymes are summarized elsewhere in this chapter Other considerations about mitochondrial purity are noted.

Integrity Determinations

A variety of assays can be used to determine the integrity of the membranes of plant mitochondrial samples, thus providing information on the structural damage (to these organelles) caused by the isolation procedure. Outer membrane integrity is often assessed by the degree of impermeability of the outer membrane to added Cyt c. The ratio of Cyt c oxidase activity before and after addition of a nonionic detergent (0.05% w/v Triton X-100) gives an estimate of the proportion of ruptured mitochondria in a sample. Two classic assays of its activity can easily be performed.

Rate of O_2 Consumption

Using an O_2 electrode, the rate of O_2 consumption in the presence of a reduced Cyt c regenerating system (to maintain a constant substrate concentration) can be measured.

1. Add ascorbate (20 mM) first to get rate of O2 consumption in the presence of this reductant, then add Cyt c (25 μM), then Triton X-100 (0.05%), and lastly KCN (1 mM) to inhibit the rate fully.
2. Cytochrome c oxidase activity is then defined as the rate in the presence of Cyt c and Triton minus the rate after the addition of KCN (this residue is also often the same rate as the ascorbate alone rate calculated before Cyt c addition).
3. The ratio "minus Triton" over "plus Triton" gives the proportion of broken mitochondria.

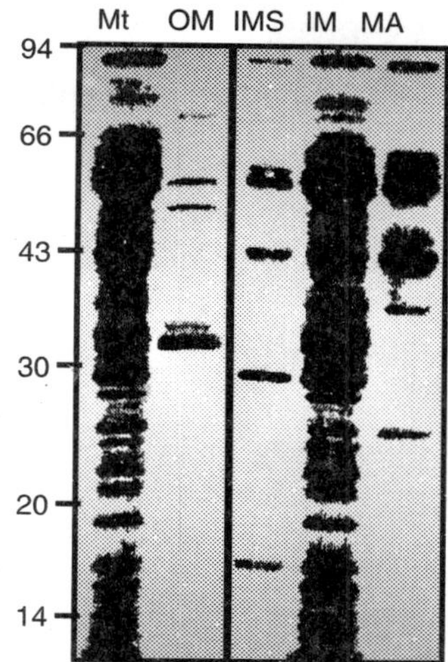

Figure.15.1: SDS -PAGE separation of the protein profile of the four subfractions of potato mitochondria visualized by Coomassie staining. Mt, mitochondria; OM, outer membrane; IMS, intermembrane space; IM, inner membrane; MA, matrix space. Proteins solubilized in Laemmli buffer prior to electrophoresis.

Oxidation of Reduced Cyt c

Using a spectrophotometer, the oxidation of reduced Cyt c can be measured at 550 nm.

1. Add mitochondrial protein to 0.1 M Tris-HCl solution containing 25 μM Cyt c (reduced). Follow Cyt c oxidation at 550 nm, then add Triton X-100 (0.05%), follow the rate, and lastly add KCN (1 mM) to inhibit the rate fully.
2. Cytochrome c oxidase activity is then defined as the rate in the presence of Cyt c and Triton minus the rate after the addition of KCN.
3. Again, the ratio "minus Triton" over "plus Triton" gives the proportion of broken mitochondria.

Storage

Once isolated by density gradient purification, mitochondria from most plant tissues can be kept on ice for 5 to 6 h without significant losses in membrane integrity and respiratory function. Longer term storage of functional mitochondria can be achieved by rapid-freezing small volumes of mitochondrial samples (>0.5 mL) in liquid N2 following the addition of dimethylsulfoxide (DMSO) to 5% (v/v) or ethylene glycol to *7.5%* (v/v). Frozen samples can be then kept at –80°C.

In our hands, mitochondrial samples prepared in this way from a variety of plant species can be stored for months without significant losses in respiratory rate, coupling to ATP production, or outer membrane integrity following thawing of these samples on ice.

Subfractionation of Mitochondrial Compartments

Using a combination of osmotic shock and differential centrifugation, plant mitochondrial samples can be fractionated in four components comprising the two aqueous compartments, the matrix (MA) and the intermembrane space (IMS) and the two membrane compartments, the inner

Table 15.2: Density Gradient Purification Strategies for Isolation of Plant Mitochondria by Different Researchers

Species	*Purification Tissue*	*method*
Non-green tissue		
Arabidopsis	Cell suspension	1, 3
	cultures	1, 2, 3
Barley	Seedlings	1, 3
Bean	Seedlings	1, 3
Cauliflower	Buds	1, 2
Maize	Embryos	1, 5, or 6
	Seedlings	1, 3
	Pea Seeds	1, 2, 3
Potato	Tubers	1, 3
		1, 2
		1, 2
	Stems	1, 3
Rice	Seedlings	1, 2, 4
Rye	Seedlings	1, 3
Soybean	Roots	1, 4
	Cell suspension	1
cultures		
Tobacco	Cell suspension	1
	cultures	1, 4
Wheat	Seedlings	1, 3
		1, 3
Green tissue		
Arabidopsis	Leaves	1, 3
Pea	Leaves	1, 4
		1, 2
Soybean	Cotelydons	1, 4
Spinach	Leaves	1, 4
		1, 2, or 3
Tobacco	Leaves	1, 3

mitochondrial membrane (IM) and the outer mitochondrial membrane (OM).

1. Initially carry out a salt wash to remove nonspecific associated proteins from the mitochondrial outer membrane. Resuspend 5 to 10 mg mitochondrial pellet in 4 mL of wash medium without BSA and add 2 M KCl to give a final concentration of 200 mM KCl. Divide in to 2 × 2-mL microfuge tubes, and centrifuge at 20,000g for 15 min at 4°C.
2. Carefully remove supernatant (sacrifice some at the interface if necessary) and discard. Resuspend pellets in a low osmotic buffer to a total volume of 6 mL and add to a 10-mL conical flask, and place in an ice slurry on a stirring block for 15 min. This step swells the mitochondria, and the outer membrane bursts because of the expanding inner membrane. Rupture of the OM can be supported by applying some mechanical shearing force to the sample (e.g., by a glass Dounce homogenizer).
3. Return to standard osmotic conditions by adding 0.9 mL of 2 M sucrose (approx 1/8 dilution of sucrose). Dispense into 2-mL microfuge tubes, and centrifuge at 20,000g for 15 min at 4°C.
4. Carefully take supernatant, which contains the removed outer membrane vesicles (OM) and soluble inner membrane space proteins (IMS), and freeze at –80°C.
5. Remove residual supernatant and resuspend each pellet (mitoplasts) in 2 mL of 250 mM KCl. Centrifuge at 20,000g for 15 min at 4°C.
6. Again remove residual supernatant and resuspend each pellet (mitoplasts) in 0.5 to 1 mL double-distilled (dd)H_2O, pool the samples in a Dounce homogenizer, and homogenize for approx 5 min to help burst mitoplasts.
7. Aliquot homogenate into two 2-mL Eppendorf tubes, make up each to 2 mL with ddH_2O, and centrifuge at 20,000g for 15 min at 4°C.
8. Carefully remove supernatant, which contains the matrix (MA), and freeze at –80°C.
9. The pellets are comprised of inner membranes (IM) and unbroken mitoplasts. Resuspend pellets in 4 mL ddH_2O, and freeze-thaw again in liquid N_2 two or three times. Centrifuge samples at 20,000g for 15 min at 4°C and keep pellets. Some of these pellets (IM) can be carbonate-washed (in 100 mM Na_2CO_3) to enrich for integral membrane proteins if required.
10. Ultracentrifugation is required to separate OM and IMS and to remove membrane fragments from the MA sample. Thaw at 4°C the sample of MA and OM/ IMS that has been stored at -80°C. Ultracentrifuge the OM and IMS mixture at 100,000g for 1 h at 4°C. The supernatant is the IMS fraction, and the pellet is the OM (retain and freeze). Ultracentrifuge the MA samples at 100,000g for 1 h at 4°C. Retain supernatant at the MA sample and freeze.
11. Sample purity can be assayed by measuring enzyme marker activities: cytochrome oxidase (COX) for IM, fumarase for MA, adenylate kinase for IMS, and antimycin A-insensitive NADH:cytochrome c oxidoreductase for OM. Antibodies to proteins can also be used in specific fractions. Separation of samples by sodium dodecyl sulfate-polyacrylamide gel electrophoresis (SDS-PAGE) show distinct protein patterns in each compartment.

Notes

1. The colloidal silica sol Percoll allows the formation of isoosmotic gradients and through isopycnic centrifugation facilitates a range of methods for the density purification of mitochondria. The most common method is the sigmoidal, selfgenerating gradient obtained by centrifugation of a Percoll solution in a fixedangle rotor.

   The density gradient is formed during centrifugation at >10,000g owing to the sedimentation of the poly-dispersed colloid (average particle size 29 nm diameter, average density p = 2.2 g/mL).

   The concentration of Percoll in the starting solution and the time of centrifuga-tion can be varied to optimize a particular separation.

2. Modification of the Percoll gradient is often required depending on the tissue being used. Ilsewhere in this Chapter shows references to a range of papers using different density gradient methods for plant mitochondrial isolation from a range of plant tissue types. Ilsewhere in this Chapter shows how to make such gradients.

3. Step gradients of Percoll are often used, as these aid the concentration of mitochondria fractions on a gradient at an interface between *Percoll concentrations*. Step gradients can easily be formed by setting up a series of inverted 20-mL syringes (fitted with 19-gauge needles) strapped to a flat block of wood, clamped to a retort stand over a rack at an angle of 45° containing the centrifuge tubes. The needles are lowered to touch the bevels against the inside, lower edge of the tubes. The step gradient solutions are then added (from bottom to top) to the empty inverted syringe bodies and each allowed to drain through in turn before the addition of the next step solution. Step gradients should, however, only be used once the density of mitochondria from a particular tissue and the density of contaminating components have been determined using continuous gradients. A range of density marker beads are available from GE Health for standardizing running conditions and establishing new protocols.

4. Cytochrome c (reduced) can be made by dilution of the oxidized form in 0.1 M Tris-HCl, pH 7.5, and addition of equimolar amounts of ascorbate, this reaction can be monitored spectrally at 400 to 600 nm. Alternatively, rather than monitoring the reaction by spectrometry (oxidized Cyt c is a red-brown colour when concentrated, and reduction leads to a general decrease in the colouration and a shift to red-pink), the progress of the reduction can be followed by eye. Excess ascorbate must be removed following reduction; this can be done using any standard desalting column (e.g., G25 or PD10), and reduced Cyt c can be stored at –20°C for many weeks.

5. Preparation of mitochondria should be undertaken as quickly as possible and without samples warming above 4°C or storage for extended periods between centrifugation runs. The time between homogenization and preparation of the washed crude pellet is the most critical for ensuring integrity and high yield.

6. Some researchers tend to omit steps 5 and 6 and place the first high-speed organelle pellet obtained in step 4 directly onto the density gradient. This choice largely depends on the success of the first low speed pellet in removing cell debris and the need to

concentrate organelles from step 4 further in order to place them on the density gradients.

7. Isolation procedures for mitochondria so far rely largely on the density of the organelles only, and this ultimately limits the purity of organelle fraction that can be obtained. Currently 2 to 5% of protein is routinely of nonmitochondrial origin even in the best mitochondrial preparations of the leading research groups. Additional steps for purity using different properties will be the next step to move purity to levels of 0.2 to 1.0% of nonmitochondrial protein in an organelle preparation. These approaches might use properties like surface charge of organelles or affinity capture techniques, but none have been routinely used to date by researchers.

8. The sucrose concentration required in the low osmotic buffer for mitoplast formation depends on the source of the mitochondria to be subfractionated. This *osmotic* strength can be altered to 10 to 80 mM to improve yields and purity significantly of the final fractions by maximizing OM rupture and *minimizing* MA protein release.

16

Chapter

PLANT PROTEOMICS AND GLYCOSYLATIA

In plants, as in other eukaryotes, glycosylation is mostly encountered on secreted proteins, although some forms of glycosylation can also be found on several cytosolic or nuclear proteins. It can be of two types: N- or O-glycosylation, depending on the linkage involved between oligosaccharides and the protein backbone. In plants, N-glycosylation is the most studied event.

N-Glycosylation

In plant cells, as in other eukaryotic cells, N-glycosylation occurs exclusively on proteins that enter the secretory pathway. It starts with the cotranslational transfer, in the endoplasmic reticulum (ER), of an oligosaccharide precursor, $Glc_3Man_9GlcNAc_2$, onto specific asparagine residues constitutive of N-glycosylation sequences present on the nascent protein (Asn-X-Ser/Thr, where X can be any amino acid except proline and aspartic acid).

This event takes place as soon as the newly synthesized protein enters the ER lumen. Then, further processing of the $Glc_3Man_9GlcNAc_2$ precursor occurs along the secretory pathway as the glycoprotein moves from the ER and through the Golgi apparatus to its final destination.

If the glycan side chain is accessible to the processing enzymes, such as glycosidases and glycosyltransferases present in the ER and Golgi apparatus, the precursor can be converted successively to high-mannose-type N-glycans, ranging from $Man_9GlcNAc_2$ to $Man_5GlcNAc_2$, and to complex-type N-glycans. Plant complex N-linked glycans present an α1-3-fucose attached to the proximal glucosamine residue of the core, β1-2-xylose residue linked to the core β-mannose and/or Gal β1-3(Fucα1-4)GlcNAc oligosaccharide terminal sequence, linked to their terminal antennae, called Lewis a.

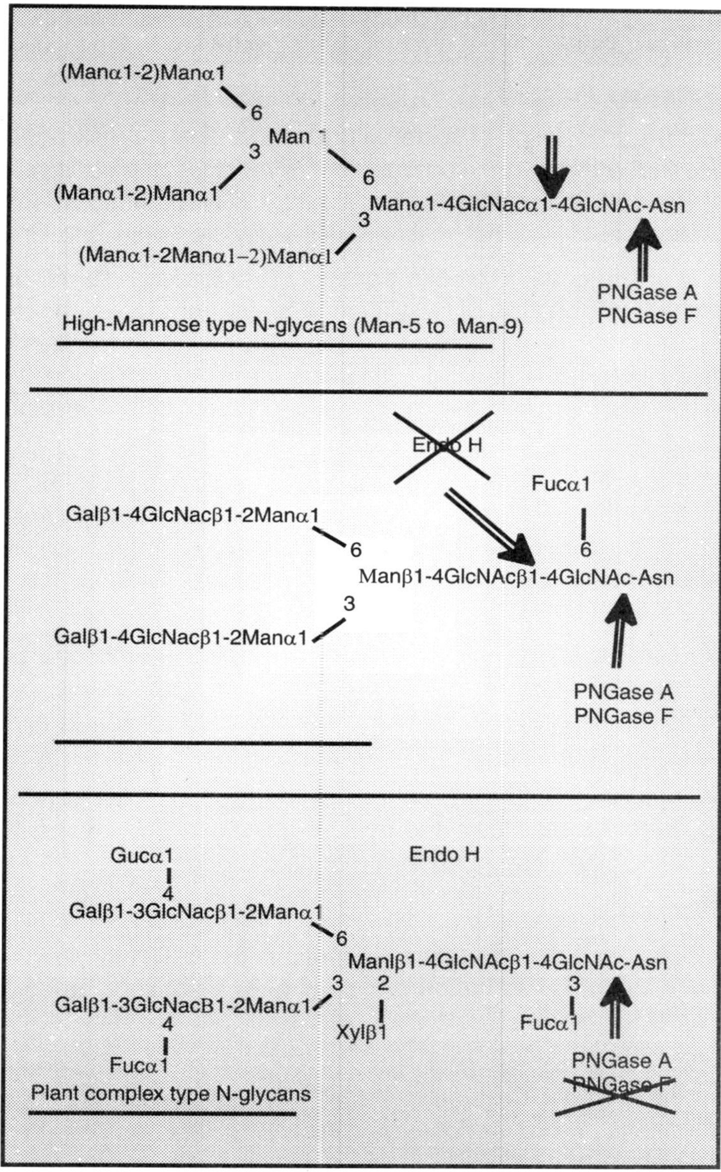

Figure 16.1: Enzymatic deglycosylation of proteins bearing N-linked glycans. High-mannose type N-glycans can be cleaved by either endoglycosidase H (Endo H), peptide N-glycosidase A (PNGase A), or peptide N-glycosidase F (PNGase F). Mammalian complex type N-glycans can be removed using PNGase A and PNGase F. Plant complex type N-glycans can only be released by PNGase A, owing to the presence of a1-3 fucose linked to the core. It should be noted that PNGase A mostly acts on glycopeptides and is not efficient on a complete glycoprotein.

O-Glycosylation

O-glycosylation can be of two types in plant cells, i.e., born by secreted proteins or by cytosolic/ nuclear proteins. The nature of the O-linked glycans differs according to the location of the proteins.

O-Glycosylation of Secreted Proteins

Little is known about 0-glycosylation of secreted proteins in plants. A few papers have reported a first class of 0-glycans that are linked to some Ser or Thr residues of plant glycoproteins. Indeed, several glycoproteins, like cell wall extensin and vacuolar sweet potato sporamin, have been described as glycoproteins containing Ser residues 0-glycosylated with one single galactose.

The characterization of rice glutelin, protein transported to the vacuole, tends to demonstrate the presence of mammalian mucin-type glycosylation. This type of 0-glycosylation in plants is still a matter of debate. The second type of secreted protein 0-glycosylation requires the hydroxylation of proline residues from glycoproteins and has been found in plants on hydroxyprolin-rich glycoproteins (HRGPs). HRGPs represent the major surface glycoproteins in plants.

These HRGPs, essentially located in the cell wall or at the outer surface of plasma membrane, represent three goups of proteins: extensins, arabinogalactan proteins (AGPs), and solanaceous lectins. Extensins and solanaceous lectins contain repeated sequences consisting of serine-(hydroxyproline)$_4$ in which the serine can be O-galactosylated, and the hydroxyprolines can be substituted by arabinosyl chains. This arabinosylation processes occur at the Golgi apparatus level.

The AGP family is composed of extracellular and plasma membrane proteins that are hyperglycosylated. Their protein moiety is rich in hydroxyproline (hyp), serine, alanine, threonine, and glycine and contains Ala-hyp repeats. Conversion of selected proline to hyp occurs either in the ER and/or in the Golgi apparatus by the action of a plant-specific prolyl-4-hydroxylase enzyme.

AGPs are Oglycosylated in a stepwise manner by addition on their hyp of polysaccharide chains consisting of (β1-3-galactose substituted by β1-6-galactose side chains, which are in turn branched with arabinose or other less abundant monosaccharide.

The O-GlcNAc Modification

Another O-glycosylation, well described in mammals, is the O-GlcNAcylation of cytosolic and nuclear proteins. This addition of O-GlcNAc on protein is quite ubiquitous among eukaryotes and occurs in a 3 linkage onto Ser or Thr.

There is no well-defined consensus sequence for O-GlcNAcylation, although the sequence Tyr-Ser-Pro-Thr*-Ser-Pro-Ser*, where the amino acids* are the glycosylated ones, represents a typical GlcNAc attachment site. O-GlcNAcylation is a highly dynamic process and often acts in a reciprocal manner to O-phosphorylation. It is driven by an O-GlcNAc transferase and an O-GlcNAcase that adds and removes, respectively, the sugar residue on/from proteins.

This glycosylation event is involved in numerous cellular processes, such as protein nuclear import, transcription, attachment of cytoskeleton to membranes, or protein synthesis, although its precise role is far from understood. In plants, little information is available concerning O-GlcNAcylation. Two enzyrnes similar to the O-GlcNActransferase, named SPINDLY and SEC, have been cloned in *Arabidopsis thaliana.*

They both present an O-GlcNAc transferase activity in vitro, but their natural substrates have not been defined yet. On the other hand, some nuclear proteins have been described as glycoproteins bearing O-linked glycans with terminal GlcNAc. Whether the latter are the substrates of SPINDLY and SEC is a question that remains to be answered.

Heterogeneity of Glycosylation

The N-glycosylation pattern of a mature secreted glycoprotein in eukaryotic cells results both from cotranslational transfer in the ER of an oligosaccharide and from its processing in the ER and Golgi apparatus. Factors like polypeptide folding can influence the availability of potential glycosylation sites in the ER.

As a consequence, when multiple N-glycosylation sites are present in a protein sequence, some sites might be glycosylated inefficiently and some others not used at all. In parallel, the processing of glycan side chains is related to their accessibility to the maturation enzymes. Glycans that are located on the protein surface will mature into complex-type N-glycans, whereas oligosaccharides that are buried in the protein structure remain unmodified high-manose glycans.

This is also true for O-linked glycans. Hence, identification of the glycosylation of a plant glycoprotein not only requires the identification of the glycan-protein linkage and the elucidation of the oligosaccharide structures attached to the glycoprotein, but also necessitates determination of their distribution/location on potential glycosylation sites of the protein backbone. In this context, the aim of this chapter is to introduce the plant proteomist to several tools and methods for determining whether the proteins identified are glycoproteins, to which category they belong, and, eventually, which part(s) of the protein backbone is (are) wearing the glycan side chain.

The last paragraph of this chapter will be dedicated to the specific identification of glycoproteins in plant glycoproteomics.

MATERIALS

Is My Protein a Glycoprotein?

1. DIG Glycan Detection Kit (Roche Applied Science, Mannheim, Germany) (store at 2-8°C).

How Is My Protein Glycosylated?

Detection of Glycans by Western Blotting

AFFINODETECTION WITH LECTINS

1. TTBS buffer: 20 mM Tris-HC1, pH 7.4, containing 0.5 M NaCl and 0.1% Tween-20.
2. Lectin-biotin: GNA *(Galanthus nivalis* agglutinin)-biotin (Vector Labs/Abcys, France) (store at 2-8°C), and WGA *(Triticum vulgaris* agglutinin)- biotin (Sigma Aldrich, Lyon, France) (store at 2-8°C).
3. Streptavidin-peroxidase conjugate (Amersham Biosciences/GE, Healthcare, Uppsala, Sweden) (store at 2-8°C).
4. TBS buffer : 20 mM Tris-HC1, pH 7.4, containing 0.5 M NaCl.
5. 4-Chloro-1-naphtol (Bio-Rad, Hercules, CA) (store at -20°C).

6. Ribonuclease B from bovine pancreas (Sigma Aldrich) (store at 2-8°C).
7. Methyl a-D-mannopyranoside (Sigma Aldrich) (store at room temperature).
8. a-A-crystallin from bovine lens (Sigma-Aldrich) (store at -70°C).
9. Ovalbumin (Sigma-Aldrich) (store at 2-8°C).
10. TTBS* buffer: 20 mM Tris-HC1, pH 7.4, containing 0.5 M NaCl, 1 mM $CaCl_2$, 1 mM $MgC1_2$, and 0.1% Tween-20.0
11. Concanavalin A (Sigma-Aldrich) (store at 5-8°C).
12. Horseradish peroxidase (HRP; Sigma-Aldrich) (store at 2-8°C).
13. Soybean agglutinin (Sigma-Aldrich) (store at 2-8°C).

Immunodetection with Specific Antibodies

1. TBS buffer : 20 mM Tris-HC1, pH 7.4, containing 0.5 M NaCl.
2. Gelatin (Bio-Rad).
3. Anti-HRP rabbit immunserum (Sigma Aldrich) (store at 2-8°C).
4. Anti-Apis *mellifera* venom (honeybee venom; Sigma Aldrich) (store at 2-8°C).
5. Mouse anti-human Lewis a monoclonal antibody (Calbiochem, Meudon, France) (store at -20°C).
6. TTBS buffer: 20 mM Tris-HC1, pH 7.4, containing 0.5 M NaCl and 0.1% Tween20.
7. HRP-conjugated goat antibodies directed at mouse polyvalent immunoglobulins (Sigma Aldrich) (store at -20°C).
8. HRP-conjugated goat antibodies directed at rabbit immunoglobulins (Bio-Rad) (store at 2-8°C).
9. Phospholipase A2 from honeybee venom (Sigma Aldrich) (store at 2-8°C).
10. Bean phytohemagglutinin L (PHA-L; Sigma Aldrich) (store at 2-8°C).

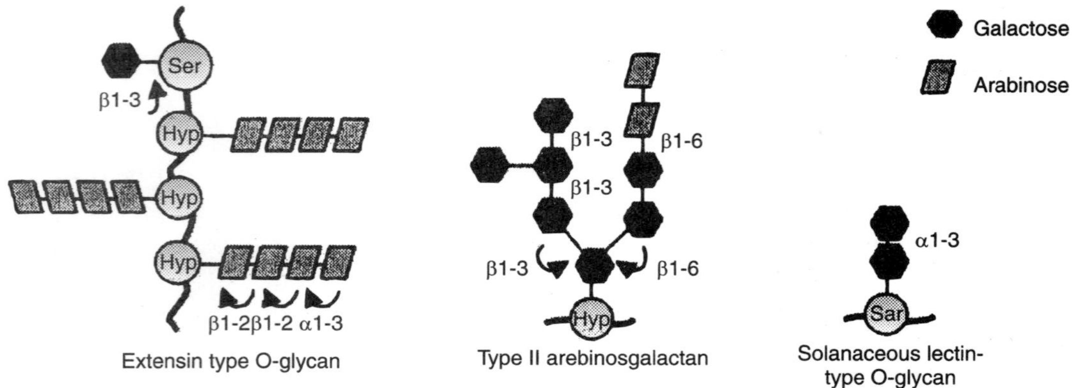

Figure16.2: Structures of O-glycans linked to plant glycoproteins.

11. Recombinant avidin expressed in maize (Sigma Aldrich) (store at 2-8°C).

Glycan Analysis by Is Monosaccharide Composition

1. Teflon-faced screw cap: kit flacons 2 mL (Interchim, Montlucon, France).
2. *2 mM* Inositol solution in water (store at 4°C).
3. Methanolic-HC1 3 *N* Kit (Supelco, Bellefonte, PA) (store at 2-8°C). Open one ampule of 3 *N* methanolic-HC1 and dilute it with methanol, in order to have a *1 N* methanol-HC1 solution.
4. Acetic acid, anhydrous 98% A.C.S. reagent (Sigma-Aldrich, Aldrich product) (store at room temperature).
5. Pyridine, anhydrous 99.8% (Sigma-Aldrich) (store at room temperature).
6. Silylation reagent: HMDS + TMCS + pyridine, 3:1:9 (Sylon HTP) kit (Supelco).
7. Cyclohexane Rectapur (VWR, Fontenay sous Bois, France) (store at room temperature).
8. Gas chromatography apparatus equipped with a silica column CP-SIL 5 CB (25 m × 0.25 mm).

Analysis of the Protein after Glycan Release

Chemical Treatment for Glycan Release

1. Glass reaction tube (Pyrex culture tubes) equipped with Teflon-faced screw cap (Bibby Sterilin Stone, Staffs, England).
2. Sodium borohydride solution: sodium borohydride freshly prepared at 40 mg/mL, in a 100 mM NaOH solution.
3. Acetic acid, anhydrous 98% A.C.S. reagent (Sigma-Aldrich, Aldrich product) (store at room temperature).

Enzymatic Treatment by N- or O-Glycosidases

1. 1% Sodium dodecyl sulfate (SDS).
2. 150 mM Sodium acetate buffer, pH 5.7.
3. Endo H: endoglycosidase H from *Streptomyces plicatus,* recombinant from E. *coli* (Roche Applied Science) (store at 2-8°C; freezing inactivates the protein).
4. 0.1 M Tris-HC1 buffer, pH 7.5, containing 0.1% SDS.
5. 0.1 M Tris-HC1 buffer, pH 7.5, containing 0.5% Nonidet P40.
6. Nonidet P40 (called now IGEPAL; Sigma-Aldrich).
7. PNGase F: peptide-N-glycosidase F from *Flavobacterium meningosepticum* (Roche Applied Science) (store between -15°C and -25°C).
8. 10 mM HCl solution, pH 2.2.
9. C 18 column: Bond Elut C 18 column (Varian, Palo Alto, CA) (store at room temperature).
10. Acetonitrile, minimum 99.5% (Sigma-Aldrich).

Table 16.1: Lectins Used for N- and O-Linked Glycan Detection in Plant Cell Extracts in Western Blotting.

Lectin	*Specificity*	*Detected oligosaccharides in plants*	*Suggested positive controls*	*Sugar inhibitor*
Concanavalin A from *Canavalia ensiformis*	αD: mannose αD: glucose	Precursor structures: M9G3 to M9G1 High-mannose structures: M9 to M5	Soybean agglutinin Ovalbumin Ribonuclease B	Methyl α-D-mannopyranoside
Galanthus nivalis agglutinin (GNA)	Terminal α1-3 mannose	High-mannose structures: M8 to M5	Ribonuclease B	Methyl α-D-mannopyranoside
Wheat germ agglutinin (WGA) from *Triticum aestivum*	Terminal GlcNAc Internal chitobiose units Ovalbumin (N-glycans)	All N-linked glycans O-linked glycans with terminal GlcNAc	α-A-crystalline from bovine eye lens (O-GlcNAc)	GlcNAc Chitotriose Chitobiose

11. 100 mM Sodium acetate buffer, pH 5.5.
12. Pepsin A from porcin stomach mucosa (Sigma-Aldrich).
13. PNGase A: peptide-N-glycosidase A from sweet almond (Roche Applied Science) (store between -15°C and -25°C).
14. Enzymatic deglycosylation kit (Prozyme, San Leandro, CA) (store at 4°C).

Where Is My Protein Glycosylated?

1. 50 mM Ammonium bicarbonate, pH 8.0.
2. Trypsin: TPCK-treated trypsin from bovine pancreas (Sigma-Aldrich, Fluka product) (store at -20°C).
3. Chymotrypsin: TLCK-treated chymotrypsin from bovine pancreas (Sigma Aldrich) (store at -20°C).
4. C18 column: Merck C18 Spherisorb ODS 2 column (5 µm, 4.6 × 250 mm, VWR)
5. Solvent A: 90:10 water/acetonitrile containing 0.1% trifluoroacetic acid (TFA).
6. Solvent B: 10:90 water/acetonitrile containing 0.1% TFA.

How Can the Whole Glycoproteome Be Analyzed?

Identification of Glycoproteins with High-Mannose N-Glycans

1. TBS buffer: 20 mM Tris-HC1, pH 7.4, containing 0.5 M NaCl.
2. Bradford Assay: Bio-Rad protein assay (store at 4°C).
3. 0.20-µm Membrane: Filtropur S 0.2 (Sarstedt, Numbrecht, Germany).
4. Concanavalin A-Sepharose: Con A Sepharose™ 4B (Amersham Biosciences) (store at 4°C).
5. TBS*: 20 mM Tris-HC1 buffer, pH 7.4, containing 0.5 M NaCl, 1 mM $CaCl_2$, and 1 mM $MgC1_2$.
6. Poly-Prep® column (Bio-Rad).
7. TTBS*: 20 mM Tris-HC1 buffer, pH 7.4, containing 0.5 M NaCl, 1 mM CaCl2, 1 $_{MM}$ $MgC1_2$, and 0.1% Tween-20.
8. Methyl α-D-mannopyranoside (Sigma Aldrich) (store at room temperature).

Identification of O-GlcNAcylated Glycoproteins

1. Miracloth (Calbiochem, Meudon, France).
2. WGA buffer: 20 mM Tris-HC1, pH 7.8, containing 150 mM KC1, 2 mM $CaCl_2$, 10 *MM* $MgCl_2$, 1 mM ditiothreitol (DTT), and protease inhibitors.
3. Protease inhibitor cocktail tablets (Roche Applied Science) (store at 4°C).
4. Amicon Bioseparation YM-10 (Millipore, Saint Quentin en Yvelines, France).
5. Nonidet P40 (Sigma-Aldrich).
5. PNGase F: peptide-N-glycosidase F from *Flavobacterium meningosepticum* (Roche Applied

Science) (store between –15°C and –25°C). 6. WGA-agarose (Sigma-Aldrich) (store at 2–8°C).

7. GlcNAc (Sigma-Aldrich).

METHODS

Is My Protein a Glycoprotein?

Only one approach is available to answer the question globally: is my protein a glycoprotein? It necessitates the use of a general sugar detection kit that allows the detection and quantification of glycoproteins on blot.

All the reagents necessary for such a determination are provided in a detection kit, the DIG Glycan Detection Kit commercialized by Roche Applied Science, which also includes the manufacturer's instructions. It should be noted that the only information the experimenter will get using this kit is that the proteins do bear at least one oligosaccharide moiety. No information will be provided about the type of glycan involved or its linkage to the protein backbone.

How Is My Protein Glycosylated?

Several strategies can be initiated to determine the nature of the glycan borne by a glycoprotein. The first one involves the use of probes specific for glycoprotein oligosaccharide moieties in Western blotting experiments. These probes are of two types, lectins and antibodies.

This strategy requires neither glycoprotein sample purification nor glycan cleavage or separation. Another strategy is to analyze the linked glycan directly by its monosaccharide composition, after the glycoprotein isolation.

Alternatively, the glycan can be selectively cleaved from the purified glycoprotein by chemical or enzymatical treatment. The glycoprotein is then analyzed by electrophoresis or mass spectrometry prior to and after deglycosylation elsewhere in this Chapter.

The selective cleavage will inform the experimenter about the nature of the linked glycans. In parallel, the protein mass difference observed by mass spectrometry will give information on the structure of the linked glycans.

Detection of Glycans by Western Blotting

The detection of glycans borne by glycoproteins can be performed after Western blotting of either 1D or 2D electrophoresis gels. The probes used for this type of detection are mostly specific for N-linked glycans.

Affinodetection with Lectins

Lectins are plant proteins able to bind oligosaccharide moieties specifically. They have been characterized on the basis of their affinity for mammalian glycoproteins. Only a few of them are available for the detection of plant glycans. They are presented in Table elsewhere in this chapter.

These lectins are available commercially in a biotinylated form. The recognition event between the lectin and the oligosaccharide is visualized on a blot using a streptavidin-peroxidase conjugate. An exception is the lectin concanavalin A, which is able to bind directly to peroxidase. The two protocols are detailed here.

The lectin-biotin/streptavidin-peroxidase procedure:

1. Separate the proteins to be studied by 1D or 2D electrophoresis and transfer onto a nitrocellulose membrane.
2. Saturate the blot with TTBS buffer for 1 h *(see* **Note 3).**
3. Incubate the blot in TTBS containing the lectin-biotin complex (0.1 mg/20 mL) for 2 h at room temperature.
4. Wash four times in TTBS, for 15 min each.
5. Incubate the blot with a streptavidin-peroxidase conjugate diluted 1:1000 in TTBS for 1 h at room temperature.
6. Wash the blot in TTBS (4 × 15 min) and once in TBS prior to development.
7. Prepare extemporaneously the peroxidase development mixture by dissolving, in a beaker, 30 mg of 4-chloro-1-naphtol in 10 mL cold methanol (–20°C), and mixing in another beaker 30 µL $_{H2O2}$ with 50 mL TBS. Mix the two solutions just prior to pouring onto the membrane, and shake gently to optimize the development reaction.
8. Stop the reaction by discarding the development mixture and rinsing the blot several times with distilled water. Dry the membrane and store between 2 sheets of filter paper.
9. Suggested controls for the experiment are presented elsewhere in this chapter.

The concanavalin A (ConA)-peroxidase procedure:

1. Separate the proteins to be studied by 1D or 2D electrophoresis, and transfer onto a nitrocellulose membrane.
2. Saturate the blot for 1 h in TTBS at room temperature.
3. Incubate the membrane in TTBS containing ConA (25 µg/mL) for 2 h at room temperature.
4. Wash the membrane four times for 15 min each in TTBS.
5. Incubate the membrane in TTBS containing horseradish peroxidase (HRP, 50 µg/mL) for 1 h at room temperature.
6. Rinse the blot four times for 15 min each in TTBS and once in TBS for 15 min.
7. Develop the peroxidase reaction as described in the lectin-biotin/streptavidinperoxidase procedure.
8. Controls have to be performed as presented elsewhere in this chapter.

Immunodetection with Specific Antibodies

We have shown that the β1-2 xylose and the a1-3 fucose epitopes of plant N-linked complex glycans are highly immunogenic in rabbits. As a consequence, sera prepared against glycoproteins containing complex glycans generally contain antibodies directed at the β1-2 xylose and/or the α1-3 fucose.

Some commercially available immunsera present the same characteristics as our home-made probes. For instance, immunsera directed against HRP can be used as a probe specific for plant complex N-glycans with α1-3 fucose and β1-2 xylose residues.

More specific is the immunserum raised against honeybee venom proteins, which can be used as a specific probe for α1-3 fucose-containing plant complex glycans. We also raised in our laboratory rabbit antibodies specific for the terminal antennae of complex N-glycans, the trisaccharide Lewis a. Anti-Lewis a monoclonal antibodies are commercially available, although they are very expensive.

AGPs and extensins can be detected using monoclonal antibodies. As we do not use these antibodies in our lab, the experimenter should refer to the initial publications describing the specificities of these antibodies for further details.

1. Separate the proteins by 1D or 2D electrophoresis and transfer onto a nitrocellulose membrane.
2. Saturate the blot with 3% gelatin prepared in TBS for 1 h at room temperature.
3. Incubate the blot in TBS containing 1% gelatin and the immunserum at a convenient dilution for 2 h at room temperature. We used the commercial immunsera described elsewhere in this chapter at the following dilutions:
   a. Anti-HRP rabbit immunserum, 1:300.
   b. Anti-honeybee venom rabbit immunserum, 1:200.
   c. Anti-Lewis a mouse monoclonal antibody, 1:100.
4. Wash the blot with TTBS buffer four times for 15 min each.
5. Incubate the blot in TBS containing 1% gelatin and the suitable conjugate, i.e., a goat anti-rabbit IgG conjugate coupled to HRP at a dilution of 1:2000, or HRP conjugated goat antibodies directed at mouse polyvalent immunoglobulins at a dilution of 1:500, for 1 h at room temperature.
6. Wash the blot four times with TTBS for 15 min each and once in TBS prior to development.
7. Develop the peroxidase reaction as described for the lectin/biotin-streptavidin/ peroxidase procedure.

Glycan Analysis by its Monosaccharide Composition

The analysis of the glycan monosaccharide composition is carried out by hydrolysis with methanol-HCl, followed by derivatization and analysis of the resulting monosaccharide(s) using gas chromatography. The monosaccharide composition obtained from this experiment gives preliminary data on the type of sugar linked to the protein (O- and/or N-linked oligosaccharide).

1. Freeze-dry the sample containing 1 mg of purified protein into a glass reaction tube equipped with a Teflon-faced screw cap.
2. Add 5 to 10 μL of a 2 mM inositol stock solution to the protein under analysis. Then freeze-dry the sample again prior to methanolysis.
3. Add 500 μL of 1 N methanol-HCl to the sample, close the tube tightly with the Teflon-faced screw cap, and place the tube overnight at 80°C.

4. Dry the sample under nitrogen atmosphere after methanolysis while heating the sample at 40°C. **Caution:** Work in a hooc because of the methanol toxicity.
5. Wash with 250 µL of methanol. D·y under nitrogen atmosphere as described elsewhere in this chapter and repeat this washing step twice.
6. Resuspend the sample in 250 µL of methanol. Then add 25 µL of acetic acid, 25 µL of pyridine, homogenize, and incubate for 6 h at room temperature. Dry under nitrogen atmosphere as elsewhere in this chapter.
7. Add 250 µL of silylation reagent and incubate at 80°C during 20 min. Air-dry.
8. Wash with 1 mL of cyclohexane. Air-dry and resuspend in 200 tL of cyclohexane, homogenize, and centrifuge befcre transferring 100 µL of the derivatized sample in a reaction vial with screw cap.
9. Set the gas chromatography apparatus to 1.4 bars with helium. Prior to injection, heat the injector and FID detector to 250°C and 280°C, respectively. Set the helium pressure to 20 psi and its flow rate to 3 mL/min. Equilibrate the column temperature to 120°C prior to the injection of 5 µL of the derivatized sample (equivalent to 25 tg of protein). Eluate the derivatized monosaccharides by a temperature gradient: (1) +10°C per minute until 160°C; (2) +1.5°C per minute until 235°C; (3) 235°C for 4 minutes; (4) +20°C per minute until 280°C; and, finally, (5) 280°C for 3 min.

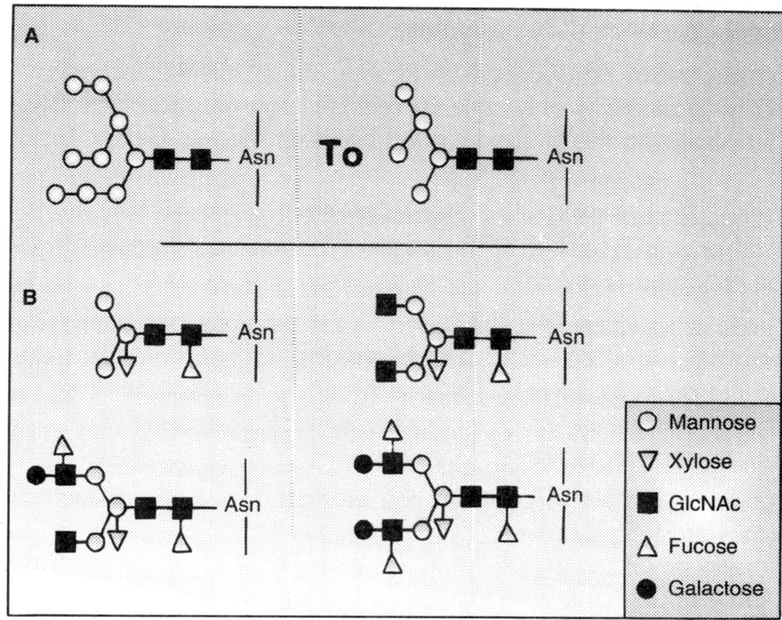

Figure 16.3: Structures of N-linked oligosaccharides in plants. High-mannose type Nglycans are composed of the core $Man_3GlcNAc_2$ substituted by two to six mannose residues (A). Complex type N-glycans present a (31-2 xylose and and/or a1-3 fucose linked to the core. Some of them also carry terminal antennae consisting of a single GlcNAc or a Gal(31-3(Fuca1-4)GlcNAc Lewis a trisaccharide (B).

Analysis of the Protein after Glycan Release

Chemical Treatment for Glycan Release

O-linked glycans can be selectively cleaved from glycoproteins by reductive amination. The deglycosylated protein will be then analyzed by 1D SDS-PAGE and its migration will be compared with the migration of its glycosylated form. An increase in its electrophoretical mobility is the proof of O-glycan occurrence on the protein.

1. Freeze-dry the purified protein in a glass reaction tube equipped with a Teflonfaced screw cap.
2. Dissolve it into 500 μL of a sodium borohydride solution. Close the glass tube tightly and incubate it overnight at 37°C.
3. Add acetic acid drop by drop to stop the reductive amination reaction until the gas production finishes. Add then 500 μL of 10% acetic acid dissolved in methanol. Air-dry under a hood. Repeat the washing step three times. This will coevaporate excess of borates.
4. After treatment, the deglycosylated protein is precipitated overnight at –20°C, by 4 vol of ethanol, and dissolved in a suitable buffer for 1D SDS-PAGE. In parallel, the glycans contained in the ethanol supernatant can be recovered and analyzed for monosaccharide composition.

Enzymatic Treatment by N- or O-Glycosidases

Further information can be deduced by treating the purified glycoprotein with glycosidases. Endoglycosidase H (Endo H) is only able to release highmannose type N-glycans from plant glycoproteins by hydrolyzing the glycosidic bond between the two GlcNAc residues on the core of the N-glycan.

Peptide N-glycosidases (PNGases) hydrolyze the bond between the Asn of the peptide backbone and the proximal GlcNAc of the oligosaccharide part, for both high-mannose type N-glycans and complex type N-glycans.

However, PNGase F, which is widely used in the analysis of mammalian glycoproteins, is active on high-mannose and complex plant N-glycans, except those presenting an α1-3 fucose residue linked to the proximal GlcNAc. PNGase A is able to release all types of plant N-glycans, but it is almost only efficient on glycopeptides and necessitates the proteolytic digestion of the glycoprotein prior to deglycosylation .

The deglycosylation protein or peptides can be analyzed for (1) an increased electrophoretic mobility, (2) the loss of glycoprotein reactivity on blots with glycan-specific probes after Endo H or PNGase F treatment, or (3) mass spectrometry analysis after Endo H, PNGase F, or PNGase A treatment. The same approach and strategies can be performed by using O-glycosidases but are rarely used in our lab and will not be detailed here.

Deglycosylation with endoglycosidase H:

1. Prior to enzyme deglycosylation with Endo H, denature the purified protein by heating it to 100°C for 5 min in the presence of 1% SDS *(w/v).*

2. Dilute the sample fivefold with 150 mM sodium acetate, pH 5.7, and add then 10 mU Endo H. Incubate the mixture at 37°C for at least 6 h.
3. After the deglycosylation reaction, add an equal volume of twice concentrated electrophoresis sample buffer if the result of the digestion is followed by electrophoresis gel, affino-, or immunodetection. The sample can also be desalted and analyzed by mass spectrometry.

Deglycosylation with PNGase F:

1. Dissolve the purified protein to be digested in 0.1 mM Tris-HCl, pH 7.5, containing 0.1% SDS.
2. Heat the sample for 5 min at 100°C to denature the protein. Let it cool down at room temperature and add an equal volume of 0.1 mM Tris-HCl, pH 7.5, containing 0.5% Nonidet P-40.
3. Incubate the sample with PNGase F (1U enzyme for 100 µg protein) at 37°C during 24 h.
4. After digestion, precipitate the deglycosylated protein overnight at –20°C, by 4 vol of ethanol.
5. Centrifuge the sample. Recover the deglycosylated protein in the pellet and dissolve it in a suitable buffer for gel electrophoresis or mass spectrometry analysis.

Deglycosylation with PNGase A:

1. Dissolve 100 µg of purified protein in 500 µL of 10 mM HCl, pH 2.2, and add 10 µg of pepsin. Incubate the sample for 24 h at 37°C and add again the same amount of pepsin. Continue the digestion for additional 24 h. Stop the reaction by heating for 5 min at 100°C.
2. Cool the sample down and take 10% of the solution to purify the peptide and glycopeptide mixture on a C18 column according to the following steps.
3. First rinse the C18 column with 5 mL of acetonitrile followed by 5 mL of water. Apply the diluted sample to 1 mL final volume with distilled water. Rinse the column with 5 mL of water to remove salts from the sample, and elute the peptides that bind to the column by using 5 mL of acetonitrile. Concentrate the peptide sample by evaporating the acetonitrile. Analyze the mass of the peptides and glycopeptides by MALDI-TOF mass spectrometry, to get the glycopeptide masses prior to PNGase A treatment.
4. Freeze-dry the sample left over and dissolve it in 500 µL of 100 mM sodium acetate, pH 5.5.
5. Incubate at 37°C for 18 h with 0.1 mU PNGase A. After digestion, freeze-dry the sample and separate the peptides from the released oligosaccharides using a C18 column, as explained above in this chapter. Analyze the deglycosylated peptides by MALDI-TOF mass spectrometry. The comparison of the two spectra, obtained before and after the deglycosylation treatment, provides a mass difference for some ions owing to the removal of N-linked glycans. This mass difference, in addition to the enzyme used for the deglycosylation procedure and the knowledge of plant glycosylation, indicate which type of glycan is N-linked to the protein and also give an idea of the structure of this glycan.

All the reagents necessary for such a determination are included in a detection kit, the Enzymatic deglycosylation Kit commercialized by Prozyme. It should be noted that, using this kit according to the manufacturer's conditions, the experimenter can obtain information about the type of glycan (N- and O-linked) borne by the glycoprotein under study.

Where Is My Protein Glycosylated?

This approach will allow the experimenter to get information about both the distribution of the glycans on the protein backbone and the structures of the glycans themselves. These experiments can be performed on a purified glycoprotein or on a protein isolated from a 2D or 1D electrophoresis gel.

First, the protein is digested by endoproteases, and the resulting peptide and glycopeptide mixture is separated by high-performance liquid chromatography (HPLC). The presence of glycans within the collected fractions is analyzed by sugar composition. The glycopeptide fractions are then analyzed by MALDI-TOF mass spectrometry prior to and after PNGase A digestion.

1. Dissolve 1 mg of purified protein in 500 µL of 50 mM ammonium bicarbonate, pH 8.0, and heat for 3 min at 100°C.
2. Add 50 µg of TPCK-treated trypsin from bovine pancreas and carry out the digestion for 2 h, at 37°C.
3. Add a second aliquot of 50 µg of the enzyme and incubate for additional 2 h at 37°C.
4. To perform a double digestion, dissolve 50 µg of TLCK-treated chymotrypsin in 50 mM ammonium bicarbonate, pH 8.0, and add them to the trypsin-digested sample. Incubate the sample at 37°C for 2 h.
5. Incubate for 5 min at 100°C to stop the protease digestion.
6. Separate peptides and glycopeptides generated by the trypsin/chymotrypsin digestion by reverse-phase HPLC (C18 column) using a 60-min linear gradient from 0 to 60% of solvent B and a flow rate of 1 mL/min with eluant peaks monitored by UV at 214 nm. Solvents are A: 90:10 water/acetonitrile containing 0.1% TFA, and B: 10/90 water/acetonitrile containing 0.1% TFA. Collect 2-mL fractions and freeze-dry them.
7. Analyze an aliquot (10%) of each collected fraction for sugar composition and select the fractions containing oligosaccharides.
8. Analyze the HPLC fractions containing the glycopeptides by MALDI-TOF to obtain the glycopeptide masses.
9. Digest the glycopeptides with PNGase A as previously described. Separate the peptides from their oligosaccharide moiety on a C18 column as described above.
10. Analyze the resulting deglycosylated peptides by MALDI-TOF mass spectrometry to determine their masses. These masses will allow the experimenter to identify the occupied glycosylation site(s) of the studied protein. The mass difference observed between glycosylated and deglycosylated peptides will provide structural information about the glycans borne by the protein.

This procedure can also be performed for proteins with unknown sequences. In this case, the deglycosylated peptide analysis will be done by LC-MS/MS, to identify its amino acids sequence. Alternatively, the released glycans can be purified and analyzed by MALDI-TOF using specific procedures described before.

How Can the Whole Glycoproteome Be Analyzed?

The identification of glycoproteins contained within a plant extract implies the determination of (1) which genes encode the glycoproteins; (2) which sites, among their potential glycosylation sites, are actually glycosylated; and (3) what is the nature or the structure of the glycans borne by these glycoproteins. All these steps are referred now to as *glycoproteomics*. Such identifications have already been realized in animal cells, but nothing has been published yet for plant glycoproteome identifications.

The current strategy developed in our laboratory to gain access to plant glycoproteomes relies on their purification on immobilized lectins, their separation by 1D or 2D electrophoresis, and their identification by mass spectrometry. Glycoproteome isolation will be illustrated through two protocols: the purification of secreted glycoproteins bearing high-mannose type N-glycans and the purification of glycoproteins bearing O-GlcNAc.

The protocols described here will only focus on the selection of the glycoproteins and will not detail their identification by electrophoresis and mass spectrometry. For more information on these techniques, see other chapters of this book.

Identification of Glycoproteins with High-Mannose N-Glycans

This protocol was developed in our laboratory for rapeseed, but it can easily be adapted to other plant material or species.

1. Extract the proteins by grinding 6 g of plant material in 50 mL of cold TBS buffer with a mortar and a pestle kept at 4°C Centrifuge the extract successively at 10,000g for 30 min, to remove the insoluble material, and then at 150,000g for 1 h, to pellet the membranes. Estimate the protein concentration of the supernatant, representing the soluble protein fraction, by a Bradford assay. Divide the sample into fractions containing 20 mg of proteins and freeze them until use.
2. Thaw a fraction of 20 mg proteins by placing it at 4°C. Centrifuge it at 10,000g for 30 min and filter on a 0.20-μm membrane to remove all insoluble material. Adjust the final volume to 35 mL with cold TBS, and add CaCl2 and MgCl2 to a final concentration of 1 mM.
3. Resuspend the equivalent of 1 mL of Concanavalin A-Sepharose resin in 50 mL of TBS*. Wash the resin twice with 50 mL of TBS*. Recover the resin after each washing step by settlement or by centrifugation (3500g for 10 min).
4. Incubate the washed resin with the 35 mL sample for 2 h at 4°C on a rotary shaker. Eliminate the nonretained proteins by pouring the resin into a Poly-Prep® column and discarding the flowthrough.
5. Resuspend the resin in 50 mL of TTBS*, before washing it five times with 50 mL of TTBS* and five times with 50 mL of TBS* successively, as described elsewhere in this chapter.
6. Pour the resin in a Poly-Prep° column and eliminate the remaining TBS *. Resuspend

then the resin within the column with 10 mL of 0.3 *M* a-methyl-mannose in TBS*. Keep the preparation at 4°C on a rotary shaker for 1 h.

7. Elute the released glycoproteins and wash the resin with 15 vol of 0.3 *M* α-methylmannose in TBS*. Pool the two fractions. The resulting mixture represents the purified glycoproteins bearing high-mannose N-linked glycans.
8. After an estimation of protein concentration by a Bradford Assay, precipitate the glycoprotein sample overnight in the presence of 12.5% trichloroacetic acid (final concentration) at 4°C. Pellet the proteins by centrifugation at 10,000g for 15 min, wash the pellet with acetone/water (9:1 [v/v]), and centrifuge again. Repeat the washing step two more times, and solubilize the proteins in a suitable sample buffer for 2D electrophoresis analysis.
9. Separate the glycoprotein sample in a 2D gel, and stain it by colloidal blue staining. Collect each observable spot individually. Digest the collected spots by trypsin. Analyze these digests by LC-MS/MS. Submit the obtained peptide sequences to a blast search for "short, nearly exact matches" against the nonredundant database of NCBI, with a limitation to the organism used as a possible option.

Once identified, the glycoproteins should be checked for two points: (1) are they potentially secreted, i.e., do they present a potential signal peptide? and (2) do they bear at least one potential N-glycosylation site? The putative subcellular location of the glycoproteins can be estimated in silico using softwares such as Predotar, Target P, or iPSORT. The occurence of N-glycosylation sites within a protein can be determined by matching this protein against Prosite.

Identification O-GlcNAcylated Glycoproteins

The strategy involved in the identification of the O-G1cNAc glycoproteome requires glycoprotein purification by affinity chromatography on immobilized WGA. Previous experiments have shown that WGA is able to bind both O-GlcNAcylated proteins and proteins with N-linked glycans.

Therefore, a preliminary elimination of N-linked glycans is an absolute necessity to purify 0-GlcNAcylated glycoproteins selectively on the affinity column. This deglycosylation step is performed by PNGase F, an enzyme able to remove high-mannose and complex N-glycans.

However, its effect is limited on plant complex N-linked glycans, as explained above, owing to the presence of an αl-3 fucose linked to the proximal G1cNAc of the N-glycan core. To overcome this restriction, it is necessary to work on plant material in which N-glycans are kept in a high-mannose form, sensitive to PNGase F.

This protocol has been developed in our laboratory using cultured cells of the A. *thaliana cgl* mutant, in which N-glycans are matured up to the $Man_5GlcNAc_2$ structure only. This protocol should be applicable to other plant materials, as long as their N-glycosylation is restricted to high-mannose N-linked glycans.

1. Filter 150 mL of cultured cells on Miracloth before incubating them in 150 mL of 0.5 M NaCl for 30 min at 4°C, to solubilize the proteins ionically bound to the cell wall.
2. Filter the cells again on Miracloth and discard the supernatant. Grind the cells in 30 mL of 20 mM Tris-HC1, pH 7.8, containing 150 mM KC1, 2 mM $CaC1_2$, 10 mM $MgC1_2$, 1 mM DTT and protease inhibitors (WGA buffer).

3. Eliminate the insoluble material by centrifuging the extract at 5500g for 15 min at 4°C and at 25,000g for 30 min at 4°C, successively. Aliquot in 5-mL fractions.
4. Concentrate a 5-mL fraction using Centricon (Amicon Bioseparation YM-10) to a 0.5 mL final volume.
5. After concentration, adjust the sample to 0.1% SDS and incubate at 100°C for 5 min to denature the glycoproteins. Add Nonidet P40 to a final concentration of 0.5% to complex the SDS. After denaturation, deglycosylate the proteins by adding PNGase F (5 U) and incubate for 24 h at 37°C under agitation. During this step, only the N-linked glycans will be eliminated.
6. Dilute the deglycosylated sample 20 times in WGA buffer and incubate it with 100 tL of WGA-agarose for 4 h at 4°C on a rotary shaker. Discard the nonretained proteins and wash the resin with 20 column volumes of WGA buffer.
7. To elute the bound material, incubate the resin with 5 to 10 vol of WGA buffer containing 0.5 M G1cNAc for 30 min at 4°C. After collection of the eluate, repeat the same treatment once more. Pool the two fractions eluted from the column. This glycoprotein fraction only contains O-G1cNAcylated proteins
8. Precipitate the glycoproteins in the presence of 12.5% trichloroacetic acid, final concentration, overnight at 4°C. Pellet the proteins by centrifugation for 15 min at 10,000g. Wash the protein pellet three times with acetone/water (9:1 [v/v]) and resuspend it in a suitable buffer for 1D electrophoresis.
9. Separate the O-GlcNAcylated proteins in 1D SDS-PAGE gel. After Coomassie blue staining, collect all the bands vizualized in the gel. After digestion by trypsin, identify the proteins by MALDI-TOF or LC-MS/MS.

CONCLUSIONS AND PERSPECTIVES

Using the methods outlined in this chapter, information about the presence, the linkage, and the composition of glycans that are constitutive of a plant glycoprotein can be obtained. The strategies discussed here are already efficient to identify N-linked glycans, whereas analysis of O-linked glycans calls for further efforts. Even when the physicochemical characterization of a plant glycan is deciphered, it is difficult to predict a priori the function(s) mediated by a given oligosaccharide or a given glycoprotein.

Indeed, the same oligosaccharide sequence may present different functions when it is carried by different glycoproteins, or found at different locations within a plant, or at different times of the plant life cycle. Therefore, the still emerging era of glycoproteomics opens new avenues that will shed lights on the roles of N- and O-linked glycans at an increasing pace. Closing gaps in this field will expand our understanding of plant physiology and also offer wider perspectives in biotechnology and medicine.

Notes

1. Use fresh ampules of 3 N methanolic-HCl and silylation reagent for each experiment.
2. The experimenter should be careful in the blot interpretation, as WGA recognizes GlcNAc

in N-linked glycans, as well as O-linked GlcNAc.

3. The solution used for blocking binding sites on the nitrocellulose needs to be devoid of glycoproteins. This is why we recommend using Tween-20 to coat the nitrocellulose (30) in this procedure.
4. Controls for specificity:
   a. It might be advisable to blot the proteins cited in Table elsewhere in this chapter on the membrane ("Suggested positive controls"), to get positive controls for the affinodetection.
   b. Lectin binding specificity must also be checked by running affinodetection in the presence of 0.3 M inhibitory sugar.
5. The available anti-O-linked glycan antibodies are: anti-AGP antibodies: LM2, JIM4, JIM13, and JIM15, JIM8, JIM14, JIM15, and JIM16; anti-extensin antibodies: LM1, JIM11, JIM12, and JIM20, JIM 19.
6. Control for N-glycan specificity of the immunodetection: it might be wise to check for the specificity of sera toward N-glycans attached to the protein studied. This can be done by performing a mild periodate oxidation on the blot prior to immunodetection. Mild periodate treatment oxidizes glycans and abolishes any recognition of the glycoprotein by the anti-glycan antibodies. Any remaining signal will be the consequence of a protein backbone antibody/recognition.
   a. After saturation with gelatin, incubate the blot in a 100 mM sodium acetate buffer, pH 4.5, containing 100 mM sodium metaperiodate for 1 h in the dark at room temperature, changing the incubation solution after 30 min.
   b. Incubate the blot in PBS containing 50 mM sodium borohydride for 30 min at room temperature.
   c. Rinse the blot with TBS, saturate for 15 min with TBS containing 1% gelatin, and perform the immunodetection, as described elsewhere in this chapter.
7. Controls for fucose or xylose specificity: some proteins can be used as positive controls for the N-glycan immunodetection. Phospholipase A2 from honeybee venom contains the α1-3 fucose residue, and is devoid of β1-2xylose. PHA-L and recombinant avidin produced in maize are glycoproteins containing both β1-2 xylose and α1-3 fucose.
8. The protocol described calls for 1 mg of protein but can be adapted to a smaller amount.
9. Inositol is used as an *internal* standard.
10. Because reductive amination is a drastic treatment, some modifications of the protein may occur. For this reason, it is sometimes better to eliminate O-linked *glycans* by *enzymatic* digestion.
11. In general in our lab, we do not use chemical treatments to release N-linked glycans from glycoproteins.
12. The role of Nonidet P40 is to complex free SDS.
13. Sixty to 80 mg of soluble rapeseed proteins were necessary, as a starting material, to

prepare enough glycoproteins for the loading of one 2D gel.

14. α-Methyl mannose is a ligand of concanavalin A and will displace the bound glycoproteins from the immobilized lectin.
15. We observed an important release of concanavalin A while eluting the affinity chromatography column. This release polluted the glycoprotein preparation and obliged us to run, in parallel to the analytical 2D gel, another 2D gel loaded with concanavalin A alone. After staining, we selected the spots present only in the analytical gel and discarded the ones resolved simultaneously in the two 2D gels.
16. One hundred and fifty milliliters of 6-day-old *cgl A. thaliana* cultured cells roughly represent 10 g of plant material.
17. The elimination of cell wall-bound proteins represents a first purification step, as these proteins do not bear O-GlcNAc. This *purification* step should be included in the *protocol* whenever possible.
18. Free GlcNAc is a ligand of WGA and will displace the bound *glycoproteins* from the *immobilized lectin.*

Chapter 17

INSECTICIDES

Resistance to pesticides has evolved, not only in insects, but also in acarines, weeds, fungi, bacteria, rodents, and other pests. The phenomenon of resistance is due to Darwinian evolution at a rate that is accelerated by the intensity of selection from pesticides. Insects present many difficult challenges related to resistance in that they are especially well adapted for dispersal through flight, they have a high reproductive capacity with multiple generations per year, and they are often difficult to detect in products traded around the world, such as grain, flowers, or even used tires.

Resistance to insecticides is defined as a significant reduction in the response to an insecticide in a population of one insect species. Although more difficult to document, many consider that this reduction in response should be accompanied by a significant loss of practical control of the pest in the field. The reduction in response should be proven with susceptibility tests conducted with samples of the population in question and compared statistically with a standard colony, with baseline susceptibility values, or with other populations that are controlled with normal response.

The situation often arises that a grower notices an apparent lack of control from an insecticide and reports this to a crop advisor or government extension agent. The problem field is inspected, information is gathered on the application of the insecticide, and samples of the insects are taken for susceptibility testing. Resistance cannot be documented from only observation of control in the field because there are many variables other than resistance that can reduce performance of the insecticide, e.g., equipment main function, unusual weather, decomposition of the insecticide or formulation, etc.

It is a combination of inadequate control plus statistically significant reduction in susceptibility that is required for the documentation of resistance in a practical sense. A recent case history of such a situation is the documentation of pyrethroid insecticide resistance in the coton bollworm, *Helicoverpa zea*, in South Carolina. Inadequate control from repeated applications of cyhalothrin and deltamethrin was reported by a crop consultant to an extension agent who made inspections of the suspect fields of bollworms.

Samples from the problem area were collected, cultured, and subjected to detailed susceptibility testing. Results were compared with a standard colony and with archived baseline data to demonstrate 35-fold resistance to cyhalothrin. Subsequent surveilance exposing pheromone-trapped males to discriminating doses revealed pyrethroid resistance in three locations, including the original problem area. Growers were advised of the resistance, and alternative methods of control were recommended.

BACKGROUND LITERATURE

Various edited books with multiple contributors have addressed the topic of resistance to insecticides, the most recent and comprehensive of which are.

Status

Resistant species include pests of agriculture and of public health in approximately equal numbers. It has been estimated that over 600 species of insects and other arthropods have developed resistance to insecticides.

This count, which was reported about twice a decade, was based primarily on published records of susceptibility test results, some of which probably lacked collaborative documentation of field failures. A more conservative estimate, in which resistance had been documented repeatedly and was widespread for each species, placed the count at about 50 species.

By either account, it seems clear that resistance to insecticides has been increasing. By the more liberal standard, it should be expected that a saturation in the number of resistant species will be reached; however, species that are already listed continue to evolve resistance to new classes of insecticides.

Biochemical and physiological factors leading to resistance with one insecticide often apply to resistance toward other insecticides. Certain major pest species have evolved resistance more quickly than replacement insecticides have been invented and developed. Among agricultural pests, notoriously resistant pests include spider mites, aphids, whiteflies, pear psylla, diamondback moth, Colorado potato beetle, red flour beetle, cotton bollworm, and tobacco budworm.

The latter two pests, proven to be difficult to manage in cotton, are among the heliothine Lepidoptera, for which a comprehensive review on resistance worldwide is available. It is common that the principal pest of a crop has evolved resistance to at least one insecticide. Most resistant among pests of health concern are mosquitoes, house fly, sheep blow fly, horn fly, cattle tick, German cockroach, head louse, cat flea, and dog flea.

On the other hand, fortunately, there seem to be a few major pests that have little capacity to become resistant, e.g., boll weevil to azinphos methyl and pecan weevil to carbaryl, but this list

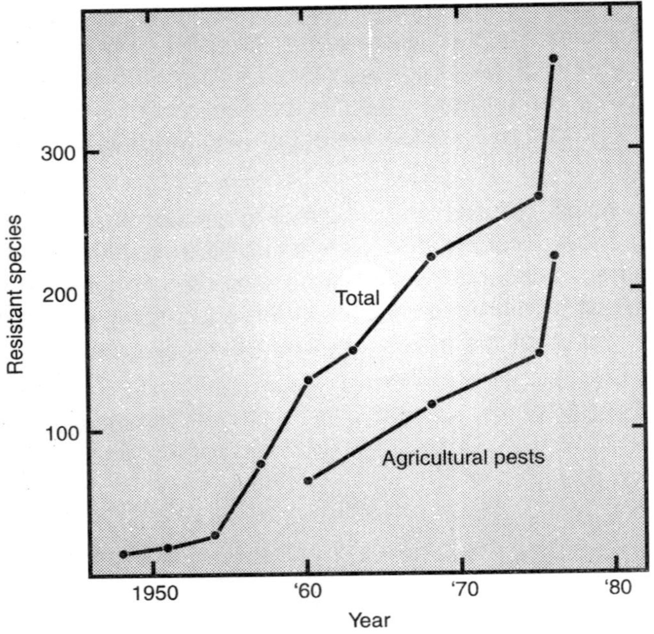

Figure 17.1: Accumulated number of arthropod species having evolved resistance to insecticides and acaricides; solid circles represent the total number resistant species, including agricultural pests and pests of human health; open circles represent the number of resistant species of agricultural and veterinary pests.

is very limited. Species that have evolved resistance are more numerous among the orders Diptera, Lepidoptera, Coleoptera, and the class Acarina, than among other taxa of insects. These are some of the more highly evolved and successful orders.

Conspicuously few species are resistant within the very highly evolved order Hymenoptera, which includes many social species. This may be due in part to the relative isolation of the reproductive individuals of the social insects; mortality to the more vulnerable worker and forager castes does not result in selection unless it results in indirect impact on colony reproduction.

Also, with the exception of many ant species, most Hymenoptera have not been exposed to many insecticides because they are not pests. Resistance has developed to nearly every chemical class of insecticide, some more rapidly than others. Especially among the Diptera, cyclod-iene insecticides induced a very rapid evolution of resistance.

A similar fate befell DDT and similar organochlorine insecticides, whereas resistance to organophosphorus and carbamate insecticides developed more gradually in general. To some degree, relative persistence in the environment was a likely contributor to the rate of resistance, in that DDT and cyclodienes were very persi-stent, as well as remarkably insecticidal, so that one application could select repeated generations of the targeted pest, whereas organophosphrus and carbamate insecticides were reactive and degraded to noninsecticidal products in the environment.

Of course, chemical persistence and toxicity are factors that must be coupled to the genetic diversity of the selected population because rare alleles are the likely source of evolution, as will be discussed. Each new class of synthetic organic insecticides has been met by practical resistance within 10 to 30 years of introduction. In succession, cyclodienes, DDT and its analogs, organophosphorus insecticides, carbamate insecticides, and pyrethroid insecticides have all been reduced in efficacy due to the evolution of resistance in major target species.

Within each class, more potent chemistry has been brought forth to meet this phenomenon, but eventually, that class had to be replaced. Currently, it is the synthetic pyrethroids that are being are reducing pyrethroid selection, but pyrethroids continue to be used versus alternatives based on short-term economic decisions with not enough regard for their long-term value.

The current focus of regulatory agencies is directed " plant pesticides," which are toxins of *Bacillus* thuringienesis expressed in engineered crops transformed with a modified bacterial gene. Since introduction in May 1995 in potato, *B. thuringiensis* toxin has replaced a significant portion of conventional insecticide application, especially in cotton and maize.

Of course, this new technology of insecticidal proteins in crops exerts selection against the pest population and it should be anticipated that evolution of resistance will follow, just as it has followed the introduction of previous chemical insecticides.

A monitoring program is in progress, and a resistance management plan was established and implimented along with the introduction of *B. thuringiensis* toxin-transformed ("Bt") crops (details of resistance management plan are discussed under Resistance Management Plans).

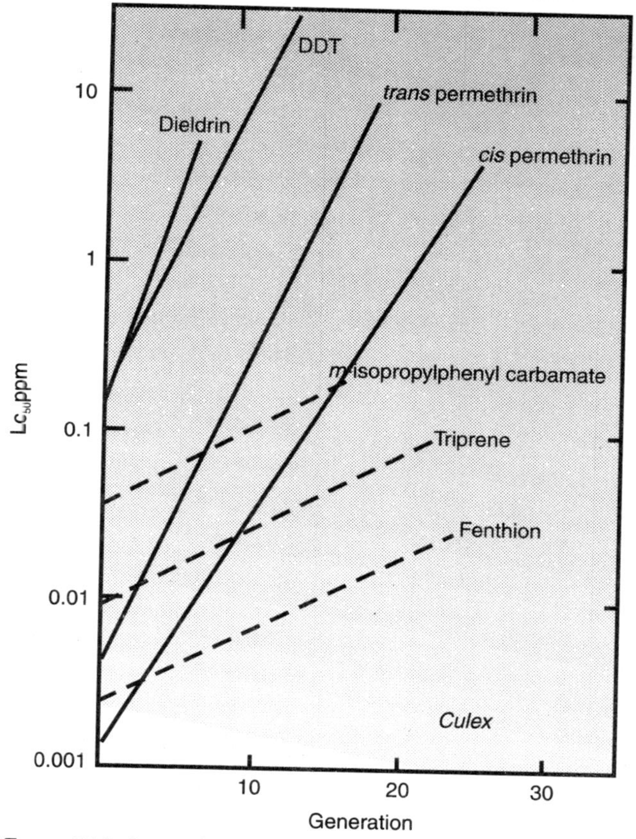

Figure 17.2: Rate of induction of resistance by experimental selection of Culex spp. larvae with insecticides; permethrin from Preiester (1978); triprene from Brown et al. (1978); other insecticides from Tadano and Brown (1966).

Resistance has evolved to conventionally applied *B. thuringiensis* toxin, particularly in the diamondback lost and replaced by a variety of less broad spectrum new chemistries, natural products, and, most importantly, transgenic insecticidal crops. However, many insecticides are lost for economic or regulatory reasons, rather than because of resistance.

The progressive loss of synthetic pyrethroids, unless it can be reversed, may be the most damaging to the technology of insect control. These mimics of natural pyrethrins possess the insecticidal potency and broad spectrum of activity of the earlier classes of synthetic organic insecticides without the detrimental environmental persistence of the organochlorines and without the extreme mammalian toxicity of the anti-acetylcholinesterases (organophosphorus and carbamate chemistry). The present status of increasing resistance to the extraordinary pyrethroid insecticides should stimulate new tactics to preserve them rather than a resignation to their loss, as was the experience with the organochlorines.

Alternatives are available, but careful management would be needed to rescue the pyrethroid class, even if it is not already too late. Basically, the alternatives must be thrust forward to spare

Table 17.1: Arthropod Pests in Which Evolved Resistance Has Led to Serious Difficulty in Control in Many Areas

Order	Species	Common name
Acarina	*Amblyomma* spp.	Ticks
	Boophilus spp.	Cattle ticks
	Panonychus citri	Citrus red mite
	Panonychus ulmi	European red mite
	Rhipicephalus spp.	Ticks
	Tetranychus spp.	Spider mites
	Anoplura *Pediculus capitis*	Head louse
Coleoptera	*Leptinotarsa decemlineata*	Colorado potato beetle*
	Oryzaephilus surinamensis	Saw-toothed grain beetle
	Sitophilus oryzae	Rice weevil
	Tribolium castaneum	Red flour beetle
Diptera	*Aedes aegypti*	Yellowfever mosquito
	Anopheles spp.	Malaria mosquitoes
	Culex quinquefasciatus	Southern house mosquito
	Haemotobia irritans	Horn fly*
	Lucilia cuprina	Sheep blow fly
	Musca domestica	House fly*
	Simulium damnosum	Black fly
	Stomoxys calcitrans	Stable fly
Homoptera	*Aonidiella aurantii*	California red scale
	Bemisia tabaci	Sweet potato whitefly
	Myzus persicae	Peach-potato aphid*
	Nephotettix cincticeps	Green leafhopper
	Nilaparvata lugens	Brown planthopper
	Psylla pyricola	Pear psylla*
	Trialeurodes vaporariorum	Greenhouse whitefly*
Lepidoptera	*Heliothis armigera*	Cotton bollworm*
	Liriomyza spp.	Serpentine leafminers*
	Plutella xylostella	Diamondback moth*
	Sitotroga cerealella	Angoumois grain moth
	Spodoptera exigua	Beet armyworm
Orthoptera	*Blattella germanica*	German cockroach

the pyrethroids in situations where resistance is developing. To some degree, transgenic crops moth on vegetables and the Indian meal moth in stored grains.

Regarding the most critical pests of transgenic field crops, cotton bollworm appeared to be naturally tolerant to *B. thuringiensis* toxin compared with tobacco budworm, on cotton. Control of tobacco budworm by Bt-cotton has been spectacular, but control of large populations of cotton bollworm has been marginal in some situations, resulting in reenforcing applications of conventional insecticides, often pyrethroids.

In Bt-maize, European corn borer is the major target. This species has been selected experimentally for resistance to Dipel, a formulation of *B. thuringiensis* toxin, but the Diperesistant strain did not survive on Bt-maize, which expressed a very high dose of the toxin protein.

Testing

In order to know the status of resistance and to confirm resistance when suspected by growers, accurate and reliable susceptibility tests must be developed. When intended for widespread surveillance and adoption by crop consultants and extension personnel, these tests must also be convenient and easy to employ in a variety of settings.

Ideally, the same basic test should be employed in both laboratory and on site; with proper planning of the test, this has been achieved for noctuid pests of cotton using insecticide deposits in glass scintillation vials and for lygus bugs using insecticide deposits in glass vials or plastic bags.

Many susceptibility tests have been developed or adopted by the World Health Organization and the Food and Agriculture Organization of the United Nations as standardized methods by which populations of a species of interest can be compared worldwide. A susceptibility test is an experiment designed to estimate the statistical median lethal dose (LD_{50}) of an insecticide and distribution of response (slope of the logio dose versus probit mortality curve) for a population.

Insects are exposed to uniform, random samples of the population to a serial dilution of doses (dosages) under defined environmental conditions for a determined period of time, after which mortality is defined using an easily recognized criterion. If a large number of dosages is employed, mortality is scored at a precise endpoint, then a distribution can be computed from the results and statistics estimated directly from that distribution.

If fewer dosages are employed (there should be at least five dosages resulting in intermediate levels of response), then a theoretical distribution model is applied and the observations tested for goodness of fit to that model, e.g., a probit model, is often employed, which assumes that a normal distribution of responses would result from the exposure.

Confidence limits are computed for the stimated median lethal dose, and that value is compared with that of a standard susceptible colony, to a baseline susceptibility value, or that of other populations, which were controlled. A statistically significant decrease in susceptibility of the test population versus the control has been used as the criterion to declare resistance. It should be recognized that resistance rarely, if ever, results as a pure, new distribution when it arises in the field.

In fact, it usually requires intensive selection in the laboratory to purify resistance to a distribution unique from the control. The assumption of a normal distribution is basically a convenience for analysis, whereas the actual resistant population most likely consists of two or multiple distributions of types of individuals regarding their susceptibilities. (It is extremely difficult to estimate the susceptibility of an individual insect.) It is very obvious whenever biochemical assays are available that, even in the case of a single gene for resistance, populations in nature consist of homozygous-susceptible, homozygous-resistant, and heterozygous individuals.

Detection of just two types in a sample requires a very large number of doses to define a "shoulder" on the calculated distribution or to indicate a deflection of the transformed log10 dose versus the probit mortality curve. Practical considerations of handling many samples of insects, changing doses many times, avoiding the introduction of error as the insects grow during the experiment, and other complications ad to the difficulty in estimating a distribution in responses with high resolution from many doses; therefore, most experiments have been done with five or six dosages and fit to one or more assumed models.

From this method, it is equivocal whether or not a flattening of the loglo versus probit mortality curve is due to the presence of a subpopulation of resistant individuals, and this means that our tests are not optimally sensitive to detect resistance while they serve to detect larger shifts in susceptibility in a rather crude fashion. Another point that might seem straightforward, but in practice can be difficult, is the determination of mortality.

When is an insect dead? At higher dosages, most respectable insecticides applied directly to the insects for a period of 24 h will result in some individuals that are unequivocally dead; i.e., they are immobile, oxidizing, and deteriorating; however, many individuals will be in various stages of progressing symptomology but still alive. Conventionally, a criterion for "death" that is not actually death is sometimes referred to as moribund.

What is important is that this condition, the actual endpoint of the assay, be clearly and unambiguously defined so that any one reading the description of moribund can pronounce the subject "dead" and that the next observer would make the same pronouncement. For lepidopteran caterpillars, moribund has been defined as the inability to translocate across the diet when probed. Actual movement across the medium is a clearer endpoint to score than is movement of a leg, because the act of probing can cause a leg to move in what is an artifact induced by the observer.

This problem can be very serious when attempting to score mortality from *B. thuringiensis* toxin because growth is reduced but death is very slow; determining weight at a set time and setting an endpoint as a threshold weight is a more practical method. Susceptibility tests can be categorized by the method of exposure. Topical application of an insecticide in solution is a very accurate method to ensure that each specimen receives the same dosage.

This method is often used for lepidopteran larvae and many other pests, and it has been the method for surveillance of resistance of *Helicoverpa armigera* in Australian cotton for many years. Various self-dosing methods have been developed in which the insect is placed on a residual deposit of the insecticide. Such is the method of exposure in the adult vial test in which insecticide is deposited in 20-mL glass vials (designed for liquid scintillation counting of radioactivity).

This test was adapted for standardization by the Pyrethroid Efficacy Group in 1978 and later by the Insecticide Resistance Action Committee. One complication of self-dosing methods is that the subject insect might avoid contact with the insecticide by resting on an untreated surface within the apparatus such as a piece of diet, a screened lid, or a piece of cotton wool; this avoidance has been documented and even selected in the case of screened end-resting onion maggots.

It is a practical difficulty to eliminate this possibility in a self-dosing container, but it seems from the precision of results that the subject insect probably moves about in the apparatus sufficiently to acquire a dose and that the acquired dose is more likely a function of the deposited amount than the frequency of movement of the insect.

Once fully developed in the laboratory, a discriminating dose for resistance can be determined as the dose killing 99.99% of a susceptible population by extrapolation of the observed log10 dose versus the probit mortality curve. It is assumed that survivors of this dose are resistant the insecticide because only 1/10,000 of the susceptible population can survive this dose.

This dose, and perhaps one or two other doses based on the knowledge of both the susceptible and resistant population responses, are employed in surveillance programs to gain a more rapid understanding of the situation in the field during the season. Such tests can be coupled with pheromone trapping to accelerate the process, and, in general, the more convenient the method, the more likely a sufficient database can be acquired during the season when collaborators are busy with many other programs.

In Texas and Louisiana, this approach has been very successful for many years with approximately 30,000 moths sampled in each state per year. Baseline susceptibility values are essential as a foundation for resistance management programs. Consulting a database of this information can save valuable time and effort when investigating a case of suspected resistance.

Insecticide susceptibility for mosquitoes and other vectors of human diseases was documented in a comprehensive monograph for the World Health Organization, but there has been no similar single document for agricultural pes with the exception of a review of methods and baseline values for pests of stored grain.

Physiological Basis of Resistance

When the physiological basis of insecticide resistance has been analyzed by comparing various parameters related to insecticide poisoning between resistant and susceptible strains, it was found that most cases of resistance are associated with one or more physiological mechanisms enhancing survival.

In many cases, one resistant strain has possessed two or more physiological mechanisms that may synergize each other to multiply the level of resistance. Behavioural resistance is the avoidance of poisoning due to a behavioural change that reduces exposure to the insecticide, thereby increasing the apparent dose needed to kill.

In methoprene-resistant *Culex pipiens*, uptake of radiolabeled insecticide was reduced due to the delay in resuming filter feeding activity after being disturbed, the labial mouthbrushes remaining still while susceptible larvae resumed a rapid beating of the mouthbrushes to make a current to move particles into the mouth.

In earlier generations of the same selected strain, a lower level resistance was associated with a significant reduction in the frequency of mouthbrush pulsation. Onion maggot adults were resistant to dieldrin due to avoidance of the insecticide, which was detected in laboratory susceptibility tests when the flies rested on the untreated screen enclosures on the ends of the testing cylinders rather than on the dieldrin deposits on the cylinder walls.

Pyrethroid-resistant horn flies in Louisiana avoided insecticide-treated legs on cattle and rested on the belly, unlike susceptible flies. There is little understanding of the genetic control of behaviour, although it appeared that altered behaviour was selected in the examples cited. *Pharmacokinetic mechanisms* of resistance are those mechanisms controlling the movement of the insecticide following exposure, including penetration, distribution, and elimination.

Elimination is considered to be the combined processes of biotransformation of a chemical and excretion of the parent chemical and its biotransformation products. Reduced penetration is a pharmacokinetic change, poorly understood, but apparently under genetic control, which slows entry of the insecticide through the integument, tracheae, or gut.

In some strains of mosquitoes, the insecticide load in the gut was eliminated more efficiently in resistant strains by the sloughing of the peritrophic membrane. In methoprene-resistant *Culex pipiens*, uptake of radioactively labeled methoprene or dieldrin from water through the cuticle was reduced in fourth instars. Reduced rates of penetration have been observed in tobacco budworm and in cattle ticks when radioactively labeled insecticides were applied topically and the cuticle subsequently analyzed by washing off the unpenetrated residual insecticide from the cuticle with solvent. The most commonly documented mechanism of insecticide resistance has been an enhanced detoxication, in which resistant strains transformed insecticides to less insecticidal products at an increased rate compared with detoxication in susceptible strains.

Detoxication is the pharmacokinetic process of biotransformation of the chemical, usually catalyzed by an enzyme, of the active insecticide to a less insecticidal product. This process takes place in the cuticle, fat body, or gut, thereby limiting the amount of insecticide that reaches the physiological target, which is often a protein in the nervous system.

This mechanism has been identified in many resistant strains by analysis of the products formed when a radioactively labeled insecticide is administered. Typically, the resistant strain detoxified the parent insecticide to more polar and less toxic products, which were excreted, and accomplished this at a faster rate than the susceptible strain.

There are many examples of enhanced detoxication among resistant species of insects. In the house fly, it was demonstrated that reduced penetration and enhanced detoxication could be combined by genetic experiments to result in a synergistic elevation of the resistance to diazinon. It is interesting to note that enhanced detoxication as a general physiological resistance mechanism, although very common in insects from the early cases to the present day, has been considered a minor physiological resistance mechanism among herbicide-resistant weeds, in which the altered target site was more commonly found, at least in the earlier cases.

In the case of many organophosphorothioate insecticides, chlorfenapyr and some others, these pharmacoknetics included an initial intoxication in which the parent pro-insecticide was transformed to a more insecticidal compound, which is the bioactivated insecticide. Subsequently, this product was detoxified and excreted.

Reduced intoication was identified as a resistance mechanism in one strain of tobacco budworm from North Carolina, which produced significantly less methyl paraoxon, the bioactivated insecticide, from the application of methyl parathion, the organophosphorothioate pro-insecticide. In this species, there are several known mechanisms of resistance to methyl parathion, reduced intoxication being one of them.

Pharmacodynamic mechanisms of resistance are those mechanisms involved with the interaction of the insecticide with the physiological target, which causes poisoning in the insect, i.e., mechanisms affecting the mode of action of the insecticide.

The majority of conventional insecticides, and some of the newer chemical classes as well (cyclodienes, DDT and its analogs, *organophosphorus insecticides, carbamate insecticides, pyrethroids, chloronicotinyls, spinosads,* and some others) kill by interfering with the chemical or electrical transmission of impuls in the nervous system.

Other physiological systems are targets of a minority of the chemical classes of insecticides, but these include *B. thuringiensis* toxin, which acts in the alimentary tract; a growing group of mitochondrial poisons, which affect respiration in muscles and other tissues; and the insect developmental inhibitors, which are directed against the endocrine system affecting growth and metamorphosis.

An early discovery of target site resistance, the reduced sensitivity of the target protein to the insecticide in resistant strains, began with the observation that some strains of house flies with resistance to DDT we not knocked down when exposed to deposits of DDT, whereas other resistant strains were knocked down, but later recovered.

These particular strains we said to possess knockdown resistance. Subsequently, it was discovered that the mechanism responsible for this resistance was a reduced physiological response in the nerve axon detected as fewer bursts of action potentials elicited by DDT in the knockdown resistant strains.

Following years of pioneering genetic and molecular experimentation, knockdown resistance was associated with a specific mutation in the voltage-gated sodium ion channel protein, which is the molecular target of DDT.

Similarly, it appears that cyclodiene resistance was due in many species to target site resistance, the target being the gamma-aminobutyric acid receptor chloride ion channel, a protein involved in inhibitory post-synaptic signaling.

Another prominent example of target site resistance is altered acetylcholinesterase, the target of bioactivated organophosphorus insecticides and carbamate insecticides, which was discovered in parathion-resistant spider mites.

This enzyme catalyzes the inactivation of acetylcholine in the ganglionic synapses of the insect central nervous system. Organophosphorus and carbomate insecticides react rapidly with this enzyme, resulting in noncatalytic phosphorylated and carbamylated acetylcholinesterase, respectively. These products are slow to reactivate, causing acetylcholine to accumulate with subsequent spontaneous firing of the neuron and eventually death.

Now altered acetylcholinesterase is a widespread physiological mechanism of insecticide resistance and is a factor in resistance in most major pests, including tobacco budworm, Colorado potato beetle, mosquitoes, and many others.

MOLECULAR BIOLOGY RESISTANCE

The biochemical basis of reduced penetration is unknown. A recent report has correlated expression of p-glycoprotein, a type of ATP-binding cassette transporter protein, with insecticide resistance in tobacco budworm, but transport of insecticides was not measured. Resistance due to enhanced detoxication has been attributed to several biochemical mechanisms in which enzymatic detoxication is increased in the resistant strain.

Carboxylester hydrolase is an enzyme that catalyzes the reaction of malathion and water to produce malathion mono- or diacid and ethanol. Catalytic hydrolysis is specific for malathion and due to a very specific hydrolase as purified from resistant aphids and from resistant mosquitoes. In the case of the green peach aphid, a form of carboxylester hydrolase, E4, not only hydrolyzes malathion, but also reacts with non-carboxylester organophosphorus insecticides to yield the phosphorylated E4.

Reactivation of carboxylesterase activity through dephosphorylation of the active site results in turnover of the non-carboxylester insecticide, although this turnover is much slower than for the malathion carboxylester hydrolase reaction. The quantity of carboxylester hydrolase is increased in resistant strains through the mechanism of gene amplification, the presence of hundreds of extra copies of the gene in genomic DNA from the resistant mosquitoes and aphids.

The quantity of hydrolase is increased to such a high concentration that organophosphorus, carbamate, and pyrethroid insecticides can be sequestered in a mechanism that does not require rapid catalysis for resistance. Cytochrome P450 monooxygenase catalyzes a wide variety of reactions of insecticides with oxygen, e.g., armatic hydroxylation, in which a hydroxyl group is added to the diphenyl ether moiety of permethrin and the product is more easily conjugated with a carbohydrate for enhanced excretion.

This is another enzyme for detoxication of insecticides that is often overexpressed in resistant strains. In resistant house flies of the Rutgers strain, there was no increase in the gene *CYP6;* however, resistant flies contained much more messenger RNA, which was likely due to increased transcription from the gene.

The *CYP6* gene was located on chromosome V, whereas increased expression was linked to chromosome II. In Learn-resistant house flies, a different CYP*6* gene was located on chromosome I, whereas increased expression was linked to chromosome II.

Analysis of the *CYP6* genes reveals no mutation compared with the gene in susceptible house flies; however, in the case of Learn, there was an insert of 15 nucleotides of DNA slightly upstream of the *CYP6* gene. Glutathione S-transferases catalyze various reactions with insecticides, in which a chemical group is transferred to the tripeptide glutathione. This enzyme is especially efficient in the O-demethylation of organophorus esters. Increased detoxication via glutathione Stransferase is another common biochemical mechanism of resistance.

As for cytochrome P450, resistance has not been associated with a specific mutation in the protein; therefore, it is likely that increased expression of the gene, or perhaps amplification, may be the cause of resistance. Unlike detoxicative mechanisms, target site resistance has been associated with point mutations in various genes of interest.

In the DDT-resistant house fly, the gene kd*r* for knockdown resistance was proven to cosegregate with a point mutation of leucine to phenylalanine in the IIS6 transmembrane subdomain of the sodium ion channel protein, whereas an additional point mutation was found in the IIS4-5 linker region in the superkdr strain that was more intensely resistant.

In tobacco budworm, the precisely homologous IIS6 codon was mutated in a Louisiana pyrethroid-resistant strain, but the alteration was leucine to histidine rather than phenylalanine. Strain Pyrethroid-R, which had been selected for several years, was fixed for the mutation valine to methionine, but it was in the analogous position in the IS6 subdomain (V421M).

Patch clamp analysis of isolated neurons from Pyrethroid-R moths indicated that there were two physiological differences compared with a susceptible strain: voltage gating was altered in the absence of pyrethroid with the resistant neurons less sensitive to voltage change, and a 10-fold increase in pyrethroid concentration was required to modify the response of the channels to voltage, indicating resistance on the molecular level. Similarly, point mutations have been detected in the gamma-aminobutyric acid receptor chloride ion channel.

The principal mutation seems to be A302G, as discovered in dieldrin-resistant fruit flies and observed in house flies, aphids, and other species; however, this codon was Q in Australian cotton bollworm and was not mutated in resistant strains. A copy of the resistance allele of the gamma-aminobutyric acid receptor and chloride ion channel was introduced into fruit flies on a transposable element to test whether or not this allele caused resistance.

This result was that the transgenic fruit flies we slightly resistant to a series of doses of dieldrin, but was not proven that the result was statistically significant. In the methyl–parathion-resistant strain Woodrow 83 of tobacco budworm, acetylcholinesterase was 36-fold less reactive with methyl paraoxon; however, this altered target was more reactive with certain other insecticides, including monocrotophos.

Five point mutations have been detected among various resistant colonies of fruit flies; unfortunately, the original published sequence from *Anopheles stephensi* mosquitoes, upon which much of the subsequent genetic analyses were based, appears to have been from a second acetylcholinesterase gene that was not genetically linked to insecticide resistance.

Evolution and Dispersal

The evolution of resistance is a special case of Darwinian evolution in which natural selection is greatly intensified by pesticides. The resistant population arises from a susceptible population due to the concentration of a preexisting gene (or genes), which was initially very rare (<1 in 10^6).

This resistance gene may be inherited as a recessive or dominant trait, meaning that the heterozygous individual may be nearly fully susceptible or nearly fully resistant or have some intermediate level of resistance when compared with the homozygous-resistant insect. In a very widespread interbreeding population of insects, when only a portion is under selection, resistance might be slow to evolve due to dilution of the resistance allele upon mating with homozygous-susceptible individuals from the untreated portion of the population.

Conversely, resistance has evolved rapidly in populations of insects restricted by geography. House flies developed resistance to organophosphorus insecticides very rapidly on Mackinaw Island, Michigan, a resort island on which only horse-drawn vehicles were used and horse manure was collected and centrally treated intensively. Similarly, a case of resistance to multiple insecticides evolved on Umenoshima (Dream Island), Japan, which was a garbage dump in which insecticides were intensively applied for house fly control. The first documented case of resistance to pyrethroid insecticides in diamondback mot$_f$ was on the northern tip of the island of Taiwan where vegetables are intensively farmed.

Resistance had developed rapidly in greenhouse populations of whiteflies and aphids and in populations of stored product pests in grain bins. These cases are circumstantial evidence for the importance of dilution of resistance through mating with untreated portions of a population. Resistance in a more extensive population has appeared to evolve initially in one location and subsequently spread outward into the susceptible populations.

This phenomenon appears to be accelerated when the insecticide of interest is used as a widespread treatment, thereo$_f$ conferring a strong selective advantage to the resistant individuals on the interface with susceptible populations. An example was the spread of dieldrin resistance from one county in Nebraska into adjoining states in the western corn rootworm, which was spreading upon soil into which this very persistent insecticide was nearly universally incorporated, a situation that allowed for little dilution of resistance. It now seems clear that once lost to resistance, an insecticide cannot be recovered; i.e., evolved resistance persists so that returning to a resisted insecticide becomes impractical.

This means that the pace of developing alternative insecticides must be accelerated in order to retain effective control. A major point to be considered in the stability of resistance is the relative fitness of resistant individuals.

In theory, an allele for resistance that also confers selective advantage in the presence of the insecticide, but imposes a fitness deficit in the absence of the insecticide, should decline to the initial very rare frequency in the absence of the insecticide.

In the case of resistant acetylcholinesterase in tobacco budworm, *Heliothis virescens*, a survey of populations across the southeastern United States indicated that the frequency of the Aceln-R allele remained at approximately 10% long after withdrawal of the primary selecting agent, methyl parathion, and its substitution by synthetic pyrethroids was assumed to have no selection for this allele.

It was unclear how this allele could have declined, but then have reached an equilibrium assumed to be much greater than the original frequency prior to any selection. Pyrethroid resistance in the tobacco budworm, *Heliothis virescens*, was documented first in California in 1980, then Texas in 1984, and later seemed to spread eastward in the midsouth states of Louisiana and Mississippi, but no further until a serious failure was suffered in Alabama in 1995.

Now pyrethroid resistance in this species seems to have dispersed across the entire Cotton Belt of the United States. Fortunately, this species is very susceptible to transgenic Bt-cotton, which is heavily planted in areas known for tobacco budworm outbreaks. It appears that earlier resistance was correlated with the mutation M421V in the sodium ion channel target of pyrethroids, whereas a second mutation, L1029H, succeeded in about 1990 as far east as Louisiana. L1029H was found in Dalzell, South Carolina, in 1998 and might have been the reason for failures in Alabama in 1995.

Although this information in this case is fragmentary at best, it demonstrates the potential for enhancing our understanding of the dispersal of resistance by tracking specific mutations in resistance genes.

Transgenic Insecticides

Transgenic insecticides are insecticidal proteins produced in crop plants due to the artificial insertion of a foreign gene that encodes that insecticidal protein. Presently, nearly all practical transgenic insecticides (also calle "plant/pesticides"), are forms of the *Bacillus thuringiens* toxin (Bt-toxin) expressed from a single gene, which was isolated from the bacterium *Bacillus thuringiensis*, modified to resemble a plant gene by changing some silent nucleotides, genetically engineered into various crops, and further modified to increase expression of the protein the crop.

The first commercial use of this technology in the United States was in potatoes in 1995, and this was followed by important introductions of genetically modified cotton, corn, and many other crops. In the development of transgenic insecticides in crops, the threat of resistance evolving in the targeted pests has been a significant consideration.

Resistance could shorten the lifetime of efficacious application of transgenic insecticides just as it threatens the utility of conventional insecticides, insect growth inhibitors, and other insecticidal agents that result in insect mortality and thereby selection of populations. Furthermore, resistance evolved to the transgenic B-toxin would likely depreciate the activity of foliar application of the natural or otherwise modified Bt-toxin in the same pest species. Transgenic insecticides potentially present nearly continuous selection when they are expressed constitutively in the modified crop.

This could result in selection for pest resistance even when pest populations occur at subthreshold levels, i.e., at levels that would not trigger application of a conventionally applied insecticide. There are a few examples of resistance to folia applied Bt-toxin (see also Status). The cosmopolitan pest of cruciferous vegetables, the diamondback moth, *Plutella xylostella*, has developed resistance in many locations when selected for many years with foliar applications of Bt-toxin.

Another widespread pest, the Indian meal moth, *Plodia interpunctella*, has developed resistance to Bt-toxin applied for the protection of stored products. Other pests, including the tobacco budworm, *Heliothis virescens*, and the European corn borer have been selected experimentally for resistance in the laboratory. Monitoring programs are established with the objective of early detection of resistance to Bt-toxin in pests exposed to transgenic insecticides in cotton.

An Australian program has determined baseline susceptibility to Bt-toxin in the cotton bollworm, *Helicoverpa armigera*, whereas tobacco budworm, *Heliothis virescens*, and cotton bollworm, Helcoverpa *zea*, have been monitored for resistance by the United States Department of Agriculture.

Resistance Management Plans

The goal of research into insecticide resistance is to find means by which resistance to newly introduced insecticides can by prevented, i.e., prophylactic measures, and to discover methods for reducing the impact of resistance once it has evolved, i.e., remedial countermeasures.

These measures are practiced in the implementation of resistance management plans, which are a set of recommended integrated pest management approaches and methods related to the use of an insecticide(s) to control a particular pest or to protect a specific crop or commodity from damage by insects that will reduce the risk that resistance will develop to any of the insecticides or other pest management tools included in the plan.

A resistance management plan should be scientifically based both on integrated pest management strategies to reduce reliance on insecticides and on resistance management strategies to minimize selection by a particular insecticide. Prophylactic measures can take many forms, ranging from abstinence at one extreme to the application of the highest possible dose on the other extreme; paradoxically, both strategies are intended to introduce no selection pressure on a population of the pest of interest.

In the absence of selection, it can be assumed that evolution of resistance could not occur. Generally, it has been assumed that the application of an insecticide will introduce some selection and that evolution will proceed, but that the rate of evolution of resistance can be significantly reduced with proper management of the insecticide.

This general assumption has been supported by mathematical models and laboratory experiments; however, there is little practical evidence that the rate of evolution has been controlled by any strategy applied to date. A resistance management plan, whether voluntary or legally mandated, must be implemented over a large area with high compliance by local growers in order to be effective because the plan applies to interbreeding pest populations without regard to crops, farms, or borders.

For this reason, some of the most successful plans have been applied where relatively few participants were involved. In Denmark, resistance management of insecticides for the control of house flies on farms could be developed with ease because it involved just a few government officials who were knowledgeable in the problem, and it was implemented successfully because a relatively small number of farms allowed a high level of personal contact to persuade farmers to comply with the plan.

This plan was intended to slow the rate of evolution of insecticide resistance by monitoring the susceptibility of house flies to each insecticide introduced sequentially and changing upon the detection of resistance. A component of this plan was to spare natural pyrethrins, which were considered very valuable to dairy farms, by denying the registration of synthetic pyrethroids based on their presumed risk of resistance due to increased persistence *vis-a-vis* the natural products.

In Australia, the first detailed resistance management plan for cotton bollworm, *Helicoverpa armigera,* was developed by early involvement of growers who were relatively few in number and who were very compliant. The key strategy of the plan was to limit synthetic pyrethroid insecticides to a window in time during the middle of the growing season when their use would have the most impact and to apply alternative measures around this window; there was nearly full voluntary compliance to this strategy. Another aspect was to restrict pyrethroid use to cotton, avoiding applications for the same pest species on other, less-valuable crops in the subject zone.

Although resistance evolved over two decades even under this plan, the rate of evolution was considered to be reduced when compared with Thailand, where resistance was rapidly selected in this pest in the absence of a similar plan. In response to the creeping increase of resistance, the Australian plan was modified to reduce the pyrethroid window of use and to focus control on elimination of pupae. Similar strategies were adopted in the United States for cotton with surveillance of pyrethroid resistance in California, Arizona, Texas, Louisiana, Arkansas, Missisippi, Alabama, Georgia, South Carolina, and North Carolina.

These state university programs were coordinated, and they also cooperated in standardization of succeptibility test methods with insecticide manufacturers and commodity organizations through, first, the Pesticide Efficacy Group, and, later, the Insecticide Resistance Action Committee. The key development of this co-operation was the agreement on a standardized susceptibiity test in which adult males of *Heliothis virescens* and

Helicoverpa zea were captured in pheromone traps and tested by contact with permethrin and later cypermethrin deposited inside 20-mL glass vials, the "adult vial test." This method was applied to surveillance of hundreds of thousands of moths for pyrethroid susceptibiity. In Louisiana, a gradual evolution of resistance was observed over two decades in the tobacco budworm, *Helithis virescens.* Another detailed resistance management plan has been implemented in Arizona for the white fly on cotton and is similar to the plan in Denmark for house fly in that sequential introduction of insecticides is carefully monitored for resistance evolution in the University of Arizona Extension Arthropod Resistance Management Laboratory.

Growers are advised to preserve an array of diverse compounds and not to rely on a single insecticide, even when a new, impressive agent is introduced. The U.S. Environmental Protection Agency has notified manufacturers of a voluntary labeling of pesticide products according to target site of action with a group number assigned to each product.

Products are to be labeled with recommendations for delaying resistance to include avoiding the consecutive use of the registered product or others within the same group by site of action against the one species. Recently, the introduction of transgenic insecticidal crops has added complexity to resistance management plans in that substitution of the transgenic insecticide for the conventional foliar insecticide often alters the pest spectrum of the crop. Also, little is known about the interactions of conventional and transgenic insecticides in relation to resistance.

Due to the advanced technology of transgenic crops and their value, the controversy over genetically modified foods, and the potential risk of Bt-cotton to the efficacy of conventional Bt-toxin, which is used by producers of "organic" foods, the U.S. Environmental Protection Agency has decided to regulate transgenic insecticides as "plant/pesticides" and to request a resistance management plan as a component of the registration application for these pesticides. Potential registrants formed the Bt Working Group to solicit expert opinion and sponsor research that was used in developing resistance management plans. Two major issues relating to resistance management plans for transgenic insecticides are the mode of inheritance of resistance to Bt-toxin in the pest species and the population ecology of the pest.

A plan implemented by Monsanto employs a strategy in which a high dose of the expressed Bt-toxin is coupled with a mandatory refuge for susceptible individuals of the targeted pest to prevent resistance. This strategy assumes that resistance to Bt-toxin would be inherited as a single recessive gene, that a high dose of the toxin in the plant would be sufficient to kill nearly all heterozygous individuals, and that a refuge of *nontransgenic crop* would produce sufficient *susceptible* individuals to prevent *homozygous* resistant from being produced. Mathematical models support this strategy in predicting that evolution of resistance would be significantly delayed; however, such models apply the assumption of a recessive inheritance.

Should resistance arise as a dominant trait, then in order to employ the high dose plus refuge prophylactic strategy, the refuge would have to be dramatically increased and the strategy would likely be disabled due to contamination of the refuge.

Regarding the genetic inheritance of Bt-resistance, a survey of the literature indicated that the average degree of dominance is greater than that assumed in some models, which often employ a dominance value of 0.1 or less.

In an experimentally selected strain of European corn borer resistant to Dipel, the degree of dominance was approximately 0.7 (on a scale to 1.0); however, it was noted that these Dipel-resistant individuals did not survive on transgenic Bt-maize plants.

Recently, it was observed that gene *hscp* for the target of pyrethroid insecticides, the sodium ion channel, is linked to a gene for detoxication of pyrethroids, *CYP6B10* in tobacco budworm.

The fact that these factors would be inherited together sugges that the dominance of the linked resistance alleles must be tested versus the inheritance of one or the other factor. Remedial countermeasures are tactics applied after the evolution of resistance to increase the efficacy of the affected insecticide.

A typical approach is to use a mixture of insecticides or a synergist, which is a second component added to the insecticide to produce a resultant mortality that is more than the expected additive result.

Synergis include piperonyl butoxide, which is commonly applied to inhibit the P450 monooxygenase, thereby reducing the rate of oxidative detoxication of the accompanying insecticide. This remedial strategy is used most effectively when the mechanism of resistance is known so that the appropriate synergist can be applied; often, knowledge of the mechanism is based on laboratory analysis with synergists.

An example of the use of piperonyl butoxide is its mixture with rotenone to control multiple-resistant Colorado potato beetle. In the case of *Heliothis virescens*, strain Pyrethroi-R, piperonyl butoxide was ineffective; however, resistance was reduced by the alternative synergist trichlorophenyl propynyl ether.

Among the organophosphorus insecticides, the combination of EPN and methyl parathion was demonstrated to reduce resistance to methyl parathion in *Heliothis virescens* in the laboratory, whereas this mixture was commonly applied by cotton growers prior to the laboratory experiments.

Resistance to malathion, a phosphorodithioate insecticide, was reduced in several species by mixtures with various organophosphates, which inhibited malathion carboxylester hydrolase to reduce detoxication.

Carbamate synergists include n-propyl carbaryl, which was highly synergistic in resistant green rice leafho per, but was never applied in practice due to oncogenesis at very high doses in laboratory animals. This compound along with several organophosphorus insecticides, including monocrotophos, displayed enhanced inhibition of a resistant form of acetylcholinesterase in *Helithis virescens*.

This is also an example of negatively correlated insecticides, which are more effective against resistant strains than against normal strains; however, there are few practical examples, despite several examples demonstrated in laboratory experiments. Chlorfenapyr was more effective against pyrethroid-resistant adults of *Heliothis virescens*, but registration of this insecticide has been canceled by the U.S. Environmental Protection Agency.

Forecast

After 60 years of living with the phenomenon of resistance, efforts should be continued to understand the molecular and genetic basis of this phenomenon and to translate that information into practical resistance management plans to extend the lifetime of new insecticides.

It is unclear from just 3 years of practical experience whether the high dose plus refuge resistance management plans for B-transgenic cotton and corn will be successful; however, these efforts are a step forward in introducing growers to highly coordinated programs for resistance management and can only be positive in this regard. It is likely that insects will continue to lead the race by evolving resistance faster than humans can replace insecticides.

Should resistance evolve to Bt-crops, it will be very challenging to find additional transgenes with the simplicity and efficacy of Bt-toxin, and it will likely bring cross-resistance to conventionally applied Bt-toxin, for which a similarly effective natural insecticide will be difficult to find.

Should resistance to Bt-crops be prevented by the current high dose plus refuge strategy, it will be possible to transfer this strategy to conventional insecticides. Advances in molecular genetic analysis and insect genomics should permit the further development of rapid diagnostics for specific resistance mechanisms and, in turn, lead to a better understanding of the evolution and dispersal of specific genes for resistance.

In addition, the molecular understanding of resistant targets may provide the basis for rationally designed countermeasures such as negatively correlated insecticides to overcome resistance. Resistance management plans will be developed and implemented along with the introduction of new chemistry and new transgenic insecticides.

As these plans are improved and more is learned about resistance, the lifetime of new inventions should be gradually increased; however, this will depend on increasing the understanding of the genetics of resistance in the target pests.

INSECTICIDAL CARBAMATES

Development of the carbamate insecticides was based on a structural lead that emerged from studies of the pharmacological properties of the alkaloid, physostigmine, isolated in 1864 as the toxic principle of the beans of the plant, *Physostigma venenosum*, which had been used as an ordeal poison in West Africa.

Extensive research on the structure and activity of physostigmine showed that it was a potent inhibitor of the enzyme, cholinesterase. Cholinesterase is the enzyme responsible for degrading acetylcholine, a substance involved in neural transmission (see Acetylcholinesterase inhibitors). Synthetic carbamate analogs used medicinally are ionizable compounds, and nonionizable analogs we shown to have insecticidal activity.

In 1947, seeral N-methylcarbamates possessing significant acetycholinesterase inhibitory activity were synthesized in Switzerland by the former Ciba-Geigy Company and developed as insecticides.

Like the organophosphates, carbamates are potent inhibitors of cholinesterase. The insecticides are strong carbamylating inhibitors of acetylcholinesterase and may also have a direct action on the acetylcholine receptors because of their pronounced structural resemblance to acetylcholine.

The N,N-dimethylcarbamates of heterocyclic enols and the N-methylcarbamates of a variety of substituted phenols that possess a wide range of insecticidal activity were described in 1954. The latter became the most widely used carbamate insecticides, and N-methylcarbamates of oximes were subsequently found to be effective systemic insecticides. The carbamate insecticides represent the third principal group of synthetic organic insecticides developed after World War II.

They are highly biodegradable, and although compounds such as carbofuran and aldicarb are highly toxic to mammals by oral ingestion, their toxic action as inhibitors of acetylcholinesterase is rapidly reversible Chemically, carbamates are esters of carbamic acids NHRCOOR′. The substituent *R* in the majority of insecticidal carbamates is a methyl group. One of the best known is carbaryl (N-methylnaphthyl carbamate), in which *R′* is a 1-naphthyl group. Free carbamic acids are unstable and decarboxylate to give the parent amine and carbon dioxide.

$$NHRCOOR' + H_2O = NHRCOOH + R'OH$$

$$NHRCOOH = NHR + CO_2$$

Production

The phenylcarbamates may be synthesized by reaction of the appropriate sodium phenoxide at 10 to 50°C with phosgene in toluene. An intermediate chloroformate is formed that reacts with excess amine to give a carbamate. This method also offers a route to the dimethyl carbamates.

$$RONa + COCl_2 = ROCOCl + NaCl$$

$$ROCOCl + R'NH_2 = ROCONHR' + HCl$$

The reaction of methyl isocyanate with a phenol or alcohol offers an alternative route. Methyl isocyanate can be obtained by reaction of phosgene with methylamine.

$$CH_3NH_2 + COCl_2 = CH_3N = C = O + HCl$$

$$CH_3N = C = O + ROH = CH_3NHCOOR$$

Oxime carbamates, such as aldicarb, are produced by a modified route. Isobutene is reacted with nitrosyl chloride to form a dimer of 2-chloro-2-methyl-1-nitrosopropane. Reaction of this compound with methylmercaptan in sodium hydroxide yields methylthiopropionaldehyde oxime, which can be converted to aldicarb by reaction with methyl isocyanate.

$$(CH_3)_2C{=}CH_2 + NOCl = (CH_3)_2CClCH_2NO \quad (CH_3)_2CClCH_2NO + CH_3SH =$$

$$(CH_3)_2C(SCH_3)CH{=}NOH \quad (CH_3)_2C(SCH_3)CH{=}NOH + CH_3N{=}C{=}O$$

$$= (CH_3)_2C(SCH_3)CH{=}NOCONHCH_3$$

The mammalian toxicity of carbamates is generally lower than that of the organophosphates. This is especially apparent when dermal toxicity is considered. However, aldicarb (2-methyl-2-(methylthio) propionaldehyde methylcarbamoyloxime) is an exception, having a somewhat higher toxicity than most members of this class (LD_{50} 0.93 mg/kg acute oral in rats).

Although compounds such as carbofuran and aldicarb are highly toxic to mammals, their toxic action as inhibitors of acetylcholinesterase is rapidly reversible. Carbamates showed reduced potential for the environmental contamination that had been associated with the organochlorines because they are readily biodegradable and they were less hazardous than were the organophosphates.

As a consequence, they became widely used for foliar applications to field, vegetable, and fruit crops, as soil insecticides, as nematicides, to protect stored products, for veterinary hygiene, and with propoxur and bendiocarb as residual insecticides for malaria control in human habitations. The N-methylcarbamates generally are biodegradable and of low soil persistence with half-lives for carbaryl and aldicarb of 1–2 weeks and of carbofuran of 1–4 months.

Certain carbamates are highly toxic to birds with oral LD_{50}s e.g., for mallard, pheasant, in mg/kg: carbofuran, 0.40, 4.2; mexacarbate, 3.0, 4.5; and methomyl, 16, 15; compared with carbaryl > 2000. Fish toxicity of carbamates is generally low, but these compounds are extremely toxic to bees. In cases of human poisoning, atropine is an antidote. The carbamates are degraded in soils, but in sandy soils, aldicarb sulfoxide and aldicarb sulfone, the principal metabolites of aldicarb, leached sufficiently to enter groundwater.

Carbofuran

Carbofuran (1) [1563-66-2], 2,3-dihydro-2,2-dimethylbenzofuran-7-yl N-methylcarbamate (mp 150–152 °C), is soluble in water to the extent of 0.7 g/L. It is a cholinesterase inhibitor with systemic activity. Carbofuran is a broa-spectrum soil insecticide and nematicide.

It is stable in weak acids and is hydrolyzed to the parent phenol in basic media. Sunlight irradiation of an aqueous solution of carbofuran gave the phenols and Carbofuran is degraded by hydrolysis and oxidation in soil before ultimate mineralization.

The rate of hydrolysis in soils is slightly higher under flooded than under nonflooded conditions. Products depend on the soil type and the prevalence of aerobic or anaerabic conditions, and it was also reported that carbofuran did not degrade under anaerobic conditions.

The products, 3-hydroxycarbofuran and 3-ketocarbofuran, have been isolated from soil extracts after incubation with carbofuran. The phenol was identified as a major product in several studies. The products are further degraded and bound to soil organic matter. Enhanced degradation may follow repeated applications of carbofuran to soils, and bacterial cultures capable of rapidly degrading carbofuran have been obtained from treated soils. Carbofuran undergoes hydrolytic and oxidative processes in mammals.

In rats, about 72% of the administered dose was eliminated in the urine within 24 hours as conjugated metabolites, mainly as conjugates of the 3 ketocarbofuran phenol. The products of metabolism in plants are 3-hydroxycarbofuran and 3-ketocarbofuran phenol.

Carbosulfan

Carbosulfan (8)[55285-14-8],2,3-dihycro-2,2-dimethylbenzofuran-7 yl(dibutylaminothio) methylcarbamate (IUPAC), is an orange to brown, clear viscous liquid (bp 124–128°C), miscible with organic solvents and solubility 0.3 ppm in water(25°C).Itiscloselyrelatedstructurallytocarbofuran, and like carbofuran, it is a cholinesterase inhibitor with systemic activity. It controls many soil and foliar insects.

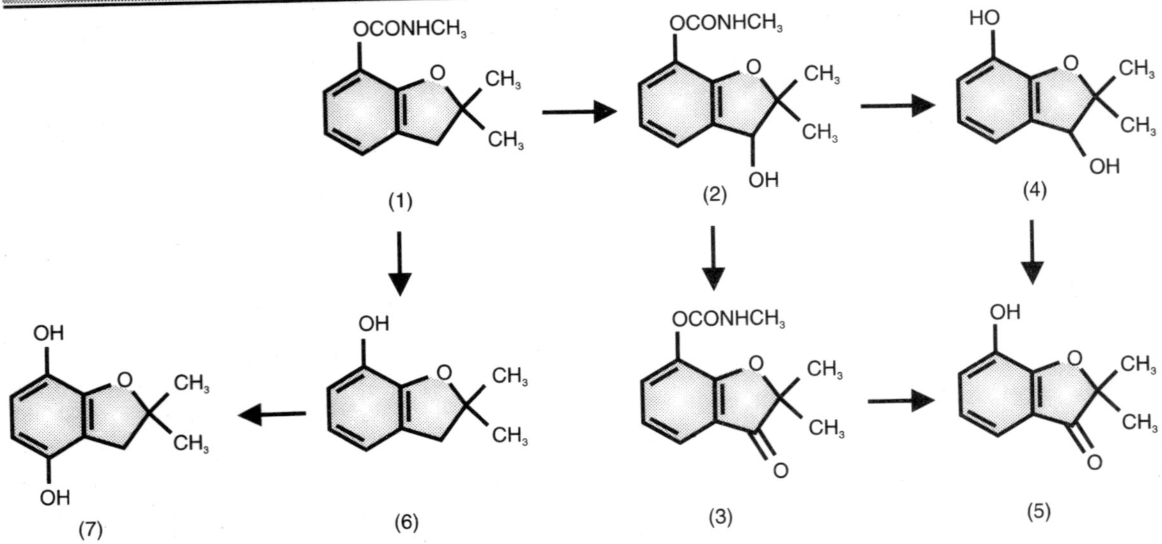

Figure 17.3: Carbofuran—environmental degradation.

The N-S bond is cleaved *in vivo* yielding carbofuran. Carbosulfan is produced by reaction of sulfenyl chloride with dibutylamine, followed by treatment with carbofuran.

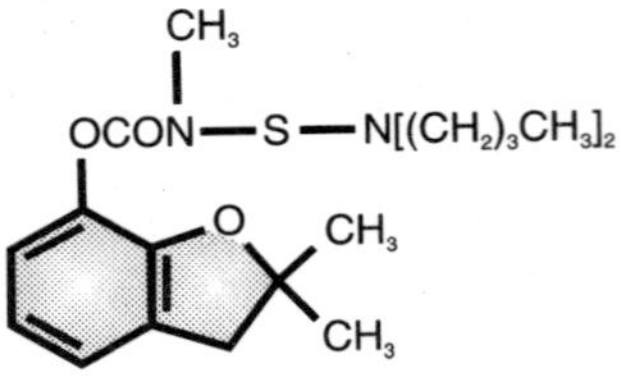

carbosulfan (8)

Its metabolic patterns are similar to those of carbofuran. In rats, it rapidly undergoes hydrolytic and oxidative processes followed by conjugation. It is not persistent in soils, with DT_{50} ca. 2–5 days, and it was rapidly degraded to carbofuran in a sandy loam soil.

Carbofuran was subsequently hydrolyzed at the carbamate ester group to form the phenol carbofuran or oxidized at the 3position. Biscarbofuran disulfide and minor products were also detected. Carbofuran was also formed in soils by nonbiological degradation processes.

In plants, carbosulfan is not translocated and its systemic activity is due to metabolites. The main metabolic routes are cleavage of the N-S bond to form carbofuran and oxidation and hydrolysis. The products of metabolism in plants are carbofuran (1) and 3-hydroxycarbofuran.

Carbaryl

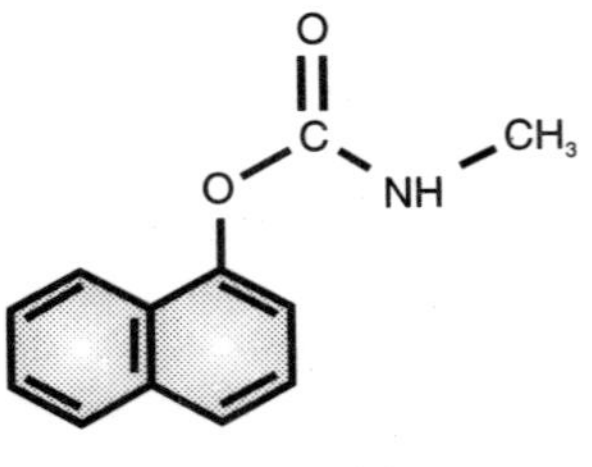

carbaryl (9)

Carbaryl (9) [63-25-2], 1-naphthyl methylcarbamate (IUPAC), $C_{12}H_{11}NO_2$, MW 201.2, mp 142°C, forms colourless to light tan crystals that are slightly soluble in water and are readily soluble in polar organic solvents. It is stable in neutral or acid media, but it is hydrolyzed under basic conditions. It is not rapidly photodegraded in the field, but it breaks down on irradiation in basic solution to form 1naphthol and 2-hydroxy-1,4-naphthoquinone.

Carbaryl is obtained by reaction of 1-naphthol with methylisocyanate in the presence of a catalytic amount of triethylamine. Carbaryl is used for the control of chewing and sucking insects in a number of crops. It is a contact and ingested insecticide. Carbaryl undergoes hydrolysis and ring oxidation in soils. The major metabolite in a number of studies was 1-naphthol. Metabolites also included 4-hydroxycarbaryl and 5-hydroxycarbaryl. In mammals, the major metabolite is 1-naphthol. This is eliminated in urine and feces, together with the glucuronic acid conjugate. Aromatic ring hydroxylation at the 3-, 4-, 5-, or 6-positions also occurs as does hydroxylation at the N-methyl group.

Bendiocarb

Bendiocarb **(10)** [22781-23-3], 2,3-isopropylidenedioxyphenyl methylcarbamate (IUPAC), $C_{11}H_{13}NO_4$, MW 223.2, mp 124.6–128.7°C, forms colourless crystals that are sparingly soluble in water and hexane and are fairly soluble in dichloromethane, chloroform, acetone, methanol, and ethyl acetate. Bendiocarb is prepared by reaction of 2,2-dimethy-1,3-benzodioxol-4-yl with methylisocyanate. Bendiocarb is used to control cockroaches, other household pests, and soil insects. It has a long residual activity and a rapid knockdown.

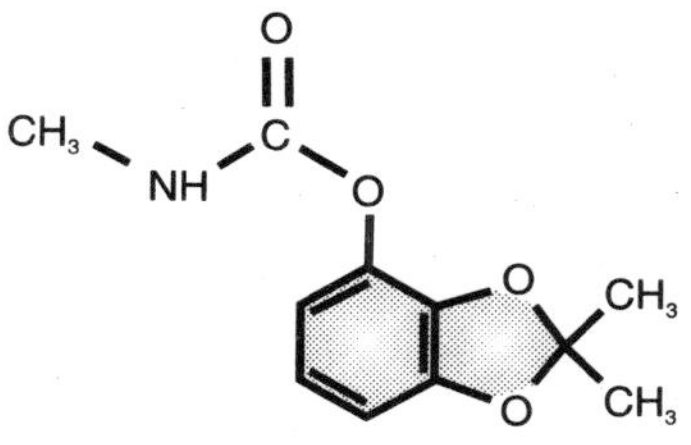

Bendiocarb (10)

Ethiofencarb

Ethiofencarb **(11)** [29973-13-5], α-ethylthio o-tolyl methylcarbamate (IUPAC), $C_{11}H_{25}NO_2S$, MW 225.3, mp 33.4°C, forms colourless crystals that are moderately soluble in water, are fairly soluble in hexane, and are readily soluble in dichloromethane, isopropanol, and toluene. Ethiofencarb is produced by reaction of 2-chloromethyl phenol with sodium ethylmercaptide to form 2-ethylthio methylphenol, which in turn is reacted with methylisocyanate. Ethiofencarb is a systemic insecticide effective against aphids. It is used on pomiferous fruit, stone and soft fruit, vegetables, ornamentals, and sugar beet.

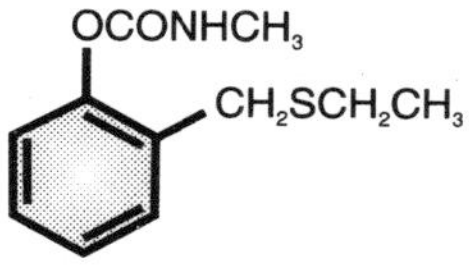

ethiofencarb (11)

Fenobucarb

Fenobucarb (BPMC) **(12)** [3766-81-2], 2-sec-butylphenyl methylcarbamate (IUPAC), $C_{12}H_{17}NO_2$, MW 207.3, mp 32°C, bp 112–113°C, is a colourless solid that is moderately soluble in water and is readily soluble in acetone, benzene, chloroform, xylene, and toluene.

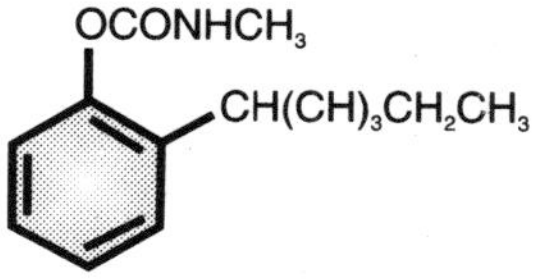

fenobucarb (12)

Fenobucarb is produced by reaction between 2-sec-butylphenol and methylisocyanate. Fenobucarb is used for the control of leafhoppers and plant hoppers on some crops and is used against bollworms and aphids on cotton.

Formetanate

Formetanate **(13)** [2259-30-9], 3-dimethylaminomethyleneaminophenyl methylcarbamate (IUPAC), $C_{11}H_{15}N_3O_2$, MW 221.3, mp 102°C, yellow crystals, solubility in water, <1 g/L at 20°C,

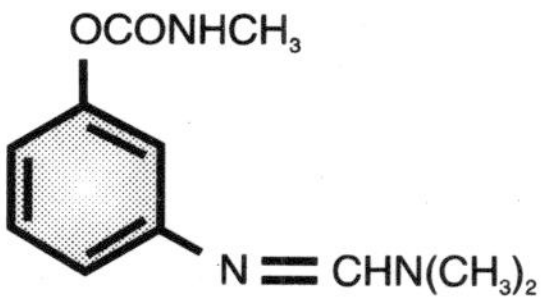

formetanate (13)

soluble in acetone, chloroform, and methanol. Formetanate is made by reacting 3-dimethylaminomethy-leneaminophenol with methyl isothiocyanate. It is active against the mobile stages of fruit tree spider mites.

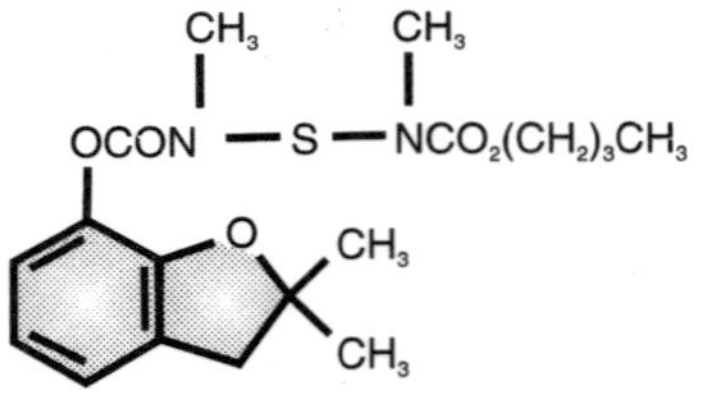

furathiocarb (14)

Furathiocarb

Furathiocarb **(14)** [65907-30-4], butyl 2,3-dihydro-2,dimethylbenzofuranyl N,N'-dimethyl-N,N'-thiodicarbmate (IUPAC), $C_{18}H_{26}N_2O_5S$, MW 382.5, bp > 250°C, is a yellow viscous liquid, which is sparingly soluble in water and is readily miscible with most organic solvents.

Furathiocarb is obtained by reaction of N-n-butyl-methylcarbamate with sulfenyl chloride, followed by treatment of the product with 2,3-dihydro-2,2-dimethyl-benzofuranyl N-methylcarbamate. Furathiocarb is applied as a foliar, soil, or seed treatment.

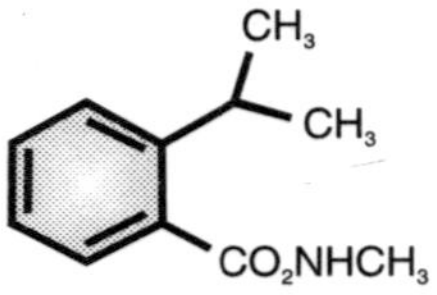

isoprocarb (15)

Isoprocarb

Isoprocarb **(15)** [2631-40-5], o-cumenyl methylcarbamate (IUPAC), $C_{11}H_{15}NO_2$, MW 193.2, mp 93–96°C, consists of colourless crystals, which are sparingly soluble in water and are readily soluble in acetone and methanol.

Isoprocarb is produced by reaction of 2-isopropylphenol with methylisocyanate. Isoprocarb is used for control of rice, cacao, sugarcane, and vegetable pests.

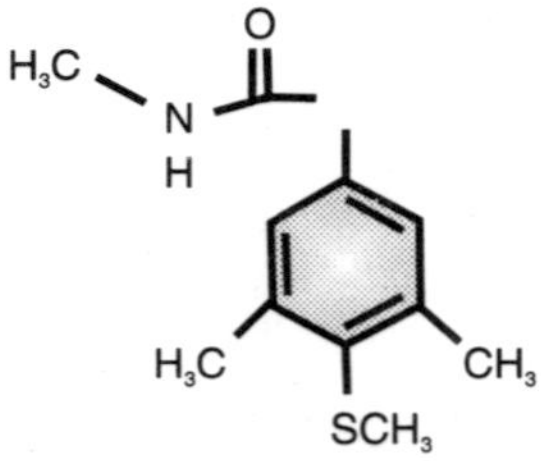

methiocarb (16)

Methiocarb

Methiocarb **(16)** [2032-65-7], 4-methylthio-3,5-xylyl methylcarbamate (IUPAC), $C_{11}H_{15}NO_2S$, MW 225.3, mp 119°C.

Colourless crystals that are sparingly soluble in water and hexane, fairly soluble in isopropanol, dichloromethane, and toluene.

Methiocarb is produced by the reaction between 3,5-dimethyl-4-methylthio-phenol and methylisocyanate. Methiocarb was first used as an insecticide.

It also has acaricidal and molluscicidal activity and has been used as a bait or seed treatment to combat slugs and snails.

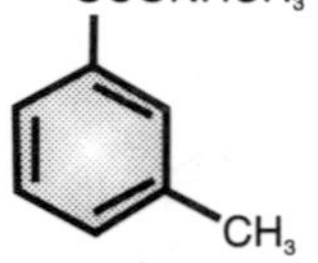

metolcarb (17)

Metolcarb

Metolcarb (17), [1129-41-5], m-tolyl methylcarbamate (IUPAC), $C_9H_{11}NO_2$, MW 165.2, mp 76–77°C, is a colour less solid that is moderately soluble in water and is readily soluble in polar organic solvents.

Metolcarb is produced by the reaction between 3 methylphenol and Metolcarb. Metolcarb is used to control leafhoppers and plant hoppers that attack rice.

Pirimicarb

Pirimicarb **(18)** [23103-98-2], 2-dimethylamino-5,6-dimethyl-pyrimidin-4-yl dimethylcarbamate (IUPAC), $C_{11}H_{18}N_4O_2$, MW 238.3, mp 90.5°C, is a colourless solid that is moderately soluble in water, acetone, ethanol, xylene, and chloroform.

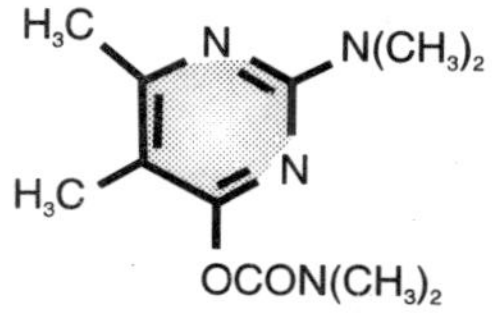

Pirimicarb is produced by reaction of 2-dimethylamino 5,6-dimethyl-4-pyrimidone with dimethylcarbamic acid chloride in the presence of a base or phosgene and dimethylamine. Pirimicarb is a systemic, selective aphicide used largely on grain crops, but also on ornamentals, cotton, fruit, and in greenhouses.

pirimicarb (18)

Propoxur

Propoxur (19), [114-26-1], 2-isopropoxyphenyl methylcarbamate (IUPAC), $C_{11}H_{15}NO_3$, MW 209.2, mp 90°C, forms colourless crystals, which are moderately soluble in most organic solvents.

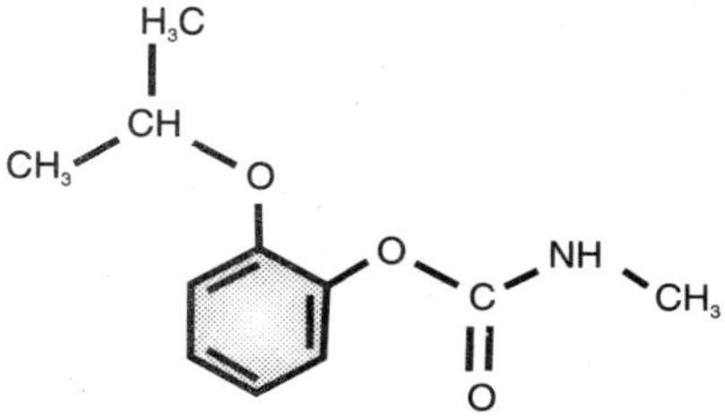

Propoxur is prepared by reaction of 2-isopropoxyphenol with methylisocyanate. Propoxur is used to control household insects such as cockroaches, bedbugs, and wasps, it is used as a replacement for DDT in malaria control, and it is applied against a number of food storage, vegetable, ornamental, and forestry pests.

propoxur (19)

Trimethacarb

Trimethacarb **(20)** [12407-86-2]. A product containing 3,4,5-trimethylphenyl methylcarbamate and 2,3,5-trimethylphenyl methylcarbamate in a ratio between 3.5 : 1 and 5 : 1, $C_{11}H_{15}NO_2$, MW 193.2, mp 105–114 °C.

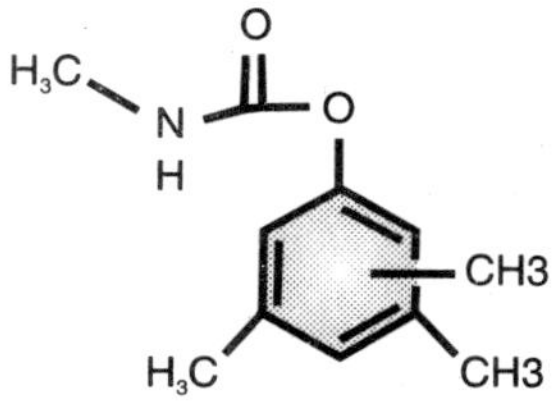

Trimethacarb is used to control a wide range of molluscan pests, and it is also a mammal and bird repellent.

trimethacarb (20)

Xylylcarb

Xylylcarb (MPMC) **(21)** [2425-10-7], 3,4-xylyl methylcarbamate (IUPAC), $C_{10}H_{13}NO_2$, MW 179.2, mp 79–80 °C, is a colourless solid, which is sparingly soluble in water and is fairly soluble in acetonitrile, cyclohexane, and xylene. Xylylcarb is obtained by treatment of 3,4-dimethyl-phenol with methylisocyanate. Xylylcarb is used to control leafhoppers and plant hoppers that attack rice, and it controls scale insects on fruit.

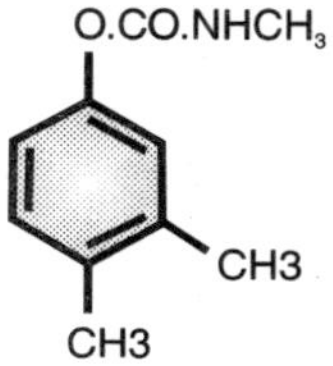

xylylcarb (21)

METHYLCARBAMATES OF OXIMES

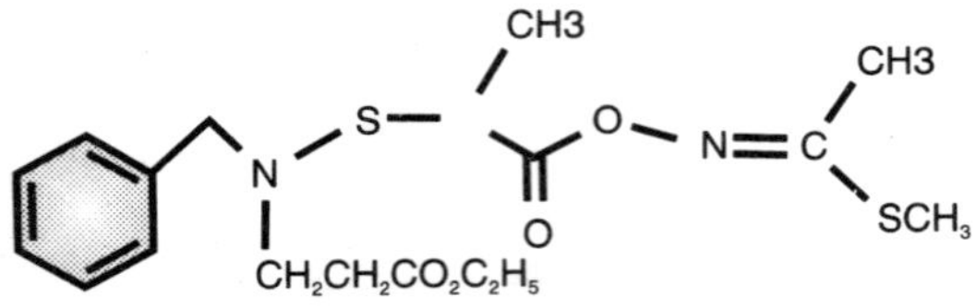

alanycarb (22)

Alanycarb

Alanycarb **(22)** [83130-01-2], ethyl (Z)-N-benzyl-N[methyl(1-methylthioethylideneamino-oxycarbonyl)amino]thiol,6-alaninate (IUPAC), $C_{17}H_{25}N_3O_4S_3$, MW 399.5, mp 46.8–47.2 °C, forms colourless crystals, which are sparingly soluble in water and are readily soluble in acetone, methanol, benzene, ethyl acetate, and dichloromethane.

It is produced by reaction of N-methyl-(carbam-oyloxy) thioacetimidate with N-benzyl-propylcarboxye-thylate sulfenyl chloride. Alanycarb (and dithiocarb) may be regarded as exerting their biological activity through biological conversion to methomyl, an inhibitor of ACHE.

It is rapidly metabolized to methomyl directly or via methomyl oxime. The ultimate products of metabolism are acetonitrile and carbon dioxide. In plants, alanycarb is rapidly metabolized to methomyl, which is further degraded via the oxime and the nitrile to carbon dioxide. In soils, degradation to methomyl may be microbial or chemical. Subsequently, methomyl oxime is formed, which is degraded to carbon dioxide.

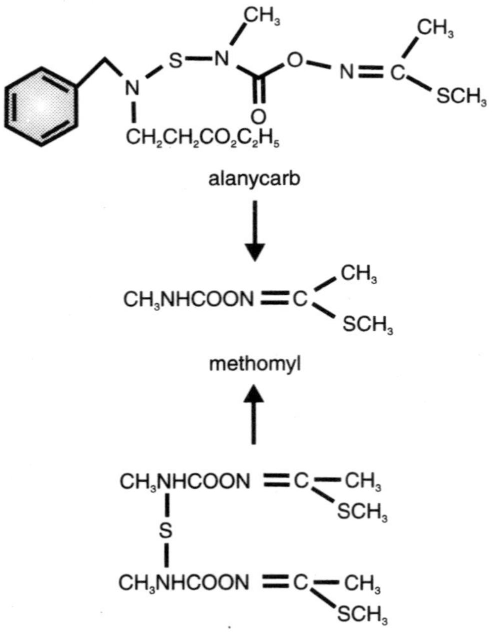

alanycarb

methomyl

thiodicarb

Figure 17.4: Alanycarb and thiodicarb as precursors of methomyl.

Alanycarb is a broad-spectrum insecticide with contact and oral toxicity to a variety of insect pests, and it has been used on fruit, citrus, tobacco, and vegetables in foliar, soil, and seed treatments.

Aldicarb

Aldicarb **(23)** [116-06-3], 2-methyl-2(methylthio)pr-opinaldehyde O-methylcarbomoyloxime (IUPAC), $C_7H_{14}NO_2S$, MW 190.3, mp 99-100°C, vp 13 mPa at 20° C, is soluble in water to 6 g/L. The rat oral LD_{50}s are 0.46-1.23 (oral) and 3.2- > 10 (dermal) mg/kg. Aldicarb acts by inhibition of cholinesterase. Aldicarb is a broad-spectrum systemic insecticide used for seed and soil treatment and as a nematicide and an acaricide.

It is stable in water with a DT_{50} of 3240 days (pH 5.5 and 15°C). The rate of hydrolysis increases under increasingly basic conditions (pH 9) with DT_{50} of ca. 75 days. Aldicarb readily oxidizes to the corresponding sulfone, aldoxycarb [1646-88-4] (mp 99 °C, vp 12 mPa at 25 °C), which is soluble in water to 10 g/L. The rat oral LD50 of aldoxycarb is 27 mg/kg, and it is a registered insecticide in its own right.

The degradation of aldicarb in soils, plants, and

mamm-als generates products that are based on the combination of hydrolytic, oxidative and hydration pathways, and the conjugation of products. Initial oxidation of the sulfide moiety is followed by oxidation to the sulfoxide and the sulfone.

Hydrolysis of the sulfoxide and the sulfone affords the corresponding oximes and nitriles. These may be further converted to amides or acids. In soil, the major pathway was oxidation to the sulfoxide, and the sulfone was also formed to a lesser extent.

These steps were followed by conversion to the oximes and nitriles, which in soils were ultimately mineralized. Pathways are shown in Figure elsewhere in this chapter. Aldicarb sulfoxide and sulfone occur as soil, mammalian, and plant metabolites.

The major metabolic pathways are similar in plants and animals, with rapid oxidation to the sulfoxide and a lesser quantity to the sulfone, but low levels of the amide and its precursor have been reported in mammals. The major products were metabolized further to the oximes and nitriles. In mammals, oximes were recovered primarily as glucuronide or sulfate conjugates.

Aldoxycarb

Aldoxycarb **(24)** [1646-88-4], 2-mesyl-2-methylpropionaldehyde O-methylcarbamoyloxime (IUPAC), $C_7H_{14}N_2O_4S$, MW 222.3, the sulfone analog of aldicarb, is also a systemic insecticide effective for control of insects (mainly sucking insect pests such as aphids, leafhoppers, etc.).

Butocarboxim

Butocarboxim **(25)** [69327-76-0], 3-(methylthio)butanone O-methylcarbamoyloxime (IUPAC), $C_7H_{14}N_2O_2S$, MW 190.3, mp 37 °C, is a brown viscous liquid that solidifies below room temperature; it is fairly soluble in water and is miscible with many organic solvents. Butocarboxim is produced by treatment of 3-methylthio-2-butanone with hydroxylamine, followed by reaction with methylisocyanate. Butocarboxim isused forthe control of aphids, mealybugs, and other sucking insects on fruit trees, vegetables, cotton, etc.

```
     CH3                    CH3
     |                      |
  CH3SCHCCH3             CH3CCHSCH3
        ||                   ||
        NOCONHCH3            NOCONHCH3
     E                      Z
```

butocarboxim **(25)**
Technical grade 85% in xylene
contains (E) and (Z) comes in ratio
(85-90 : 15-10)

Butoxycarboxim

Butoxycarboxim (26), [34681-23-7], 3-methylsulfonylbutanone O-methylcarbamoyloxime (IUPAC), $C_7H_{14}N_2O_4S$, MW 222.3, mp 85 -89°C, consists of colourless crystals which are readily soluble in water and polar organic solvents.

Butoxycarboxim is obtained by the reaction of 3-methylsulfon-2-butanone with hydroxylamine, followed by treatment with methylisocyanate [1440]. Butoxycarboxim is used on potted ornamentals against similar pests to those controlled by butocarboxim.

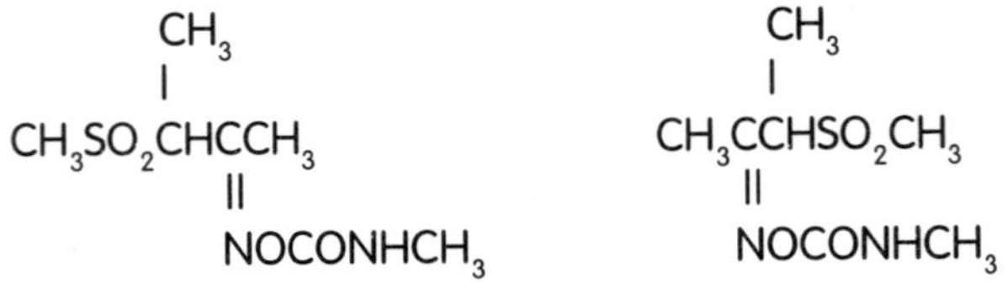

```
        CH3                          CH3
        |                            |
CH3SO2CHCCH3                 CH3CCHSO2CH3
          ||                     ||
          NOCONHCH3              NOCONHCH3
        E                            Z
```

butoxycarboxim **(26)**

Methomyl

Methomyl **(27)** [16752-77-5], S-methyl-N-(methylcarbamoyloxy)thioacetimidate (IUPAC), $C_5H_{10}N_2O_2S$, MW 162.2, mp 78–79 °C, consists of colourless crystals, which are fairly soluble in water and highly soluble in methanol, ethanol, acetone, and isopropanol. Methomyl is produced by chlorination of acetaldoxime and conversion of the resulting a-chlorooxime with sodium methylmercaptide.

Methomyl controls a wide range of insects and spider mites in fruits, vines, olives, hops,

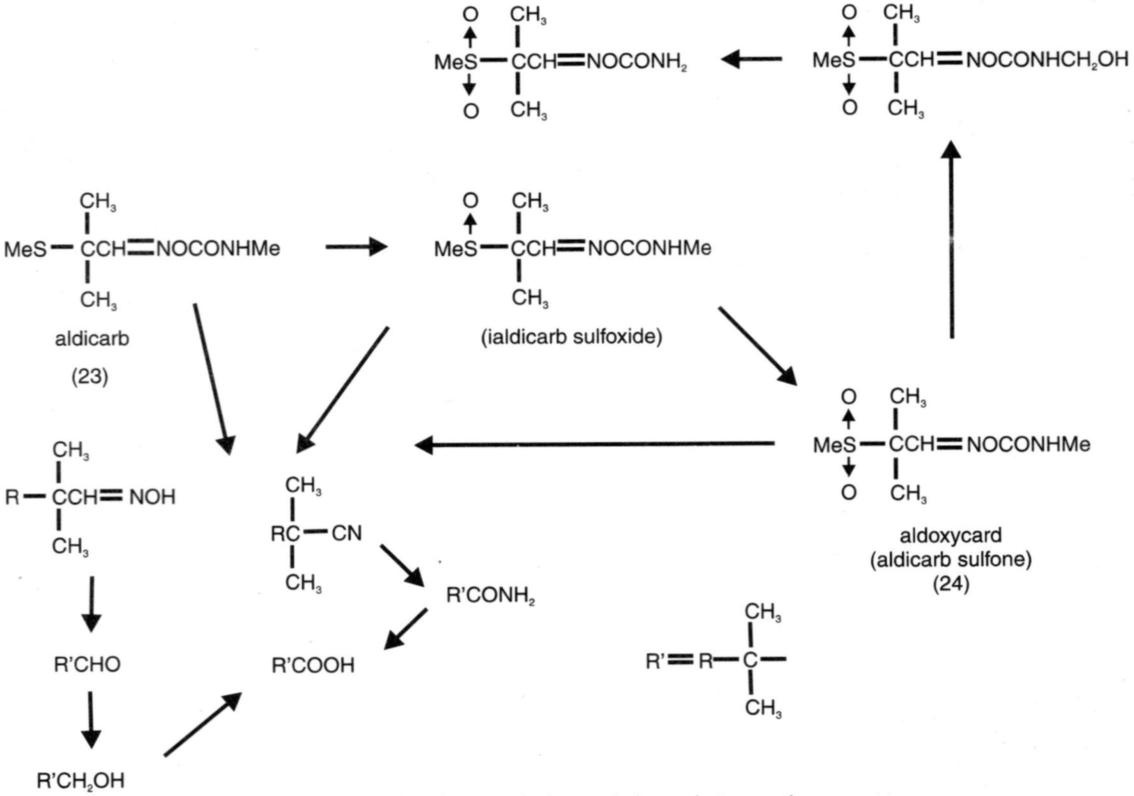

MeS—C(CH3)2CH=NOCONHMe
aldicarb
(23)

MeS(→O)—C(CH3)2CH=NOCONHMe
(ialdicarb sulfoxide)

MeS(→O)2—C(CH3)2CH=NOCONHMe
aldoxycard
(aldicarb sulfone)
(24)

MeS(→O)2—C(CH3)2CH=NOCONHCH2OH

MeS(→O)2—C(CH3)2CH=NOCONH2

R—C(CH3)2CH=NOH

RC(CH3)2—CN

R'CONH2

R'CHO

R'COOH

R'CH2OH

R'=R—C(CH3)2—

Figure 17.5: Aldicarb: metabolic and degradative pathways.

vegetables, and ornamentals.

$$CH_3NHCO_2N{=}C\begin{matrix} SCH_3 \\ CH_3 \end{matrix}$$

methomyl **(27)**

Oxamyl

Oxamyl **(28)** [23135-22-0], N,N-dimethyl-2-methylcarbamoyloxyimino-2-(methylthio)acetamide (IUPAC), $C_7H_{13}N_3O_3S$, MW 219.3, mp 100–102 °C, consists of colourless crystals, which are readily soluble in water, methanol, ethanol, acetone, and fairly soluble in toluene.

$$(CH_3)_2NCOC(SCH_3){=}NOCONHCH_3$$

oxamyl **(28)**

Oxamyl is produced by chlorination of the oxime of methylglycolate, reaction with methanethiol and alkali, and conversion to the carbamate with methyl isocyanate. Oxamyl (announced in 1968) is used for control of chewing and sucking insects, spider mites, and nematodes in ornamentals, vegetables, potatoes, and other crops.

Thiodicarb

Thiodicarb **(29)** [59669-26-0], 3,7,9,13-tetramethyl-5,11-dioxa-2,8,14-trithia-4,7,9,12-tetra-azapentadeca-3,12-diene-6,10-dione (IUPAC), $C_{10}H_{18}N_4O_4S_3$, MW 354.5, mp 173–174 °C, consists of colourless crystals, which are sparingly soluble in water, readily soluble in dichloromethane, acetone, methanol, and xylene.

$$CH_3NCO_2N{=}C(CH_3)_2 \;-\; S \;-\; CH_3NCO_2N{=}C(CH_3)_2$$

Thiodicarb (29)

Thiodicarb is produced by reaction of N,N′-thiobis(me-thylcarbamic acid fluoride) with 2-methylthioacetaldoxim in the presence of a base.

Thiodicarb is a carbamate insecticide (cholinesterase inhibition) with molluscicidal properties. It was introduced in France in 1988 and used as a pelleted bait containing 4% active ingredient at 5 kg/ha.

Thiofanox

Thiofanox **(30)** [39196-18-4], 3,3-dimethyl-1-methylthio-2-butanone O-methylcarbamoyloxime (IUPAC), $C_9H_{18}NO_2S$, MW 218.3, mp 56.5–67.5°C, is a colourless solid, which is moderately soluble in water and is readily soluble in common organic solvents. Thiofanox is obtained by reaction of 3,3-dimethyl(methylthio)-2-butanone with hydroxylamine, followed by reaction with isocyanate. Thiofanox is a systemic soil insecticide effective against a wide range of insect pes on many crops.

$$CH_3NHCO_2N{=}C(C(CH_3)_3)CH_2SCH_3$$

thiofanox (30)

MODE OF ACTION

The insecticidal carbamates are cholinergic. Poisoned insects and animals exhibit violent convulsions and other neuromuscular disturbances. These insecticides carbamylate acetylcholinesterase and may have a direct action on acetylcholine receptors. The mechanism of interaction with acetylcholinesterase is analogous to the normal three-step hydrolysis of acetylcholine.

However, the third reaction step is much slower for the carbamylated enzyme than for the acetylated one. The importance of structural complementarity of the insecticidal carbamates to the active site of acetylcholinesterase is demonstrated by the pronounced difference in activities of D-2-(sec-butylphenyl) methylcarbamate and L-2-(sec-butylphenyl) methylcarbamate (the L isomer is five times more toxic) and of the 2-3-, and 4-substituted phenyl methylcarbamates, where the 4-isomers are virtually inactive.

Detoxification of carbamate insecticides occurs *in vivo* through microsomal hydroxylation, N-demethylation of carbamyl nitrogen, side chain oxidation, and ring hydroxylation. Methylenedioxyphenyl synergists prevent oxidation largely by inhibiting the microsomal enzymes.

NEREISTOXIN PRECURSORS

Cartap

Cartap **(31)** [15263-52-2], S,S′-(2-dimethylaminotrimethylene) bis(thiocarbamate) hydrochloride (IUPAC), $C_7H_{15}ClN_3O_2S_2$, MW 273.8, mp 179–181 °C, forms colourless, slightly hygroscopic crystals that are readily soluble in water, are slightly soluble in methanol and ethanol, and are insoluble in acetone, diethyl ether, ethyl acetate, and benzene.

CH_2SCONH_2
|
$CHN(CH_3)_2$
|
CH_2SCONH_2

cartap (31)

Cartap is obtained by hydrolyzing 1,1-dithiocyanato-2-dimethylaminopropane with hydrochloric acid. Cartap is the pro-insecticide of the natural toxin nereistoxin. It is used for the control of chewing and sucking insects, at almost all stages of development, on many crops. Its structure is based on that of the naturally occurring neurotoxin, nereistoxin. Cartap is hydrolyzed in base to the dihdronereistoxin, which is oxidized to the insecticide, nerotoxin **(32)**. The conversion occurs within plants, and the monoxide **(33)** was identified as a minor metabolite. In rats, cartap was rapidly excreted in urine. It was hydrolyzed, converted to the sulfoxide, and N-demethylated.

Nereistoxin does not inhibit cholinesterase. Instead, it acts as an antagonist at the nicotinic acetylcholine receptor and blocks neural transmission.

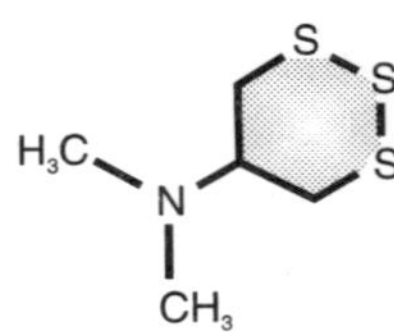

thiocyclam (34)

Thiocyclam

Thiocyclam (34) [31895-21-3], N,N′-dimethyl-1,2,3-tthian-5-ylamine (IUPAC), $C_5H_{11}NS_3$, MW 181.3, mp 125-128°C [decomp.]. Thiocyclam is a pro-insecticide of the natural toxin nereistoxin and is rapidly converted into the latter in biological media. It has limited systemic activity with stomach and contact action. It causes paralysis by ganglionic blocking action on the insect central nervous system. It is used as the hydrogen oxalate salt. This is converted to the active nereistoxin, and its oxide in soils and plants and

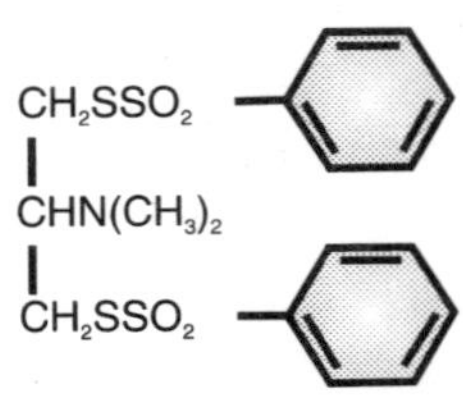

bensultap (35)

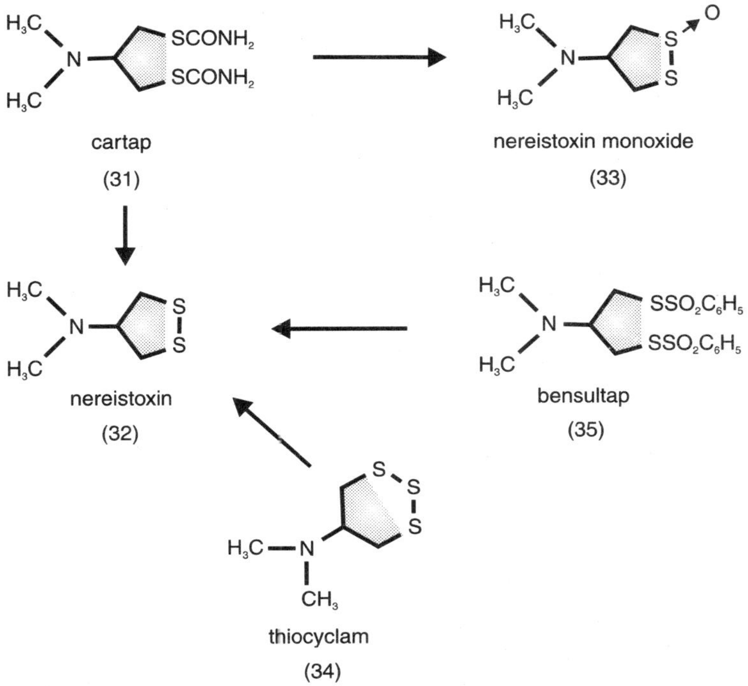

cartap (31)

nereistoxin monoxide (33)

nereistoxin (32)

bensultap (35)

thiocyclam (34)

Insecticidal precursors of nereistoxin

Figure 17.6: Nereistoxin precursors.

these are ultimately broken down into smaller molecules. DT50 in soil 1 day (pH 6.8, 22 °C, organic content 2.8%).

Bensultap

Bensultap **(35)** [17606-31-4], S,S′-2-dimethylaminotrimthylene di(benzenethiosu-lfonate) (IUPAC), M.W. 431.6, C17H21NO4S4, pale yellow crystalline powder mp 83–84 °C. Solubility in water 0.7–0.8 mg/kg (30 °C). Solubility in methanol 25 g/kg (25 °C), in acetone, acetonitrile, and N,N-dimethyl formamide >1000 g/kg. Hydrolyzed in neutral or alkaline solution. It is an insecticide with contact and stomach action for control of major insect pests. Acts a pro-insecticide or analog of nereistoxin. It inhibits the action of acetylcholine by blocking the receptor site at the post-synaptic membrane.

INSECTICIDE RESISTANCE ACTION (IRAC) COMMITTEE

This committee was formed in 1984 to provide a coordinated crop protection industry response to the developmental of resistance in insect and mite pests. It obtains information by surveys, reports, and other means to determine the extent of resistance. Its mission is to develop resistance management strategies to enable growers to use crop protection products in a way to maintain the efficacy.

There are many cases in which such a classification system may be useful, but in some cases weeds may exhibit multiple resistance across many of the groups. In such cases the key may be of limited value. The system itself is not based on resistance risk assessment, but can be used by the farmer or advisor as a tool to choose herbicides in different mode of action groups. This may be useful in selecting mixtures or rotations of active ingredients.

INSECTICIDES

Imidacloprid

The insecticide world market has long been dominated by well-established products belonging to the organophosphate and carbamate classes, which act by inhibition of acetylcholin-esterase, and pyrethroids, which act on voltage-gated sodium ion channels.

In the late 1980s, these three classes accounted for more than 87% of sales, and other insecticide classes with different modes of action were of limited economic importance. There was an increasing demand for new broad-spectrum insecticides with new modes of action, especially since the insects resistance to the major insecticides became a serious problem.

In 1986, imidacloprid (1) was discovered as a new insecticide with a unique structure and with a hitherto unrecognized insecticidal performance. Since being launched on the Japanese market in 1992, its sales potential has increased yearly, and it ranked as one of the top selling insecticides in the world in 2000. Following imidacloprid, insecticidal molecules of analogous structure have been developed as listed in Figure elsewhere in this chapter. Nitenpyram (2), acetamiprid (3), and thiacloprid (4), which also carry the "activator" 6-choloro-3-pyridinylmethyl (trivial name: 6-chloronicotinyl) group in the structure, are already on the market.

Then followed thiamethoxam (5) of oxadiazine skeleton carrying the 2-chloro-5-thiazolylmethyl group. Further, clothianidin (6) and dinotefuran (7) are likely going to be marketed soon. These new insecticides have been named chloronicotinyls or neonicotinoids, which show a similar mechanism of action. The chloronicotinyl insecticides are growing in importance compared with conventional organophosphates, carbamates, and pyrethroids (4). Imidacloprid, besides its agricultural use, is also used for the control of subterranean pests and pet ectoparasites.

Nomenclature and Physicochemical Properties

The nomenclature and physicochemical properties are listed in Table elsewhere in this chapter.

AGRICULTURAL USES

Formulations

The product is available as dustable powder, granules (including nursery box granules), seed dressing (flowable suspension, water-dispersible powder for slurry), emulsion concentrate, soluble liquid, suspension concentrate, wet-table powder, water-dispersible granule, and tablet.

Efficacy on Target Pests

Imidacloprid is highly effective for the control of hemipteran pests, i.e., bugs, aphids, leafhoppers, planthoppers, and whiteflies. The compound is also active against some species of the orders *Isoptera, Thysanoptera, Coleoptera, Diptera,* and *Lepidoptera. No* activity against

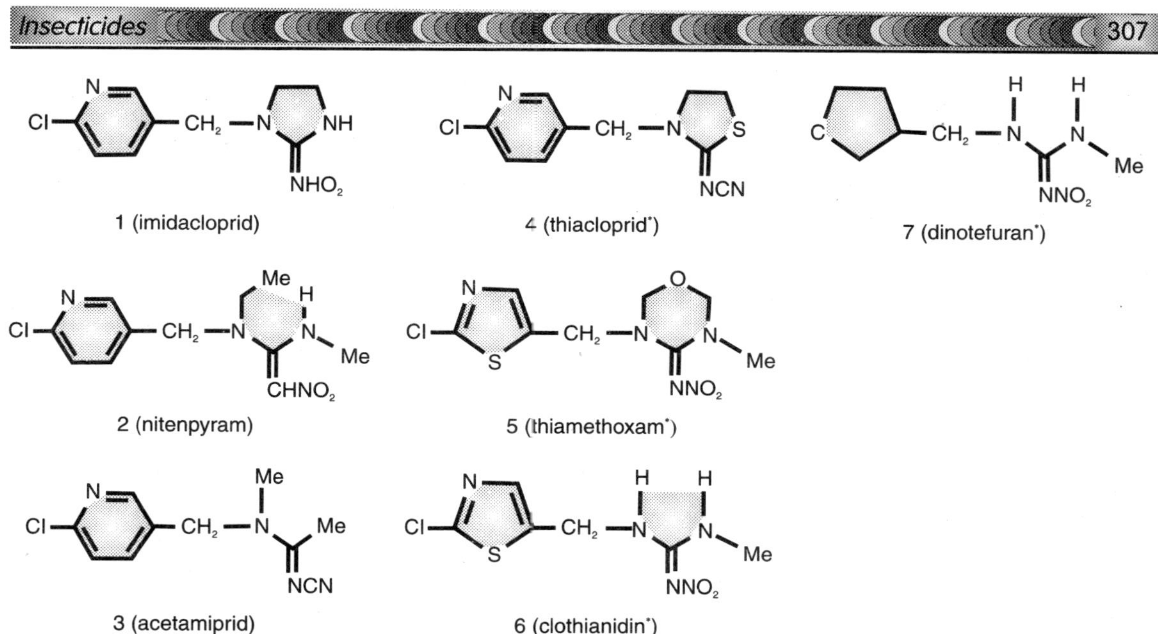

Figure 17.7: Imidacloprid and developed chloronicotinyl/neonicotinoid insectides.

nematodes and spider mites has been found. Imidacloprid has a systemic activity that makes it especially useful for seed treatment and soil application, but it is equally effective after foliar application.

Foliar Application

The spray application targets especially pests attacking crops such as cereals, maize, rice, potatoes, vegetables, sugar beet, cotton, citrus, tea, and deciduous fruits. Table 2 shows the acute activity (estimated LC_{95} in ppm a.i.) of imidacloprid against a variety of pests following foliar application (dip and spray treatment) of host plants under laboratory conditions.

Imidacloprid is very active on a wide range of aphids. Most susceptible is the damson hop aphid *Phorodon humuli* (LC_{95} = 0.32 ppm), which is often highly resistant against other classes of insecticides. Imidacloprid is highly effective against some of the most important sucking rice pests, leafhoppers and planthoppers.

The product is generally less effective against biting insects with the LC_{95} of 8 and 200 ppm against eggs, larvae, and pupae of *Chilo suppressalis, Heliothis virescens, Plutella xylostella, Spodoptera frugiperda, Lema oryzae, Leptinotarsa decemlineata,* and *Lissorhoptrus oryzophilus.*

The translaminar transport, where the ingredient moves from the treated upper side of a leaf to the lower surface, is very effective for controlling pests with a furtive lifestyle. For imidacloprid, the reinforcement of efficacy by translaminar action is observed in the tests using cabbage leaves.

Soil and Seed Treatment

A major strength of imidacloprid is based on efficacy in soil application and seed dressing due to its considerable mobility from the roots to the upper parts of plants through the xylem and

Table 17.2: Nomenclature and Physicochemical Properties of Imidacloprid.

Common Name	*Imidacloprid (ISO)*
Developing name	NTN 33893
IUPAC name	1-(6-chloro-3-pyridylmethyl)-N-nitroimid-azolidin-2-ylideneamine
Chemical abstract name	1-(6-chloro-3-pyridinylmethyl)-N nitro-2-imidazolidinimine
CAS Registration number	[138261-41-3]
Commercial names	For crop protection: Admire, Confidor, Gaucho, Provado; for termite control: Premise, Hachikusan; for animal health applications: Advantage
Structural formula	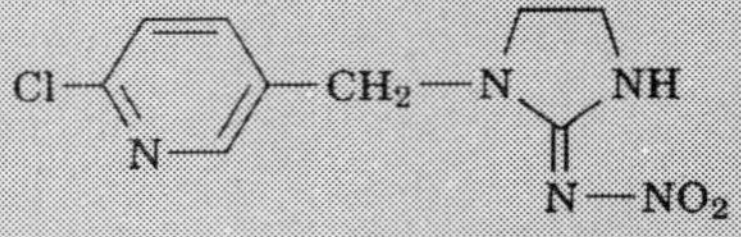
Molecular formula (Mol. Wt)	$C_9H_{10}ClN_5O_2$ (255.7)
Physical state (20°C)	colorless solid with a faint odor
Melting point	144°C
Density (20°C)	1.41 g cm^{-3}
Vapor pressure (20°C)	4×10^{-10} Pa
Hendry's constant (calcd, 20°C)	2×10^{-10} Pa m^3 mol^{-1}
Water solubility (g/L, 20°C)	0.61
Solubility in organic solvents (g/L, 20 °C)	Hexane: <0.1, Acetone: 50, Toluene: 0.69, 2-Propanol: 2.3, DMSO: >200, Acetonitrile: 50, Dichloromethane: 67, Methanol: 10
Partition coefficient *n*-octanol/water (21 °C)	log $P_{OW} = 0.57$

the adequate residual activity. These applications have the advantage over conventional methods not only in agronomic value such as uniform distribution of the active ingredient, accurate dosing, reduced dose rates, and longer application intervals, but also in their environmental benefit in that only a small fraction of the land is exposed to the insecticide.

As shown in Tables elsewhere in this chaper, various soil and foliar insects are controlled for long periods by incorporation of ingredient into the soil or by seed dressing or seed coating. It is highly effective against early season sucking pests such as *Myzus persicae* at a soil concentration of 0.31 ppm for more than 8 weeks and *Aphis fabae* at a concentration of 1.25 ppm for more than 5 weeks.

The likely long period of efficacy is shown by seed treatments. The seedling-box treatment, a method of placing granules in a nursery box before machinery transplanting to the field, made possible more than 11-week control of noxious insect pests for rice. The relation of pest control efficacy to the systemic property and the residual activity was investigated.

Studies on seeds of winter wheat treated with radioactive imidacloprid frevealed a continuous uptake of the applied radioactivity into the sprouting wheat from 1% at the first leaf stage to approximately 19% at full maturity. Concentrations as low as 0.12 mg/kg of younger leaves at the end of the shooting phase 195 days after sowing showed an insecticidal efficacy of 98% against *Rhopalosiphum padi.*

In a seedling-box treatment, a continuous movement of the ingredient from the roots to the leaves and sheaths occurs and 0.01 ppm exist in the aerial parts 80 days after application. As a result, a one-shot application of imidacloprid to the nursery box (1 g a.i. per box) can control the brown hopper for the entire season.

It can be calculated that an 85% less amount of pesticides is needed to achieve equivalent contro of pests by imidacloprid-coated seeds for sugar beet production in the United Kingdom and less than 5% of the land exposed to granules or sprays of conventional products

Antifeedant Activity Against Homopteran and Coleopteran Species

The antifeeding effect of imidacloprid at lower (sublethal doses is different from the fast-acting insecticidal efficacy at recommended field rates. The sublethal effects may contribute particularly to the biological activity of imidacloprid against plant-feeding homopteran pest species.

Low concentrations of imidacloprid have been shown to elicit behavioural changes in aphids, whiteflies, and planthoppers. Such behavioural changes include the *depression* of honeydew excretion and wandering, resulting in death due to starvation. Similar effects were also described for coleopteran pests such as black maize beetles and wireworms when feeding on stems of seed-treated maize plants.

Furthermore, sublethal concentrations have been shown to reduce the fertility of aphids, thus, demonstrating its potential to prevent the buildup or spread of populations later in the season when the con-centration of imidacloprid in systemically treated plants declined.

Control of Virus Vectors

The transmission of plant-pathogenic viruses by plant-sucking homopteran pest species is

Table 17.3: Spectrum of Activity (LC_{95} in ppm a.i.) of Imidacloprid After Foliar and Soil Application under Laboratory Conditions.

Foliar Application		
Pest Species	***Developmental Stage***	***LC_{95} (>ppm)***
Homoptera		
Aphis fabae	Mixed	8
Aphis gossypii	Mixed	1.6
Aphis craccivora	Mixed	1.6
Aphis pomi	Mixed	8
Brevicoryne brassicae	Mixed	40
Myzus nicotianae	Mixed	8
Myzus persicae	Mixed	1.6
Phorodon humuli	Mixed	0.32
Laodelphax striatellus	Larvaethird instar	1.6
Nephotettix cincticeps	Larvae third instar	0.32
Nilaparvata lugens	Larvae third instar	1.6
Sogatella furcifera	Larvae third instar	1.6
Pseudococcus comstocki	Larvae	1.6
Bemisia tabaci	Larvae second instar	8
Hercinothrips femoralis	Mixed	1.6
Lepidoptera		
Chilo suppressalis	Larvae first instar	8
Helicoverpa armigera	Larvae second instar	200
Plutella xylostella	Egg, Larvae second instar	200,200
Spodoptera frugiperda	Egg, Larvae second instar	200,40
Heliothis virescens	Egg	40
Coleoptera		
Leptinotarsa decemlineata	Larvae second instar, Adult	40,40
Lema oryzae	Adult	8
Lissorhoptrus oryzophilus	Adult	40
Phaedon cochleariae	Larvae second instar	40
Diabrotica balteata	Larvae third instar	1.6

(Table Contd.)

Soil Application		
Pest Species	**Developmental Stage**	**LC_{95} (>ppm)**
Soil insects		
Hylemyia antiqua	Larvae	5
Diabrotica balteata	Larvae	2.5
Agriotes spp.	Larvae	5
Agriotis segetum	Larvae	20
Reticulitermes flavipes	Imago	10
Foliar insects		
Phaedon cochleariae	Larvae	5
Myzus persicae	Mixed population	0.16
Aphis fabae	Mixed population	0.16
Spodoptera frugiperda	Larvae	10

one of the major threats in many cropping systems, e.g., sugar beet, cereals, and vegetables. It has been shown that imidacloprid—foliarly and systemically applied—particularly prevents the persistent transmission of circulative, phloem-restricted viruses, due to the rapid action on homopteran virus vectors.

Some successful examples include the prevention of the secondary spread of barley yellow dwarf virus transmitted *by R. padi* and S. avenae, the control of virus yellows and potato leafroll virus transmited by *M. persicae*, the leafstripe virus transmitted by *L. striatellus*, and the tomato-spotted virus transmitted by *B. tabaci*.

Activity on Resistant Insect Species

The chloronicotinyl/neonicotinoid insecticide imidacloprid acts agonistically on nicotinic acetylcholine receptors (nAChR). Because this is a new biochemical target addressed by modern insecticides, in principle, existing resistance mechanisms are not expected to confer cross resistance to imidacloprid.

Table elsewhere in this chapter shows that imidacloprid is equally effective against insecticide-susceptible and *M. persicae* resistant to organophosphates, carbamates, and pyrethroids. Many examples were reported in which the resistant strains were as susceptible to imidacloprid as were the sensitive ones: aphids, whiteflies, planthoppers, and the Colorado potato beetles.

Although no practical loss of field efficacy seems to be observed at present, selection studies in laboratories revealed that certain pest insects can be selected for lesser susceptibility. A population of small brown hopper reared under malathion and propoxur pressure showed 18-fold tolerance to imidacloprid. In another example, the continuous hydroponic laboratory selection of a population of *Bemisia argentifolii* resulted in more than 80-fold resistance after 24 generations.

Table 17.4: Residual Activity (Weeks) Against Pest Species (>95% Mortality).

Method/Dose Rate (a.i.)	*Pest Species Residual*		*Activity (Weeks)*	
Soil treatment (ppm)	*Myzus persicae* (green peach aphid)		*Aphis fabae* (black bean aphid)	
	2.5		>8	>5
	1.25		>8	5
	0.63		>8	4
	0.31		>8	3
	0.15		4 2	
Seed dressing (g/kg seed)	*Aphis gossypii* (cotton aphid)		*Aphis fabae* (black bean aphid)	
	1.0		>5	>5
	0.25		2 5	
	0.06		<2	3
	Seed coating (g/unit)		*Aphis fabae* (black bean aphid)	
	30		7.7	
	70		8.3	
	110		9.3	
Seedling-box treatment *(g/box)*	*Nephotettix cincticeps* (green rice leafhopper)	*Nilaparvata lugens* (brown planthopper)	*Sogatella furcifera* (white-backed planthopper)	
	3	>11	>11	>11
	1	>11	>11	>11

A clear indication of cross-resistance was reported between imidacloprid and other chloronicotinyls in *B. tabaci* from southern Spain. Resistance management strategies have been proposed to prevent or delay the development of resistance to imidacloprid in pest insects.

They include deploying several active ingredients of different insecticidal mechanisms in an annual or regional rotation, applying the preparations at the recommended doses and spray intervals, alternating practice with transgenic crops, combining biological pesticides and physical pest control methods, performing crop rotation to delay colonization, reducing immigrant density, and decreasing the number of generations or manipulating crop cutout to force insect survivors into refuge areas prior mating.

For applying such strategies, collecting baseline data and continuous monitoring of susceptibility of pests to imidacloprid is indispensable. It is also argued that systemic applications of persistent insecticides such as imidacloprid should be limited to minimize the dangers of continuing to select when control is no longer needed.

Agriculture Significance

Due to the novel biochemcial target, broad-spectrum, and high intrinsic activity as well as

Table 17.5: Efficacy of Imidacloprid and Conventional Insecticides (LC_{95}, ppm) Against a Susceptible (S) and Resistant (R) Greenhouse Strain of Myzus persicae (Leaf Dip Bioassay).

Insecticide	*LC_{95} (ppm) Against S/R Strains*	*RF*
Imidacloprid	8/8	1
Ethylparathion	8/>1000	>125
Pirimicarb	40/1000	25
Cypermethrin	8/1000	125

good plant compatibility, the control target of imidacloprid encompasses a large variety of insect species for various plant crops. Further, the appropriate physicochemical properties such as a good systemic activity, adequate stability in field, and the amphipathy allow imidacloprid to be applied in unusually diverse ways, e.g., as a foliar spray, or systemically as a drench or in irrigation water, as a granular soil application, as seed dressing, or as a paint-on formulation. The representative use rates and target pests are summarized in Tables elsewhere in this chapter. Further technical information is available from the product brochures.

Chemistry

Spectral Data and Dissociation Constant

Electronic absorption λ_{max} (nm)/log s(water): 269 (4.17); ^{1}H-NMR (S, CDCl3): 3.54 (m, 2H), 3.83 (m, 2H), 4.55 (s, 2H), 7.35 (d, J = 8.4, 1H), 7.70 (dd, J = 7.0/2.6, 1H), 8.32 (d, J = 2.6, 1H), 8.20 (bs, 1H); IR (KBr, v, cm^{-1}): 3155, 1580, 1565, 1300, 1280; EIMS (70eV, m/e, rel int): 209 (55%), 173(80%), 126 (100%). The crystal structure with the structure parameters is reproduced in Figure and Tables elsewhere in this chapter.

Synthesis

The first laboratory synthesis of imidacloprid is outlined in Figure elsewhere in this chapter. Reduction of 2-chloro-5-pyridinecarbonyl chloride to 2-chloro-5-hydroxymethylpyridine was carried out by excess NaBH4 in water, which was converted to the chloride **(10)** by $SOCl_2$.

Imidacloprid was obtained by the coupling reaction of **10** with 2-nitroiminoimidazolidine in acetonitrile with potassium carbonate as base. This method was successfully applied to the synthesis of [^{3}H]imidacloprid using $NaB[^3H]_4$ as the tritium source.

Technical production starts with the Tschitschibabin reaction of 3-methylpyridine giving 2-amino-5-methylpyridine, which is transformed to the chloride **(15)** by the Sand-meyer reaction in the presence of hydrogen chloride.

A successive operation of chlorination of the methyl group to 10 and the subsequent substitution of the active chloride with ethylene diamine to **16** are carried out without isolation of the intermediates. The final product is produced by ring formation with nitroguanidine. This multistep process affords the product at a purity of >95%.

Table 17.6: Use Rates for Foliar and Soil Application for Representative Crops.

Crops	Target Pest	Recommended Rates of Active Ingredient	
		Foliar Treatment (g/ha)	Soil Treatment (g/unit)
Vegetables	Aphids, thrips	50-200	100-300
Vegetables	whiteflies	100-200	100-300
Cotton	aphids	25-100	0.05-0.075 (g/m)
Cotton	whiteflies	100-350	0.05-0.075 (g/m)
Potatoes	aphids, Colorado potato beetle	50-100	180-200 (in furrow)
Tobacco	aphids, thrips	50-100	100-250 (drip/soil drench)
Pome fruits	leafminer	50-100	0.5-1.0/tree (drip/soil drench)
Pome fruits	aphids	50-100	0.25-0.5/tree (drip/soil drench)
Citrus	aphids	100-200	0.2-2.0 (drip/soil g/tree)
Rice	hoppers	25-50	100-200 (0.5-1.0 g/box)
Rice	rice water weevil		100-200 (0.75-1.0 g/box)

Table 17.7: Target Insects of Imidacloprid After Seed Treatment.

Crop	Soil-dwelling Insects	Early Leaf-feeding and Sucking Insects
Corn	Wireworms, black maize beetle, ground beetle, rootworms, seed corn maggot	Argentine stem weevil, fruit fly, aphids, jassids, Armyworm
Sorghum	wireworms, false wireworms, fire ants	Aphids, green bug, chinch bug
Cotton	cotton root weevil, wireworms, termites	Cotton aphid, thrips, jassids, white fly
Canola (oilseed rape)	Flea beetle	Aphids, thrips, cabbage root fly
Cereals	Wireworms, ground beetle	Bird cherry aphid, grain aphid, Russian wheat aphid, Hessian fly
Rice	Termites	Green leafhopper, smaller brown planthopper, brown planthopper, rice leaf beetle, rice water weevil, *Stechaenothrips* spp.
Sugar beet	Pygmy mangold beetle, wireworms, millipedes, springtails, beet root weevil, flea beetle	Green peach aphid, black bean aphid, beet leafminer, flea beetle, lygus (bugs)
Sunflower	Wireworms, ground beetle	Aphids, *Zygogramma exclamationis*
Beans	Rootworms, grubs	Black bean aphid, pea and bean aphid, whitefly, leafhopper
Potato	Wireworms	Colorado potato beetle, green peach aphid, jassids

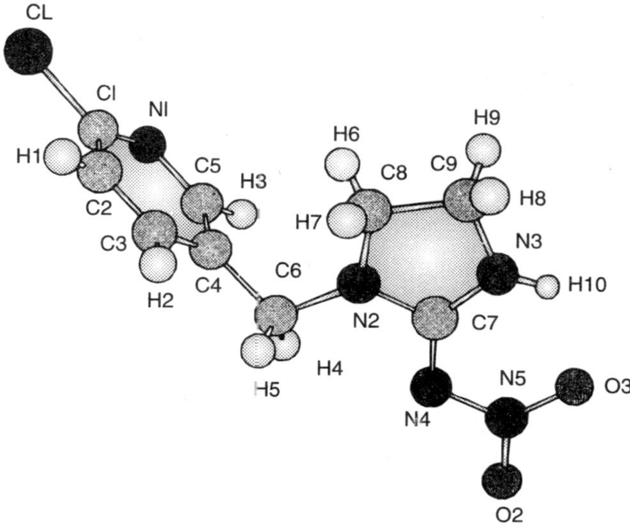

Figure 17.8: Crystal structure of imidacloprid.

Biological Activity and Mode of Action

In vivo symptomatology in American cockroach, *Periplaneta americana*, after injecting an LD_{50} dose of 1 mg a.i./insect is characterized by the following sequence: walking up and down with a loss of strength in legs within 15 min; leg tremor followed by whole body shaking and prostration with a curled abdomen 30 min later; and death.

The symptoms are similar to poisoning with nicotine, suggesting that both molecules act on a similar set of receptors in the central nervous system. Imidacloprid displaces radiolabeled α-bungarotoxin (α-BTX), a competitive antagonist of acetylcholine, from its binding site on the insect nicotinic acetylcholine receptors (nAChR) like nicotine.

For imidacloprid and nicotine, the IC_{50} values in 10^{-6} M, the concentrations to displace 50% specific binding of the probe [^{3}H] or [^{125}I]α-BTX on nAChR, are respectively, 1.95 and 1.25 in honeybee head membrane, 2.9 and 0.086 in stable fly head membrane, and 0.20 and 9.8 in American cockroach cholinergic motor neuron.

Direct evidence of the molecular target of imidacloprid is that [^{3}H]imidacloprid binds with 95% specificity to membranes from housefly, *Musca domestica*, and other insect nerve tissue, and that the binding is displaceable by acetylcholine, nicotine, and other nicotinic ligands. The high affinity binding to nAChR is also described for the aphid.

Electrophysiological results support its primary site of action at the nAChR in insects. At concentrations below 10^{-6} M, imidacloprid induces a rapid depolarization of postsynaptic neurons accompanied by the induction of action potentials and subsequently by a complete block of the nerve impulse propagation.

The induction responses are very similar to acetylcholine responses and are blocked by nicotinic receptor antagonists.

These biphasic effects are evident in patch-clamp and two-electrode voltage-clamp studies

Table 17.7: Selected Interatomic Atom Distances and Angles of Imidacloprid.

Bond Length		*Bond Angle (degree)*		*Tortional Angle (degree)*	
C4-C6	1.50	C3-C4-C6	122.2	C4-C5-C6-N2	117.8
C6-N2	1.45	C4-C6-N2	113.1	C7-N2-C6-C4	106.9
N2-C7	1.34	C6-N2-C7	125.7	C6-N2-C7-N4	-0.5
N2-C8	1.45	C6-N2-C8	122.2	N2-C7-N4-N5	176.3
C7-N3	1.32	C7-N2-C8	112.0	C7-N4-N5-O1	0.0
N3-C9	1.45	N2-C8-C9	102.7	C7-N4-N5-O2	179.4
C8-C9	1.53	C8-C9-N3	102.8	N4-C7-N2-C8	176.5
C7-N4	1.34	C7-N3-C9	112.5	N3-C7-N4-N5	-3.0
N4-N5	1.35	N2-C7-N3	109.6	C9-N3-C7-N4	178.8
N5-O1	1.23	N2-C7-N4	117.1		
N1-N2	4.35(5.45)b	N3-C7-N4	133.3	Interplane angle (degree)	
N2-O1	4.47(5.80)b	C7-N4-N5	116.7		
		N4-N5-O1	115.4	75.7	
		N4-N5-O2	122.9		

on insects, and the insecticidal potency of imidacloprid-related compounds correlates with their capacity to cause excitation in cockroach nerve cords.

In contrast to insects, the affinity to the vertebrate nAChRs is very weak. It failed to recognize the specific binding sites in brain membranes from human, dog, mouse: and chicken or electric organ of the electric eel.

The low mammalian activity was demonstrated further by the low potency as an inhibitor of [^{3}H]nicotine binding in rat brain and [^{3}H]α-BTX binding to the muscle-type nAChR from *Torpedo* spp., and the weak agonistic action in mouse NIE-115 neuroplasma and BC3H1 muscle cells, low activity in ion channel activation compared with acetylcholine with rat $\alpha_4\beta_2$ and α_7 subtypes expressed in *Xenopus oocytes,* weak or partial agonistic nature with recombinant chick $\alpha_4\beta_2$ receptor, and very low affinity to immuno-isolated nAChRs of varying subunit composition.

Table elsewhere in this chapter compares the pharmacological characteriztions with nicotine in radioligand binding sites in insects, *Torpedo* spp., and rodents. The safety factor, the ratio of the lethal doses for mammals to insects, can be estimated as 7300 from the LD_{50} of 0.062 mg/kg for *M. persicae* by topical application and an *LD50* for rats of 450 mg/kg after oral administration, which ranks imidacloprid as one of the insecticides of highest selectivity known.

The binding models of chloronicotinyl/neonicotinoid insecticides with the recognition site on the nAChR were proposed, and a comparative molecular field analysis (CoMFA) for the binding mode was performed.

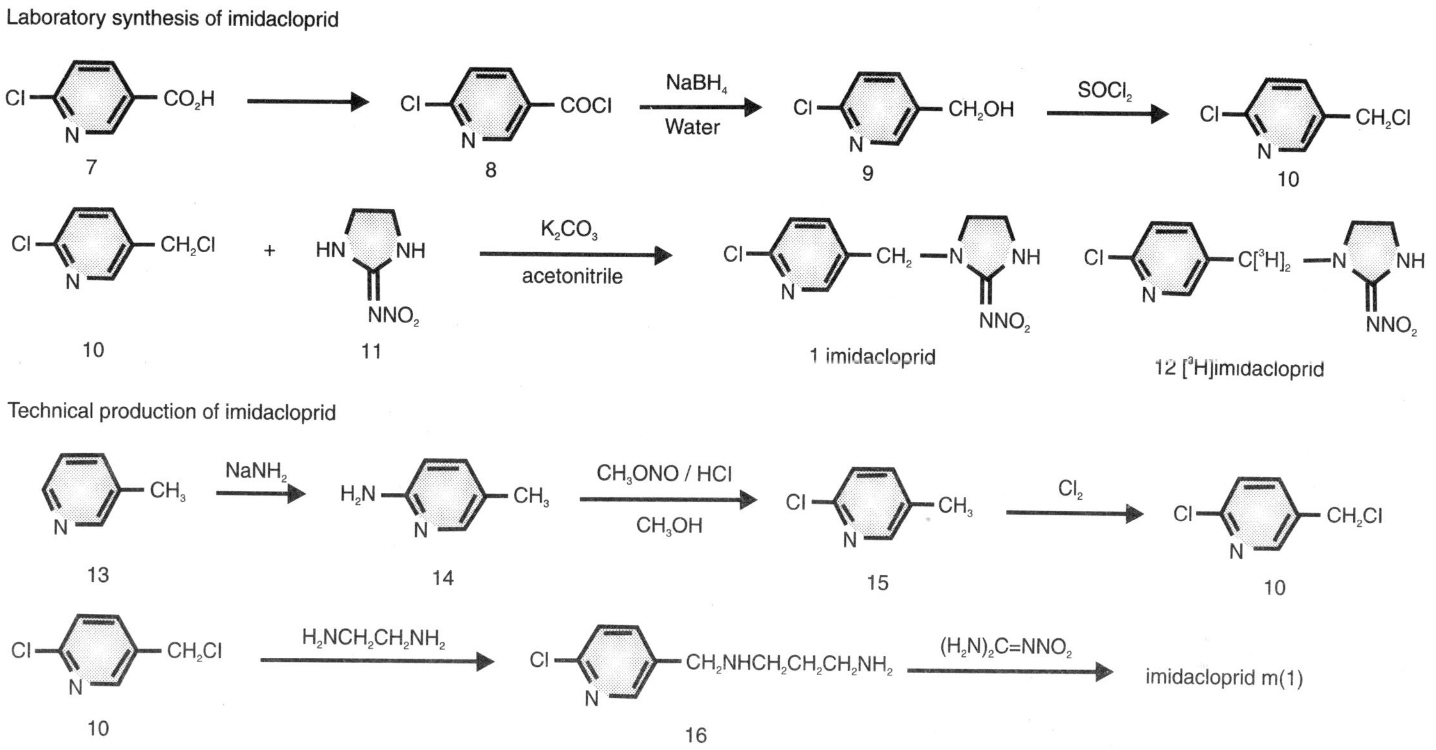

Laboratory synthesis of imidacloprid

Cl–(pyridine)–CO_2H (7) → Cl–(pyridine)–COCl (8) —$NaBH_4$ / Water→ Cl–(pyridine)–CH_2OH (9) —$SOCl_2$→ Cl–(pyridine)–CH_2Cl (10)

Cl–(pyridine)–CH_2Cl (10) + HN–(imidazolidine)–NH, =NNO_2 (11) —K_2CO_3 / acetonitrile→ Cl–(pyridine)–CH_2–N–(imidazolidine)–NH, =NNO_2 (1 imidacloprid); Cl–(pyridine)–$C[^3H]_2$–N–(imidazolidine)–NH, =NNO_2 (12 [^{3}H]imidacloprid)

Technical production of imidacloprid

(pyridine)–CH_3 (13) —$NaNH_2$→ H_2N–(pyridine)–CH_3 (14) —CH_3ONO / HCl, CH_3OH→ Cl–(pyridine)–CH_3 (15) —Cl_2→ Cl–(pyridine)–CH_2Cl (10)

Cl–(pyridine)–CH_2Cl (10) —$H_2NCH_2CH_2NH_2$→ Cl–(pyridine)–$CH_2NHCH_2CH_2CH_2NH_2$ (16) —$(H_2N)_2C{=}NNO_2$→ imidacloprid m(1)

Figure 17.9: Lbtory and thil preparation of iidloprid.

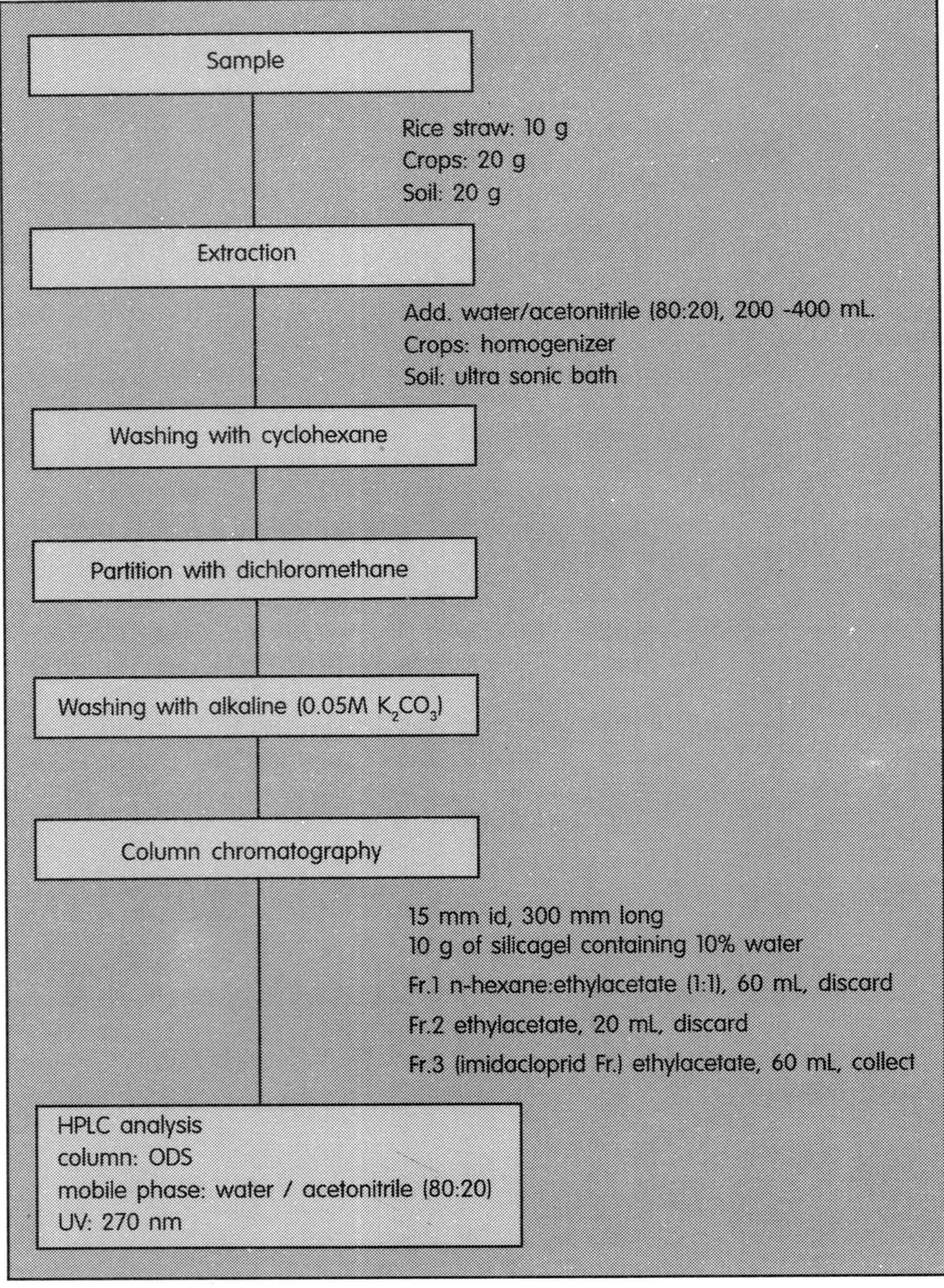

Figure 17.10: Analytical procedure to separate imidacloprid from crops and soil.

Table 17.8: Pharmacological Characterization of Imidacloprid and Nicotine in Radioligand Binding Sites in Insects, Torpedo, and Rodents.

	IC_{50} (10^{-6} M)				
Ligand/Probe	Housefly [^{3}H]IMI	Aphidb [^{3}H]IMI	Torpedo [^{3}H]α-BTX	Ratd [^{3}H]NIC	Moused [^{3}H]α-BTX
Imidacloprid	0.002	0.00076	1060	0.98	42
(–)-Nicotine	0.6	2.1	23	0.0096	1.9

ENVIRONMENTAL FATE

Stability in Storage

Formulated imidacloprid is stable at room temperature in the dark under usual storage conditions; the active ingredient is labile in alkaline media. The calculated half-life is 57 min based on the direct photolysis in water. The photostability of nitroimine with nitromethlene or cyanoimine chromophore, which functions an electron-attracting group in chloronicotinyl molecules in Figure elsewhere in this chapter, was compared on the quantum chemical basis.

Residue Analysis

The strong absorption at 270 nm attributable to the nitroimino chromophore is a reliable indication for the analysis of imidacloprid. The HPLC-UV method is standardized for the detection and quantification of the parent molecule.

The imidacloprid concentration can be measured by this method at levels >10 ng/g in different fruits and vegetables, and at 5–20 ng/g in rice plants and soil. Other detectors are also used for HPLC analysis of the parent molecule such as diodearray and pulsed reductive amperometry.

The metabolites or derivatives may be subject to an alternative determination; photochemically induced fluorimetry of 2-hydroxyimino derivative, GC-MS of 3-N-perfluoroalkanoyl derivatives (63,64), or the urea after hydrolysis. Total residue analysis based on 6-chloronicotinic acid is a well-established method. The recovery procedures from the test plant tissues and the HPLC measurement conditions including the totcl residue analysis are described in detail. Figure elsewhere in this chapter shows a representative residue analysis scheme for crops and soil.

Fate in Soil

The aerobic decomposition rates in soil are variable depending on type of soil, vegetation, and conditioning of the soil, e.g., by treatment with manure. The mean of the half-lives reported in the literature is 80 days.

This is in line with a laboratory experiment simulating groundcover when a half-life of 48 days was found. According to laboratory trials, degradation is more rapid under anaerobic than under aerobic conditions. The halflives are 27 days in a water/sediment system after it had attained

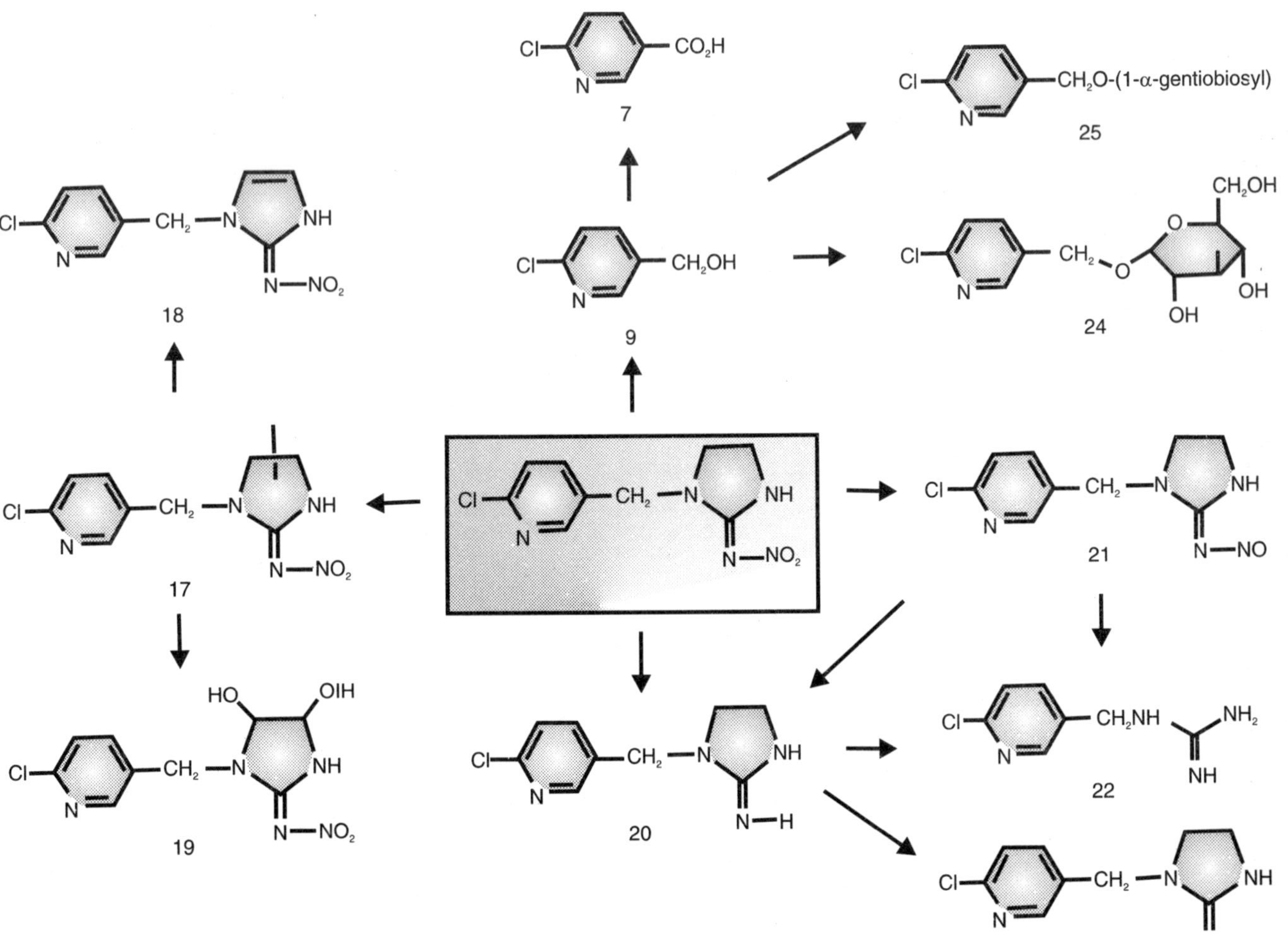

Figure 17.11: Metabolic degradation of imidacloprid (1) in Plants.

anaerobic conditions, and 53 and 69 days in two Japanese paddy soil types, volcanic and alluvial, respectively.

More than half of the amount is decomposed in less than 1 week in the presence of light. The sunlight and microbial activity of a water/sediment system are important factors for the degradation of imidacloprid.

The study in soil using [^{14}C]imidacloprid over 1 year shows that he metabolism proceeds via loss of the nitro group to form guanidine **(20)** and simultaneously via cleavage and oxidation of the imidazolidine ring to yield 6-chloronicotinic acid and finally carbon dioxide.

The main product in recovery was imidacloprid in aerobic soils or guanidine in anaerobic and paddy soils, and none of the identified metabolites was found in concentrations above 10 ppb after 100 days. According to a 5-year lysimeter study, ca. 40% of the total radioactivity is lost as CO_2 in 2 years and the radioactive residues of leachate are adsorbed within 20 cm of the soil surface.

Fate in Plants

Imidacloprid is well translocated through xylem from the roots to the upper parts of plants due to its adequate water solubility and lack of an acidic or basic moiety. The active ingredient is accumulated in younger leaves after seed treatment.

The degree of metabolism is significantly influenced by the application method. In the case of sprays or surface treatments, an unchanged parent compound is the main constituent of the residue because only a part of the applied component penetrates into the plant and is metabolized after application. Following granular application or box treatment, the active ingredient is readily taken up via the roots and the degree of metabolism is higher.

The metabolic route in plants is presented in Figure elsewhere in this chapter. The main degradation routes are:

1) hydroxylation of the imidazolidine ring leading to the mono- and dihydroxylated compounds **(17,19)** with subsequent removal of water to form the olefin metabolite,
2) transformation of the nitro group to the guanidine metabolite **(20)** possibly partly the nitrosoimine intermediate, and
3) oxidative cleavage of the methylene bridge to form 6-chloropicolyl alcohol (and the conjugates **24,25)** and further oxidation to 6-chloronicotinic acid. These metabolites are detected in variable amount in treated cultures.

The following values in parentheses describe in order the total residue in ppm and the main metabolites with the amounts at the harvesting times for representative plants after the recommended application dose (71): spray tomato (0.85, 1: 88.0%), spray potato (Leaves; 1.35, 1: 37.9%, 20: 12.6%, 17: 7.0%/Tubers; 0.009, 7: ca. 30%, 1: ca. 5%), granular potato (Leaves; 5.76, 1: 26.7%, 7: 8.3%, **20:** 8.2%, **17:** 4.6%/Tubers; 0.091, 1: 48.3%, 20: 11.3%, 7: 9.4%, 17: 8.0%), granular corn (Straw; 3.08, 1: 22.2%, **20:** 10.9%, 17: 6.0%/Grain; 0.039, **1:** 25.2%, **18:** 13.1%, 17: 9.3%), granular cotton (Leaves; 0.11, 9: 13.2%, **20:** 9.8%, 24: 6.3%), granular rice (Straw;1.47, **20:** 25.6%, 1: 11.5%/ Grain; 0.036, 1: 6.3%), nurserybox rice (Straw; 1.31, **20:** 36.2%, 1: 8.1%/Grain; 0.014, 1:13.6%). The residues in the storage organs of the plants in the crops and fruits are very low.

TOXICITY AND SAFETY ASPECT

Metabolism in Animals

After oral administration of [^{14}C]imidacloprid to rats, the radioactivity was readily absorbed and distributed to all organs within 1 hour and eliminated by 96% after 48 hours, about 75% thereof with the urine and about 21% with the feces. At the time, the radioactive residues in the body amounted to ca. 0.5% of the administered radioactivity.

Therefore, bioaccumulation of imidacloprid is low in rats. Maximum plasma concentration was reached at 1.1 and 2.5 hours. Two major routes of biotransformation are proposed. The first route includes an oxidative cleavage of the parent compound rendering 6-chloronicotinic acid and its glycine conjugate. Dechlorination of this metabolite formed 6-hydroxynicotinic acid and its mercapturic acid derivative.

The second route includes the hydroxylation to 17 followed by elimination of water of the parent compound rendering olefin **18.** There were no sex, dose, or label differences in toxicokinetics. The metabolism in hens and goats is similar to that in rats.

Acute Toxicity

The following data for acute toxicity are reported: acute oral LD_{50} for male and female rats 424-475, for mice 131-168 mg/kg; dermal LD_{50} for rats >5000 mg/kg; inhalation LC_{50} (4h) for rats 69 mg/m^3 aerosol, >5323 mg/m^3 dust; nonirritating to eyes and skin (rabbits) and not a skin sensitizer (guinea pig). The toxicity class is II on World Health Organization (WHO) (a.i.) and II on U.S. Environmental Protection Agency (EPA) (formulation).

Chronic Toxicity

The following no-observed-effect levels (NOELs) are established: rat, 5.7 mg/kg body weight per day; and dog, 15 mg/kg body weight per day. The ADI is 0.057 mg/kg body weight. The compound has no mutagenic or carcinogenic potential. In the developmental toxicity study, the maternal and fetal NOELs for rats were 10 mg/kg and 30 mg/kg body weight per day, respectively.

The corresponding NOELs for rabbits were 8 mg/kg and 24 mg/kg body weight per day. No fetal malformations were observed in rats and rabbits at any dose level. From the reproductive toxicity study in two-generation and two-litter rats, a pup reproductive NOEL of 12.5 mg/kg body weight per day was estimated.

No effect on mating behaviour, fertility, gestation, conception, litter size, or mortality was observed. Malformation did not occur. There was no primary neurotoxicity, and the NOELs for neurotoxicity were 307 mg/kg body weight per day after acute exposure and 196 mg/kg body weight per day in the 3-month study. The tolerances by EPA are listed in Table elsewhere in this chapter.

Ecotoxicology

The median tolerance limits (TLms) in milligrams/liter are 190 (48 h) for carp, 237 (96 h) for golden orfe, 211 (96 h) for rainbow trout, and 85 (48 h) for water fleas. The EC_{50} value for algae (*Scenedesmus subspicatus*) is also larger than 10 mg/L. The hydrophilic properties are responsible

Table 17.9: Tolerances in ppm for Residuesa of Imidacloprid (EPA).

Crop	Tolerance	Crop/Product	Tolerance
Apple	0.6	Sweet corn (kernel plus cob with husk removed)	0.05
Pear	0.6	Soybean (meal)	0.5
Tomato	1.0	Wheat (grain)	0.05
Peas (legume vegetables)	0.1	Rice (grain)	0.2
Sugar beet (roots)	0.05	Meat	0.3
Potato	0.5	Milk	0.1
Legume vegetables (seed)	0.3	Egg	0.02
Leafy green vegetables	3.5		

for the low fish toxicity and low bioaccumulation. The acute bird toxicity is high; oral LD_{50} is 150 mg/kg for hen, 31 mg/kg for Japanese quail, 152 mg/kg bobwhite quail, and 33 mg/kg for canary and pigeon. In practice, however, considering the very low residues in seeds and the reported repellent effect on birds, the avian hazard may be negligible. The activity of soil microorganisms is not impaired even at very high dose rates of 2000 g a.i./ha.

For earthworm (*Eisenia foetida*), the LC_{5C} is 10.7 mg/kg dry soil. Spray applications, four times overdosed, had a transient effect on the population, which had been recovered by autumn of the year of application. The coated sugar beet seeds had no effect on this worm.

The effects of imidacloprid on beneficials have been tested. It was harmless to the predator insects *Deraeocoris nebulosus* (Uhler), Olla v-nigrum (*Say*), *Chrysoperla rufilabris* (Burmeister), and a few predatory mites after foliar application at the near-recommended field rate of 127 mg a.i./L). No significant mortality was observed for the predator *Perillus bioculatus* (*F.*) after 24 h contact with potato foliage after foliar spray. On the other hand, harmful effects on nymphs and adults of predators *Podisus maculiventris* and *Orius laevigatus* were reported.

The product is harmful to honeybees by direct contact and should not be applied during the flowering period. Also, it is suspected to be harmful to silkworms. Mulberry leaves should not be fed to silkworms until 40 days after spray. Because virtually no apparent impact on most beneficial insects has been observed after seed dressing, whenever possible, all types of systemic applications such as seed dressing, soil treatment, or stem painting/injection may be recommended.

Application in Nonagricultural Fields

High insecticidal potencies together with nonvolatility and stability under storage conditions, especially under the shelter of sunlight, impart imidacloprid, a good candidate for protecting wooden structures from subterranean pests. Imidacloprid has been successfully applied as a termiticide.

Suitable toxicological properties such as high insecticidal activity, low mammalian toxicity, and absence of eye/skin irritation and skin sensitization potential allows imidacloprid to be used to control parasites on pets (cats and dogs) and in homes.

Monthly topical application of 10 mg/kg kills rapidly existing and reinfesting flea infestations on pets and breaks the flea life cycle by killing adult fleas before egg production begins. Owing to accurate, quick, and simple dosing through dose unit pipettes, which helps to ensure compliance by pet owners, the flea spot-on solution is finding a substantial veterinary market.

MODE OF ACTION

The following document was prepared by the Insecticid Resistance Action Committee. The Insecticide Resistance Action Committee was formed in 1984 to provide a coordinated crop protection industry response to the developmen of resistance in insect and mite pests. "The mission of IRAC is to develop resistance management strategies to enable growers to use crop protection products in a way to maintain the efficacy. The organization is implementing comprehensive strategies to confront resistance."

IRAC Mode of Action Classification

Label will contain a box marking the group and type of material:

CHEMICAL GROUP	IA	INSECTICIDE

Table.

Primary Target Site	*Chemical Subgroup*
1. Acetyl choline esterase inhibitors	A. carbamates
	B. organophosphates
2. GABA-gated chloride channel antagonists	A. cyclodienes
	B. fiproles
3. Sodium channel modulators	Pyrethroids and pyrethrins
4. Acetyl choline receptor agonists/antagonists	Aa. chloronicotinyls
	B[a]. nicotine,
	C[a]. cartap, bensultap
5. Acetyl choline receptor modulators	Spinosyns
6. Chloride channel activators	Avermectin, emamectin benzo-ate Milbemycin
7. Juvenile hormone mimics	Aa. methoprene, hydroprene
	B. fenoxycarb
	C. pyriproxifen
8. Compounds of unknown or nonspecific mode of action (fumigants)	Aa. methyl bromide
	B. aluminum phosphide
	C. sulfuryl fluoride

(***Table Contd.***)

9. Compounds of unknown or nonspecific mode of action (selective feeding blockers)	A. cryolite B. pymetrozine
10. Compounds of unknown or nonspecific mode of action (mite growth inhibitors)	A. clofentezine, hexythia-zox B. etoxazole
11. Microbial disrupters of insect midgut membranes (includes Transgenic B.t. crops)	A1. B.t. israelensis A2. B.t. sphaericus B1. B.t aizawai B2. B.t. kurstaki C. B.t. tenebrionis
12. Inhibition of oxidative phosphorylation, disrupters of ATP formation	A. diafenthiuron B. organotin miticides
13. Uncoupler of oxidative phosphorylation via disruption of H proton gradient	Chlorfenapyr
14. Inhibition of magnesium-stimulated ATPase	Propargite
15. Inhibit chitin biosynthesis	Benzoylureas
16. Inhibit chitin biosynthesis type 1-Homopteran	Buprofezin
17. Inhibit chitin biosynthesis type 2-Dipteran	Cyromazine
18. Ecdysone agonist/disruptor	Tebufenozide
19. Octopaminergic agonist	Amitraz
20. Site II electron transport inhibitors	Hydramethylnon, dicofol
21. Site I electron transport inhibitors	Rotenone, METI acaricides
22. Voltage-dependent sodium channel blocker	Indoxacarb

ORGANOCHLORINES

Few chlorinated organic insecticides remain in use in north America and Europe. Several have been classified as Persistent Organic Pollutants (q.v.) and proscribed globally. However, their considerable benefits to humanity in the past should not be overlooked nor should the lessons that were learned during the period when they we applied to control disease vectors and agricultural pes throughout the world.

Unfortunately, their injudicious use and, a- the time of their introduction, ignorance of processes affecting fate and transport of pesticides and their residues led to the widespread occurrence of pollutants. The problems ranged from gross contamination at manufacturing sites to low level contamination the water of lakes, rivers, and estuaries.

Organochlorine compounds were not only used as pesticides, but they also had many industrial uses. Large quantities we used in heat-exchange systems and insulators. Chlorinated organic compounds were also byproducts of a number of chemical manufacturing processes, and as a result of the careless disposal and handling of wastes, quantities we spilled into the environment. Pollution of the environment was not the only problem. Associated with the increasing use of organochlorine insecticides were questions of long-term toxicity, other effects on wildlife, and the issue of increasing insect resistance. The following sections describe the properties of the different classes of organochlorine compounds used as insecticides, miticides, or acaricides.

Table elsewhere in this chapter summarizes the nomenclature and properties of the principal organochlorine compounds that have been used in pest control. A stimulus for the investigations that led to the initial discoveries of the insecticidal activity of organochlorine compounds was the spread of World War Ii. In Europe, supplies of traditional botanical insecticides used in crop production, such as pyrethrum extract and nicotine, were limited by wartime blockades and shortages.

The critical need to protect crops from insect pests and to protect personnel in tropical areas from malaria and other insect-borne diseases accelerated the search for synthetic replacements. The insecticidal properties of *hexachlorocyclohexane* were discovered almost simultaneously in France and England in 1940. The discovery of the insecticidal activity of lindane (the gamma isomer of hexachlorocyclohexane), followed by the well-known successes of DDT for controlling vector-borne diseases, stimulated the evaluation of synthetic organic compounds as new insecticides and their *commercialization*.

During the 1950s, the application of the Diels–Alder reaction to chemical synthesis gave rise to a new group of organochlorine insecticides, the cyclodienes. The insecticidal and acaricidal properties of halogen derivatives of benzene depend on the number and type of halogen atoms and their positions of substitution in the benzene molecule.

The insecticidal activity of the fluorobenzenes is relatively weak but greater than that of benzene. Insecticidal activity of the chlorobenzenes increases with the number of chlorine substituents up to three; trichlorobenzenes are the most active. 1,4-Dichlorobenzene has been used as a mothproofing agent. Bromobenzene and dibromobenzenes are somewhat more active than are the corresponding chlorobenzenes. Toluenes containing halogen in the aromatic nucleus are similar in activity to the corresponding benzene derivatives.

Activity is considerably higher when halogen is present in the methyl substituent, but the resulting derivatives are unsuitable for practical use because they have a strong irritating effect on mucous membranes. The fungicidal activity of chlorobenzenes increases from monochlorobenzene to hexachlorobenzene. There are several classes of organochlorine insecticides, but the use of many of them was discontinued or restricted because they were found to persist in the environment and accumulated or bioconcentrated through the food chain.

Directly or indirectly they showed potential for adverse effects in humans and the environment. Their effectiveness diminished as many insect species showed resistance after repeated applications. Initially, organochlorine insecticides were heavily used, but their disadvantages soon became apparent. An advantage was that, generally, they were not expensive to manufacture.

For example, toxaphene and related insecticides were produced by chlorination of camphene to give an insecticidal product, which, although consistent in properties and composition, was a mixture of more than 175 individual products. The technical product was an effective insecticide and was used in large quantities in the United States (some cotton received each year up to 22 pounds per acre cumulatively). After withdrawal of DDT and the cyclodienes in the 1970s, toxaphene sales in the United States in 1976 continued and were higher than those of any other insecticide, but the problems of resistance and the association of toxaphene spraying with the occurrence of spine injury in fish were among factors that led to its discontinuance. Implementation of national environmental policies became, from 1970, a major driving force in the choice of molecules that were appropriate for development as pesticides.

Research on mode of action was a guide to the design of molecules that might be more effective or selective. Understanding of the mode of action also served to indicate potential resistance problems that might occur when compounds were used injudiciously.

When an insect species had developed resistance to a class of insecticides, there was the likelihood that resistance to compounds of other classes possessing the same mode of action might develop. In a particular strain of insects, such resistance due to the same mechanism is termed "cross resistance" in contrast to "multiple resistance," which is the resistance of a strain to different compounds but resulting from different mechanisms.

The organochlorine compounds affect neural transmission. The elucidation of their modes of action and the structure-toxicity relationships have demanded many years of investigation. They were succeeded by carbamate and organophosphate insecticides, which showed more acceptable environmental behaviour because they were much more readily degraded than were the majority of organochlorine pesticides, but both classes were $inhibitor_f$ of acetylcholinesterase.

Consequently, the potential for resistance was a major concern, as was high acute toxicity to mammals, and it was important to exploit other modes of action in designing new insecticides. This was addressed by introduction of the synthetic pyrethroids and many new structural types, such as imidacloprid, that acted at different target sites.

Section 1

DDT

DDT: 1,1,1-trichloro-2,2-bis(4-chlorophenyl)ethane (IUPAC); 1,1,1-trichloro-2,2-di-(4-chlorophenyl)ethane; the technical product: 1,1,1-trichloro-2,2-bis(chlorophenyl)ethane contains 30% o,p′-DDT (1,1,1-trichloro-2-(2-chlorophenyl)-2-(4-chlorophenylethane)ethane); CAS RN [50-29-3]; $C_{14}H_9Cl_5$; m.wt. 354.5 practically insoluble in water; vapor pressure 0.025 mPa (20°C, p′-DDT). DDT is produced by condensation of chlorobenzene with chloral (obtained by oxidation of ethanol with bleaching powder) in the presence of strong sulfuric acid. The product is p,p′-DDT with o,p′-DDT as a significcant impurity (20–30% depending on conditions) and a trace o,o′-DDT.

DDT was relatively inexpensive to produce on a large scale, and as it had a wide spectrum of insecticidal activity with low mammalian toxicity, it became rapidly the insecticide of choice for control of many insect-borne diseases, such as malaria. It was also used on a widespread scale for control of some forest pests, such as the gypsy moth.

Initially, its environmental stability appeared to present a considerable advantage, but following its widespread application, residues of DDT and other organochlorine insecticides were shown to be ubiquitous in environmental samples and in wildlife. DDT was widely used in many agricultural crops. The primary use of DDT currently is as a vector control for eradication of malaria-bearing mosquitoes.

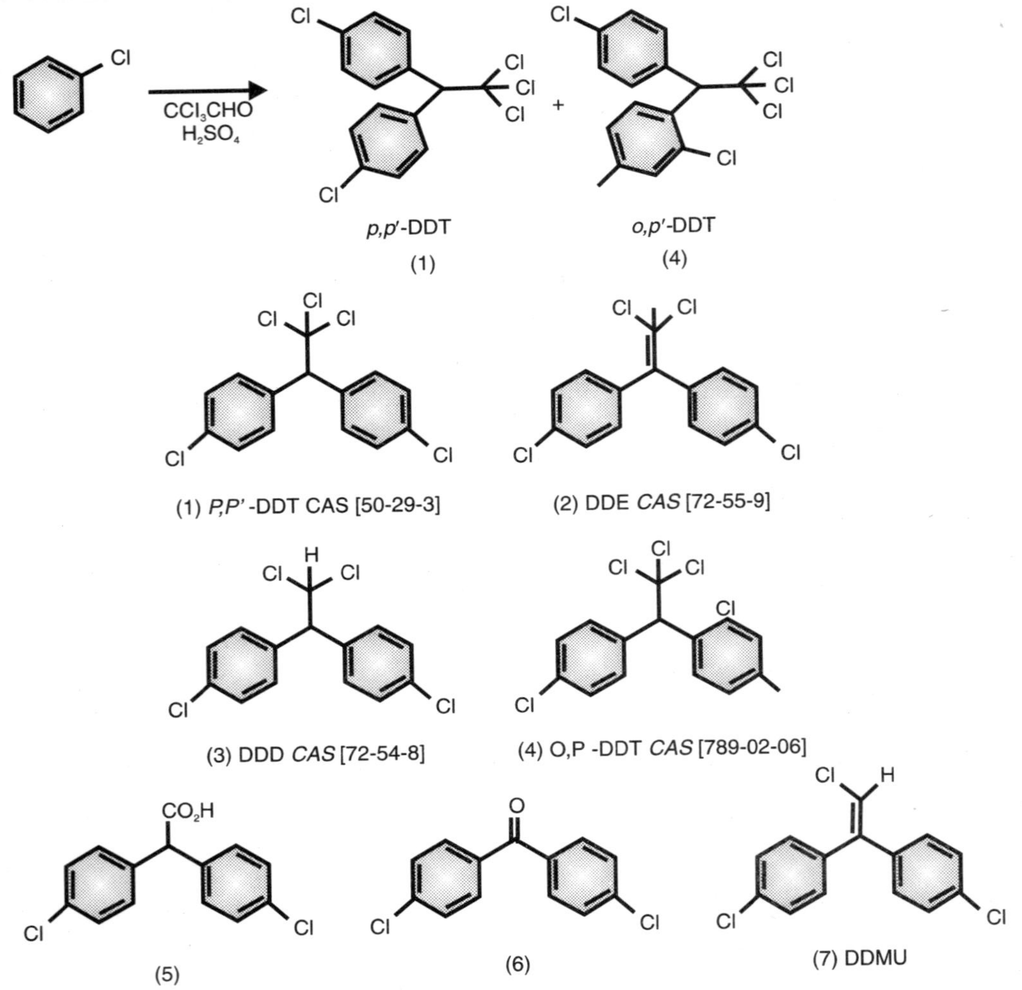

CCl_3CHO
H_2SO_4

p,p'-DDT
(1)

o,p'-DDT
(4)

(1) *P,P'* -DDT CAS [50-29-3]

(2) DDE *CAS* [72-55-9]

(3) DDD *CAS* [72-54-8]

(4) O,P -DDT *CAS* [789-02-06]

(5)

(6)

(7) DDMU

Figure 17.12: Manufacture of DDT: DDT structures 1-4; DDT structures 5-7.

Less-persistent insecticides have replaced DDT for control of insects on crops and in forests. DDT is a nerve poison that affects the sodium channel of nerve membranes. It is a nonsystemic nsecticide with contact and stomach action. The most important reactions of DDT are dehydrochlorination to DDE and reductive dechlorination to DDD.

These reactions occur abiotically, *in vivo* and in soils. The products resemble DDT in their recalcitrance toward environmental degradation. The stability of DDT and its principal metabolites DDD and DDE, in combination with their lipid solubility and resistance to biological degradation, resulted in their bioconcentration fish and other organisms exposed to extremely low levels of these compounds in water.

Although metabolism of DDT in mammals may proceed via DDD to give [4,4]dichlorodiphenylacetic acid, DDE is also formed and stored in fat. It may be slowly depleted by oxidative reactions, and ring hydroxylated derivatives have been detected in mammals and wildlife samples. Consumption of DDT residues in wildlife and fish by predators resulted in adverse effects. Although its mammalian toxicity is low, DDT is highly toxic to fish. It is only moderately toxic to birds, but DDE, which occurs as a significant environmental residue, is associated with thinning of eggshells in raptorial birds.

Decreased eggshell thickness resulted in considerable breakage of eggs with a concomitant decline in population. It has been shown by Lundholm that eggshell thinning caused by p,p′-DDE in susceptible species, such as the duck, is due to the inhibition of prostaglandin synthetase.

This leads to reduced levels of prostaglandin E_2 and lower uptake of calcium by the eggshell gland mucosa. Analogs of p,p′-DDE, such as o,p′-DDE, p,p′-DDT, o,p′-DDT and p,p′-DDD, were inactive, in both the enzyme assays and in causing eggshell thinning. Reductive dechlorination to DDD occurs readily in flooded soils.

A variety of microorganisms convert DDT to DDE or DDD. Both DDE and DDD resemble DDT in their recalcitrance toward environmental degradation, and in mammals, residues may be stored in lipids. Further oxidation gives 4,4′-dichlorodiphenylacetic acid, the predominant excretory metabolite.

DDT, DDD, and DDE occur widely as residues of DDT, but subsequent metabolism is generally slow in most organisms. In alkaline solution and at temperatures above its melting point (the technical product has m.p. 108.5–109 °C), DDT decomposes thermally with elimination of hydrogen chloride to form DDE. In solution in a proton donor solvent, such as methanol, DDT decomposes thermally (e.g., in a heated metal gas chromatographic inlet) to give products that include DDD.

Environmental Fate

DDT decomposed very slowly in sunlight, and 93% was recovered unchanged from the surface of an apple after 3 months. DDE decomposed more rapidly than DDT in sunlight. Other reports indicate that DDT was photolyzed under field conditions to give products, including DDE, 4,4′-dichlorobenzophenone, 4-chlorobenzoyl chloride, 4-chlorobenzoic acid, and 4-chlorophenyl 4-chlorobenzoate.

Irradiation of DDT at shorter wavelengths under laboratory conditions gave a variety of products that arose from reactions of photolytically generated radicals. The nature of the products and the composition of the product mixture depended on the solvent and the presence or absence of oxygen. Some of the many compounds isolated or detected after irradiation of DDT in solvents by energetic ultraviolet irradiation (less than 260 nm) are shown.

Irradiation of a methanolic solution of DDT by wavelengths around 260 nm gave a complex mixture of products. In methanol under nitrogen, major products were DDD and 1,1-bis(4-chlorophenyl)-2-chloroethylene (DDMU, 7). More than 30 components of the mixture obtained by irradiation of DDT in oxygenated methanolic solution were characterized by gas chromatography–mass spectrometry.

Many of these probably arose by the interaction of photolytically generated free radicals with oxygen or solvent. Products included DDD, DDE, 4,4′-dichlorobenzophenone, and methyl 4-chlorophenyl acetate. Reduced products were formed by hydrogen abstraction from the solvent. Bond rearrangements also generated a variety of structures. For example, expulsion of a molecule of carbon monoxide from 4,4′-dichlorobenzophenone gave 4,4′-dichlorobiphenyl.

Chlorobenzoic acid and chlorophenol may have been formed by an alternative pathway from 4,4′-dichlorobenzophenone. Under similar irradiation conditions, DDE was photooxidized to 3,6-dichlorofluorenone in approximately 10% yield. Subsequent photooxidation reactions may result in a chlorinated biphenylcarboxylic acid.

When DDT was exposed to light (253.7 nm) on quartz for 2 days, 80% of the original DDT

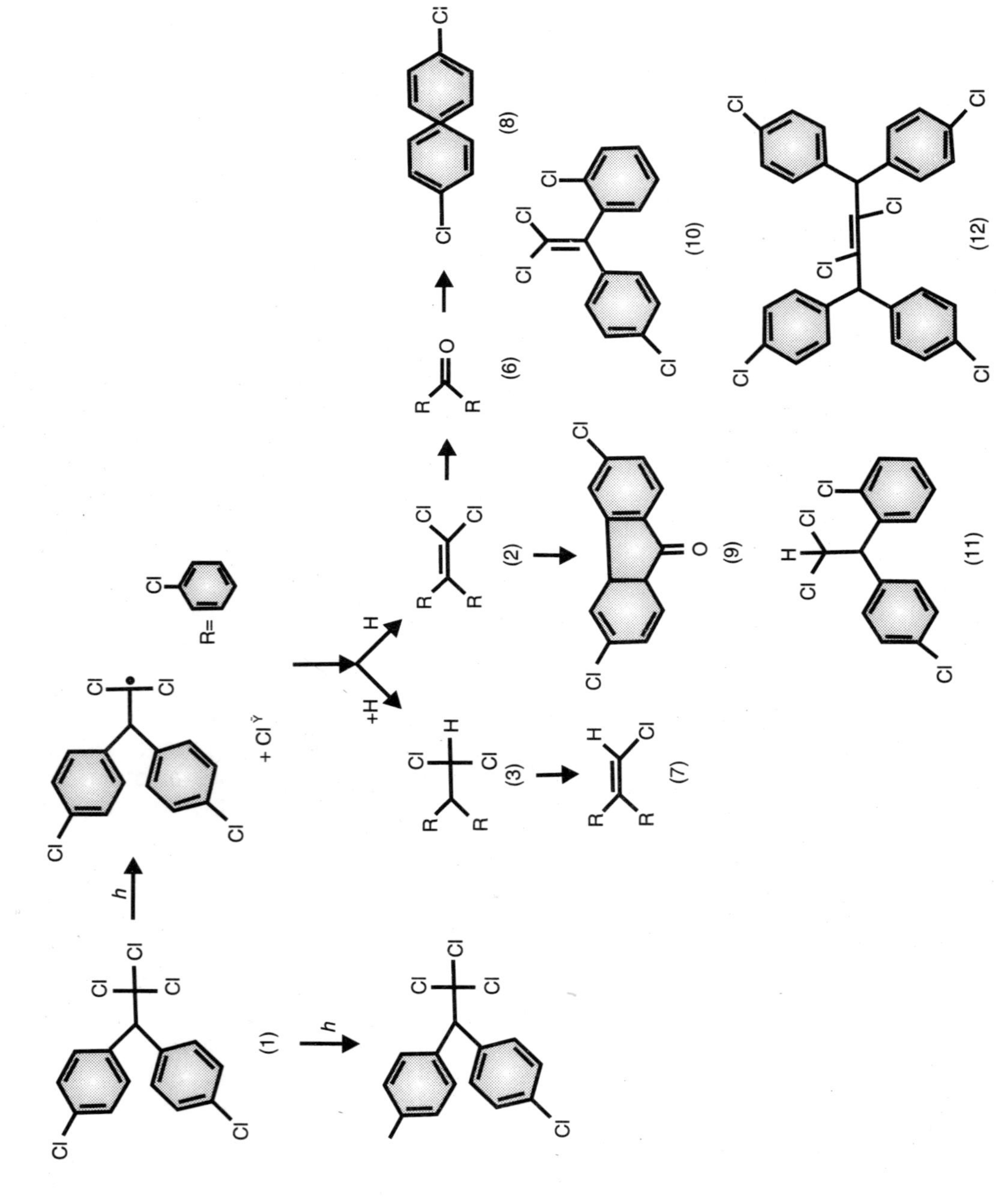

Figure 17.13: Products of DDT irradiation: DDT structures 10–12.

degraded to 4,4′-dichlorobenzophenone, DDE, and DDD. Irradiation of solid DDT moistened with benzene at 235.7 nm gave 4,4′-dichlorobenzophenone and a number of unidentified ketonic compounds. An intermediate DDT hydroperoxide was later identified.

The irradiation of DDT in water (5 mg in 100 ml) with a 1.2-kW high pressure mercury lamp with a quartz filter gave 17 photoproducts identified by gas chromatography–mass spectrometry (with electron impact, chemical ionization, and negative ion chemical ionization). The major products included DDD, DDE, and DDMU.

DDT was also converted into rearrangement products, including 1-(2-chlorophenyl) (4-chlorophenyl)-2,2-dichloroethylene (o,p′-DDE, **(10)** and 1-(2-chlorophenyl)-1-(4-chlorophenyl)-2,2-dichloroethylene (o,p′-DDE, **11).** Subsequently, DDD, DDE, and DDMU we irradiated separately. The mixture of products formed is consistent with a free radical mechanism.

Homolytic fission of C-Cl bonds gave dechlorinated radicals that may abstract hydrogen, lose hydrogen chloride, or undergo bond rearrangement. The addition of 5% acetone to the aqueous solution increased the rate of photolysis 1.5 to 2 times. A dimer, 2,3-dichloro-1,1,4,4-tetrakis (p-chloropheny-2-butene, was isolated in 10% yield after 26 hours of irradiation (360-watt mercury lamp) of DDT dissolved in ethanol with a lamp in the absence of air.

In the presence of air, 4,4′-dichlorobenzophenone was a photoproduct. The photolysis of DDT at long wavelengths was induced by the presence of an aromatic amine. DDT in cyclohexane in the presence of diethylaniline (or other aromatic amine) decomposed by irradiation at 310 nm to give the same products as those obtained by direct photolysis, including DDD, DDE, and 4,4′-dichlorobenzophenone.

Transformations in Soils and Plants

Many micro-organisms are capable of degrading DDT, but degradation in soils is very slow, DT_{50} equalling 380 days. The primary metabolites of DDT in soils are DDD and DDE. Conversion of DDT to DDD occurred rapidly under anaerobic conditions and slowly to DDE under aerobic conditions.

Minor amounts of dicofol were also detected. Soil metabolites of DDT under aerobic conditions also reported were 4,4′-dichlorobenzophenone, 4,4′-dichlorodiphenylacetic acid **14, 15, 16,** and 4-chlorophenylacetic acid. The fate of DDT in soils is influenced by water content.

Under flooded conditions, there was virtually no release of carbon dioxide and DDT was rapidly converted to DDD. Flooding significantly reduced formation of DDE from DDT, and DDD comprised almost half the extractable products after flooding. Data support the hypothesis that loss of DDT associated with increasing soil water content is partly due to the creation of anaerobic microenvironments for microbial degradation of DDT via DDE.

A proposed pathway of microbial degradation of DDT by *Aerobacter aerogenes,* based on studies with whole cells or cell free extracts, is supported by the presence of 4,4′-dichlorobenzophenone in crude extract of flooded soil.

In microcosms, mineralization of DDT and DDE was less than 1% of the added radiolabel (^{14}C-labeled DDE), consistent with half-lives observed in the field. However, in tropical soils, DDE mineralized at a considerably greater rate possibly due to the higher temperatures at which

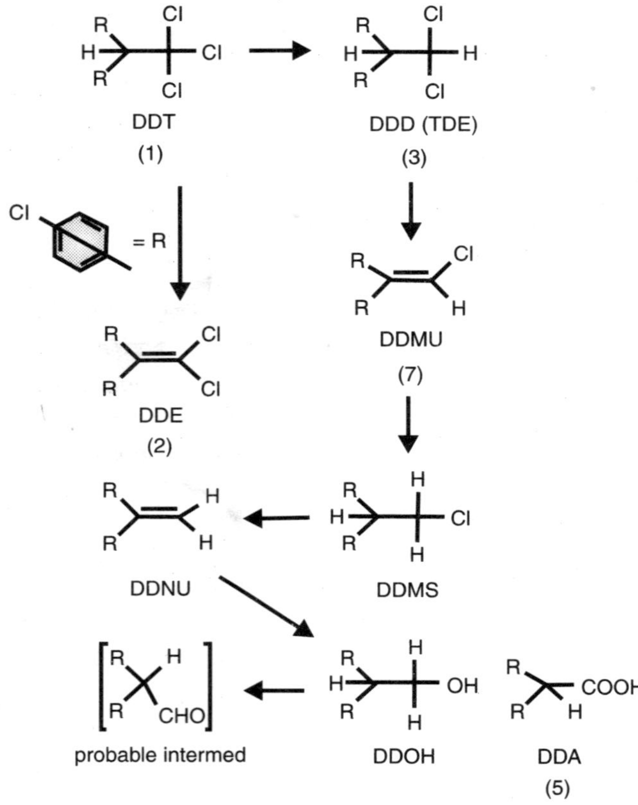

Figure 17.14: Metabolism of DDT.

experiments were conducted. Activated sludge degraded DDT with a half-life of 7 hours. DDD, 4,4′-dichlorobenzo-phenone 4, and a compound identified as bis(4-chlorophenyl)acetonitrile were formed. This compound was also found in the sediment layer of a Swedish lake and isolated from sewage sludge of a water treatment plant in Uppsala.

DDT slowly degraded after application to spinach and cabbage. Metabolites identified were DDD, DDE, 4,4′-dichloro-diphenylacetic acid 4, and conjugates of-dichlorodipheny-lacetic acid and 4,4′-dichlorobenzhydrol.

In ensiled pasture herbiage, DDT was extensively converted to DDD and DDE.

Mammalian Metabolism

Humans excreted 4,4′-dichlo diphe-nylacetic acid in urine after oral ingestion of DDT. DDD is metabolized and degrades and is rarely found a stored metabolite, whereas DDE resists breakdown to 4,4′-dichlorodiphenyl-acetic acid. The higher levels of DDE than DDT found in the general human population reflect its stability. Human embryonic lung cells incubated with ^{14}C-labeled DDT produced DDD by reductive dechlorination and gave 4,4′-dichlorodiphenylacetic acid as the only other metabolite.

The primary metabolites of DDT in the rat are DDE and DDD. The latter is converted to 4,4′-dichlorodiphenylacetic acid, which is excreted in feces or urine conjugated with glucuronic acid or amino acids.

DDE was largely excreted unchanged after oral ingestion, but about 5% of the dose was excreted in feces as metabolites in which the aromatic rings were hydroxylated. The dichloroethylene moiety remained intact, and the shift in the position of the aromatic chlorine substituent suggested that an arene oxide intermediate is involved. When ^{14}C-labeled DDT was administered orally to mice, urine samples contained DDT, DDD, DDE, 4,4′-dichlorobenzophenone, 4,4′-dichlorodiphenylacetic acid, and dicofol. Five unidentified metabolites and some conjugates were also obtained.

Metabolism of DDT in mammals may proceed by oxidation of DDE to the acid, 4,4'-dichlorodiphenylacetic acid, followed by excretion as conjugated products or oxidative attack on the aromatic rings of DDE. Tissue samples from wildlife specimens collected in the field contained

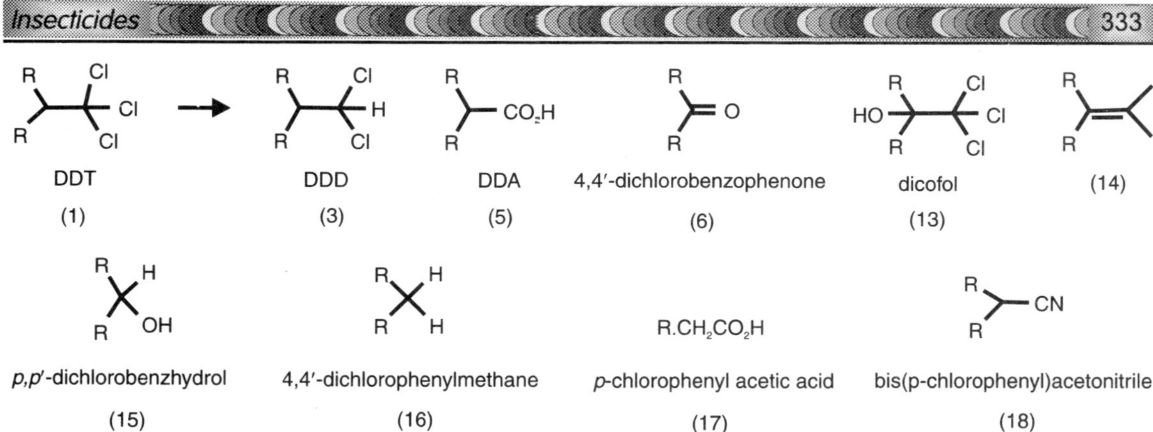

Figure 17.15: Products of DDT microbial and/or soil transformations.

(19) (20) (21)

DDT structures 19 21

many metabolites of DDT. Many of these were hydroxylated in the phenyl rings. Pooled tissue extracts from guillemots and gray seals contained two hydroxylated derivatives of DDE, identified as 1,1-dichloro-2-(4-chloro-3-hydroxyphenyl)-2-(chlorophenyl)ethylene and 1,1-dichloro-2-(3-chloro-4-hydroxyphenyl)-2-(4-chlorophenyl)ethylene, and some samples also contained 4,4′-dichlorobenzophenone and 4,4′-dichlorodiphenylacetic acid.

These two compounds and a third, identified as 1,1-dichloro-2-(4-chloro-2-hydroxyphenyl)-2-(4-chlorop-henyl)ethylene, were obtained from rats fed DDD. These structures suggest that the metabolic pathway involves an arene oxide intermediate with subsequent ring opening accompanied by a shift of the chlorine substituent. A methanesulfonyl derivative of DDE **(21)** found in seal blubber and human milk may arise in a mechanistically similar manner involving the addition of the sulfhydryl group of glutathione to an arene oxide.

A principal pathway of resistance to DDT in houseflies may be enzymic dehydrochlorination by the enzyme DDTase to the relatively less toxic DDE. There is evidence that the enzyme exists in insects in a number of forms. However, other mechanisms of resistance are important.

Mode of Action in Invertebrates

DDT has been shown to interfere with nerve axon ion channels, resulting in a prolongation of the sodium inactivation mechanism and suppression of the potassium conductance increase. The combined effect is to slow down the repolarization of the nerve membrane after an action potential, resulting in sustained depolarization and repetitive action potentials.

DDT has little or no capacity to cause nerve conduction blockage, and this is therefore

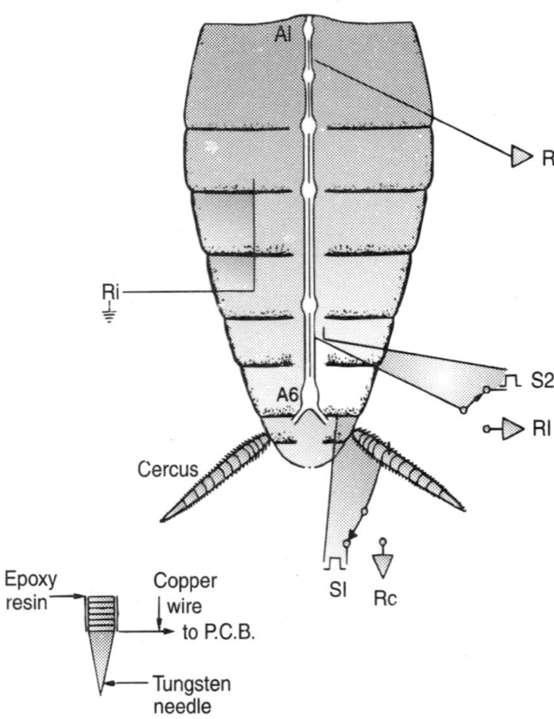

Figure 17.16: The ventral surface of a cockroach abdomen.

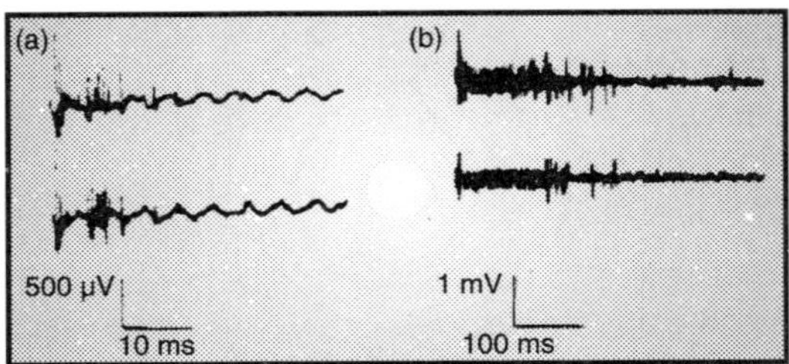

Figure 17.17:. Nerve responses of a DDT-dosed cockroach at 16.5°C: a) 6 h after treatment with DDT (5.25 µg), single electrical stimulation of a cercal nerve (S1) resulted in after-discharge of abdominal neurons, recorded by R2 (upper trace) and R1 (lower trace).

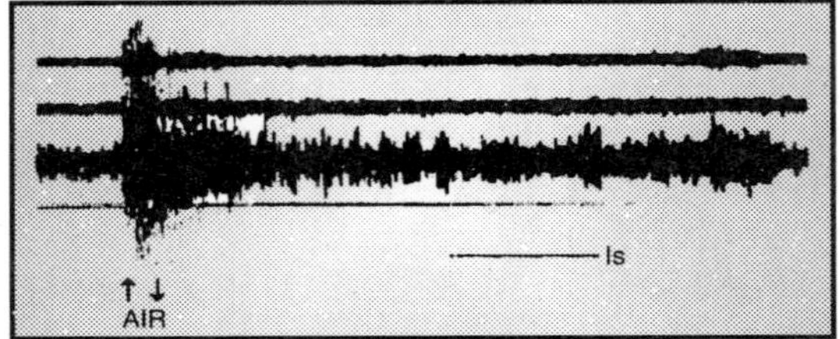

Figure 17.18: Nerve responses of a DDT-dosed cockroach at 16.5°C: Top trace R2; second trace, R1 (nerve chord); third trace, Rc (cercus); bottom trace air-puff marker. Recorded 5.5 h after treatment with DDT (5.25 µg) about 1 h before tremors developed. The cockroach was ataxic with an occasional kick of a leg.

considered to arise through secondary mechanism(s) that are initiated by the intense nerve activity in poisoned insects. Some of the proposed secondary effects may be related to one or more of the biochemical effects listed below.

However, the likely relevance of such secondary effects playing a role in poisoning *in vivo* must be addressed by the consideration of inactive isomers of DDT (e.g., o,p'-DDT) that are often significant impurities in samples of p,p'-DDT. Metabolites of DDT and less active analogs also need to be considered in the structure-activity relationship before a putative effect can be ascribed a role in the poisoning process that results in insect mortality. Another interesting feature of DDT action *in vivo* is its negative temperature coefficient of toxicity in insects (similar to Type I pyrethroids).

This is not simply a result of more rapid metabolism/excretion of DDT at higher temperatures because doses can be chosen that allow insects to go into and out of tremors, just by changing the temperature. Thus, any effect of DDT that becomes more pronounced as temperature is raised probably does not play a key role in the development of the signs leading to insect death.

In an attempt to correlate the nerve effects with the *in vivo* poisoning signs, electrode-implanted, free-walking cockroaches (Figure elsewhere in this chapter shows the electrode positions) were dosed with LD_{95} doses at three temperatures. Effects on the peripheral and central nervous systems were measured along with the stage of poisoning.

Abnormal nerve activity commenced prior to the development of poisoning signs at all three temperatures. Repetitive firing following stimulation began within 2 hours of dosing, and an example is shown in Figure 6a, recorded 6 hours after dosing at 16°C (5.25 μg/insect).

These discharges became more pronounced as poisoning progressed to the tremoring stage, shown at 19 hours. Sensory axons (in the cercus) also fired repetitively following a brief air-puff stimulus, from an early stage, shown in Figure 7 at 5.5 hours, about an hour before tremors developed, at 16°C. At 25°C, an estimated LD_{95}(20 μg/insect) caused effects that were qualitatively similar.

Examples of repetitive firring in the central nervous system (CNS) following electrical stimulation, presynaptically and postsynaptically, were recorded from the same experiment at 25°C. In the experiments at 25°C, tremoring developed at about 6.5 hours after dosing and prostration followed at 24-30 hours.

The insects at 25°C became paralyzed around 37 hours after dosing. At both 16°C and 25°C, as well as at 32°C, these after-discharges reached a peak in terms of both duration and intensity during the paralyzed stage, which was often many days after dosing.

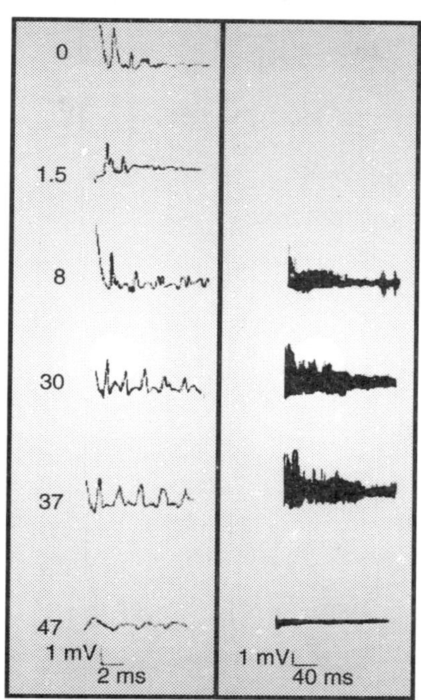

Figure 17.19: Nerve responses of a DDT-dosed cockroach at 25°C. Evoked responses to a single electrical stimulation of a cercal nerve (at S1) recorded in the nerve chord at R2 (upper) and R1 (lower) at different sweep speeds.

This was also true for the discharges in both sensory and motor neurones. The duration of the repetitive firing in the CNS was measured at each temperature. Although the nerve effects and poisoning signs were qualitatively similar at the three temperatures, the maxi-mum discharge duration was greater as temperature was lowered, despite the reduced amount of DDT required with reducing temperature.

This suggests that the effects on the CNS may be critical in causing insect mortality. However, effects on peripheral nerves were not quantified in these experiments, making this claim uncertain. It was subsequently shown, using an in vitro assay, that p,p′-DDT caused consistent repetitive firing in a cercal sensory nerve at a threshold concentration of 2 × 10-8 M.

This response was identical to that of Type I pyrethroids, for which the assay was developed. Although DDT was only about one-tenth the potency of, for example, bioallethrin, DDT is less acutely toxic than is this pyrethroid to a similar degree.

Given the intense disruptions to the peripheral and central nervous systems of the cockroach that were evident following dosing with p,p′-DDT, it is likely that there we consequent changes in neurohormone levels and other secondary factors that may have been the ultimate cause(s) of death.

It may be relevant that, just as the nerve discharges became most intense as cockroaches became paralyzed, so also did the release of diuretic hormone and plasticizing hormone in a Hemipteran (*Rhodnius prolixus*) become apparent during the paralyzed stage.

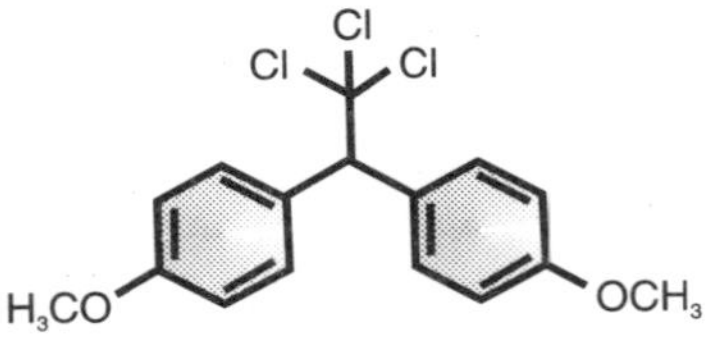

(22) methoxychlor 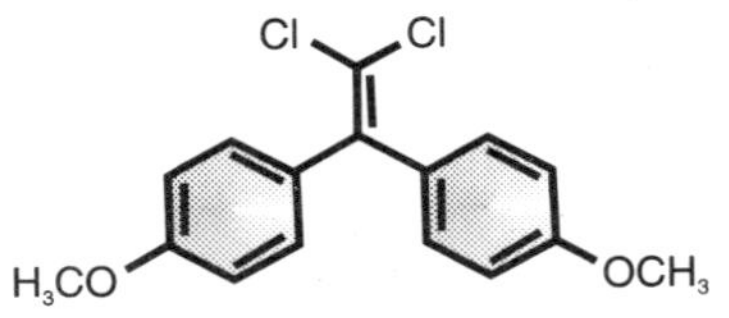 (23) 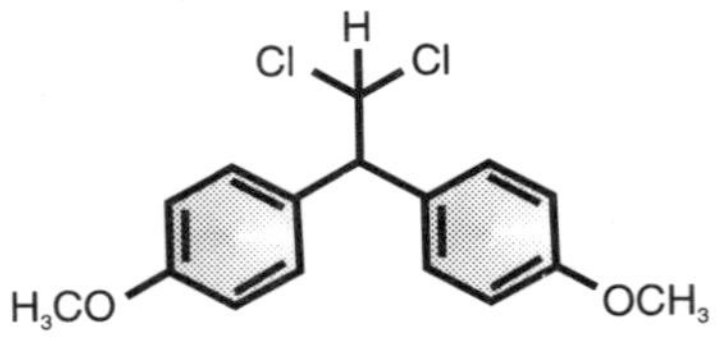 (24) 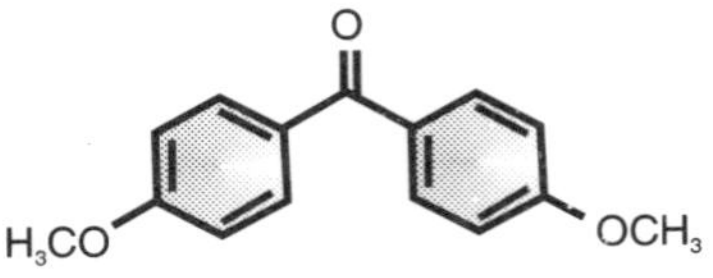 (25) 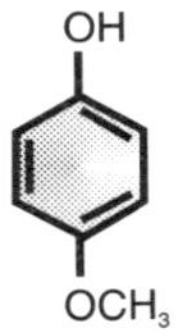 (26)  (27) 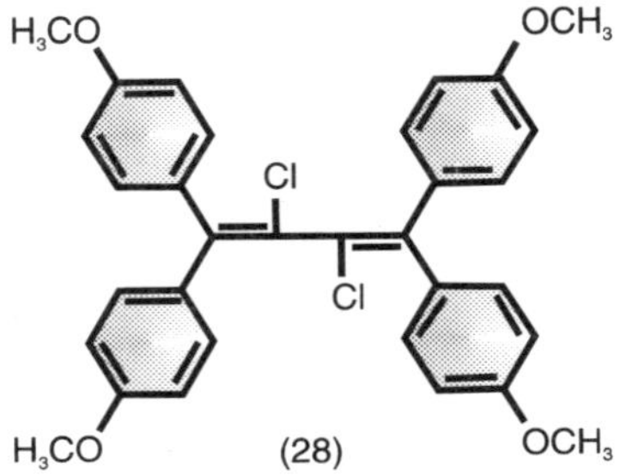 (28) 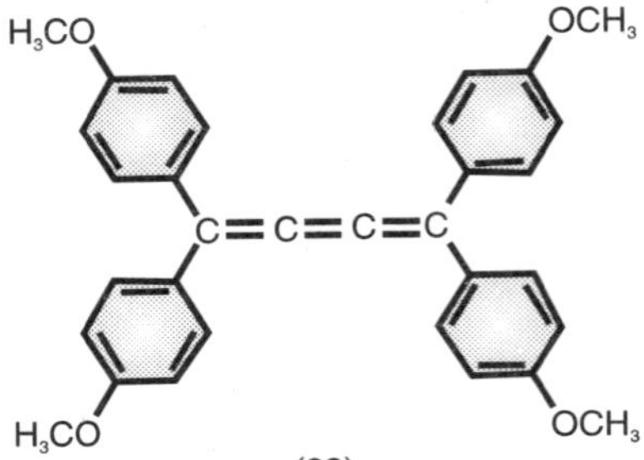 (29) 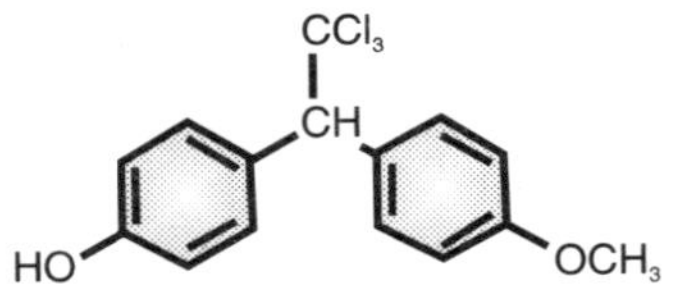 (30) 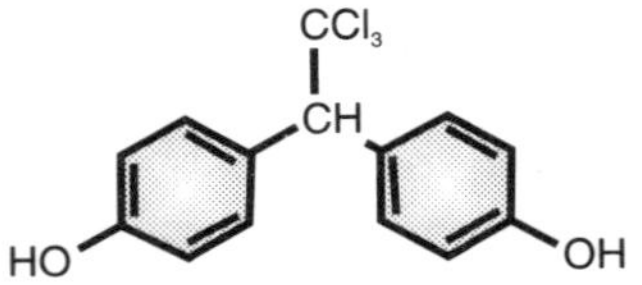 (31)

Figure 17.20: Structure 22, 23, 24, 25, and 28; structures 26, 27, and 29; structures 30, 31.

Table 17.10: Approximate Maximum Duration of Abdominal After-Discharges Following Single Electrical Stimulation of a Cercal Nerve (S1) and the Nerve Cord (S2) of DDT-Dosed Periplaneta.

Temperature (°C)	*Dose of DDT µg*	*S1 Maximum Duration (ms)*	*S2 Maximum Duration (ms)*
16	5.25	420	140
25	20.0	180	120
32	27.6	90	<50

Toxicology

The mammalian toxicology of DDT and analogs has been expertly reviewed recently by Smith (31). The effects can be categorized as follows:

Clinical Signs

Animals dosed with p,p′-DDT show an increase in susceptibility to fear, an increase in "spontaneous" movements and hyperirritability followed by a fine tremor. This develops into a coarse tremor that may last for several days, accompanied by a drop in body weight resulting from a fall in food intake. Repeated doses of DDT can result in tremors lasting for weeks or for months. In some species, tonic–clonic convulsions with episthotonos have been observed, and the overall process has been likened to the human condition of amyostatic syndrome.

Animals that survive the first few hours tend to recover completely. Many laboratory animals become cold to the touch as signs develop, and hyperthermia has often been concluded to be the cause of death in toxicology studies.

Acute Toxicity

The acute oral gavage LD_{50} of p,p′-DDT in the rat is about 150 mg/kg. The dermal toxicity is approximately 10-fold lower. The o,p′-DDT isomer is over 20-fold less toxic by oral ingestion than is p,p′-DDT. The hamster appears to be resistant to acute and chronic toxic effects.

Following a single dose, young rats are less susceptible than are adults to the toxic effects of p,p′-DDT, but this difference is not apparent with repeated doses. Among several possible explanations, it has been suggested that young rats are less susceptible than are adults to DD-induced hyperthermia.

Distribution and Storage

Animals store DDT in adipose tissue, but the metabolite DDE is generally stored to a greater extent than is the parent. Thus, as DDT intake decreases, the proportion of DDT-derived compounds extractable from a mammal comprising DDE increases. It has been suggested that the level of DDT in the rat brain at the time of death is generally greater than 25 ppm (mg/kg), whereas the survivors tend to have brain concentrations below 25 ppm.

Excretion

Following intravenous administration to rats with cannulated bile ducts, the majority (65%) of DDT and metabolites (mostly DDA) was excreted via the bile with only 2% in urine and 0.3% in feces. The excretion of DDT in milk has been studied extensively in a variety of species, including humans.

Cows dosed with substantial amounts of DDT excreted 10% or more in their milk. A large body of epidemiological data have been compiled in an attempt to correlate DDT deposits in human mammary glands with an increased incidence of breast cancer. Although some studies have claimed that DDT is a plausible cause, the most recent studies have failed to establish a link.

Biochemical Effects

Rodent Nervous System Effects. The principal mechanism of nerve excitation in mammals is probably through an interaction with voltage-sensitive ion channels in the nerve axon membrane, as described for invertebrates. Such effects probably account for the majority of reported clinical signs of DDT intoxication. Other effects have been reported, often at very low concentrations, and these may play a role in the poisoning syndrome.

The inhibition of Na/K ATPase in rabbit brain has been reported, as has Ca ATPase inhibition in rat brain. In the latter case, inhibition of this enzyme in lobster leg axon *in vitro* was shown to have an IC_{50} of *ca.* 10^{-9} M. Ihas also been suggested that changes in the metabolism of brain serotonin and norepinephrin may be responsible for hyperthermia, and ACh changes may be related to tremor and convulsions.

An accumulation of ACh and cyclic GMP in the cerebellum was found in rats exhibiting tremors and convulsions. However, it is unclear whether these effected recorded *in vivo* were a direct result of DDT or a secondary effect. There are no pathological changes in the nervous system associated with DDT.

Induction of Enzymes

DDT and related compound readily induce mixed function oxygenases in the liver. However, there is considerable variation between species in the amount of DDT needed to induce these enzymes. Maximal increases were found at tissue levels of 3 ppm DDE in the rat and 40 ppm in quail. In the female rats multiple doses of DDT induced hepatic CYP2B and CYP3A, but not CYP1A1 or CYP1E1, resulting in increased hydroxylation of testosterone at 16 and 6,6.

In the male rats CYP2B and CYP3A were induced by DDT as well as by DDE and DDD (Nims et al., 1998). In the rat renal cortex, a single dose of >100 mg/kg resulted in the induction of pyruvate carboxylase, phosphoenolpyruvate carboxykinase, fructos 1,6-diphosphatase, and glucose 6-phosphatase.

This induction may explain the hyperglycemia often found during the early stages of DDT intoxication in the rat. Histological changes in the liver that are typical of mixed function oxygenase inducers have been reported for DDT.

Cause of Death

Three main effects have been identified that may be lethal in mammals. Disruption in thermoregulation results in coldness of the skin along with hyperthermia. In rats dosed orally at a lethal dose of 600 mg/kg, a 3 °C rise in body temperature was noted. Other experiments have shown respiratory arrest to be the lethal lesion, usually following a convulsion.

Large intravenous doses of DDT in the dog were found to result in ventricular fibrillation. This was accompanied by sensitization of the heart to epinephrine. It was concluded that ventricular fibrillation could be lethal in animals that died shortly after poisoning signs first appeared.

Genotoxicity and Carcinogenicity

An extensive database has been developed describing the genetic toxicity of DDT in a variety of systems. DDT was not mutagenic in bacterial or fungal assays, with or without metabolic activation. In the majority of studies, DDT did not reveal genotoxic effects in rodent or human cells.

In chronic dietary studies in rodents, DDT caused an increased incidence of liver tumors in mice, in several studies. These tumors occurred in both sexes, were dose-related, and included both hepatomas and carcinomas (malignant).

Similarly, in the rat, there is also evidence for liver tumors. However, in four studies with hamsters, at doses of 500 ppm and 1000 ppm (*ca.* 50 or 100 mg/kg/day), there was no apparent increase in tumors, in any organ. In conclusion, it is possible that the clear evidence of carcinogenicity in the mouse and rat is secondary to liver toxicity. There is little or no epidemiological data to support a classification of DDT as a human carcinogen. IARC have classified DDT as Class 2B, a possible human carcinogen.

Reproductive and Developmental Toxicity

DDT injected subcutaneously into cockerels at 300 mg/kg/day caused a marked reduction in both testicular growth and the development of secondary sex characteristics. This effect was attributed to an estrogenic effect of DDT. However, in multigeneration feeding studies using rats, mice, and dogs, little was observed to substantiate the gonadotoxic effects in the cockerel.

Perhaps this is a function of the high cose used in the cockerel study. More recent studies have shown that DDT and DDE act as anti-androgens at the rat androgen receptor. Effects in male rats were detected following *in utero* exposure, but only at maternal doses above 10 mg/kg/day.

In female rats, technical DDT injected intraperitoneally at 5 mg/kg or o,p′-DDT at 1 mg/kg had estrogenic effects measured as an increase in uterine weight. The o,p′-DDT isomer also inhibited the uptake of estrogen binding by the uterus *in vivo*, an affect perhaps attributable to competition for binding sites.

It was subsequently shown that *in vitro*, o,p′-DDT acts as an estrogen receptor agonist, of low potency (four orders less than diethylstilbestrol), but long persistence. Injected subcutaneously into rat pups on the first 4 days after birth, o,p′-DDT resulted in earlier puberty and earlier persistent vaginal estrous, but only at doses of >ca. 80 mg/kg/day.

In cultured bovine oviduct and uterine cells, o,p′-DDT stimulated DNA synthesis at low concentrations and inhibited at high concentrations. Methoxychlor had the same effects, although to a lesser degree. It remains to be seen whether the o,p′-isomer of methoxychlor has a greater effect than does the insecticidal p,p′-isomer. At lower dosages, o,p′-DDT fed to ewes at 10 ppm for 2–9 months had no effect on reproduction.

Similarly, a dose of 40 ppm (2.1 mg/kg/day) fed to rats through two matings had no effect on reproduction or lactation. Female rats exposed to up to 200 ppm DDT in the diet (10 mg/kg/day, but 30 mg/kg/day during lactation) reproduced normally through two generations. Furthermore, mice dosed at 25 ppm (>3.33 mg/kg/day) through six generations were normal with respect to fertility, gestation, viability, lactation, and survival.

At 100 ppm, there was a slight reduction in lactation and survival in some but not all generations, but at 250 ppm, several parameters of reproductive toxicity were affected. In a dog study, animals were fed at 0, 1, 5, and 10 mg/kg/day, 5 days/week for thre generations.

The only reproductive effect recorded was an earlier onset of the first estrous (by 2 to 3 months). Developmental toxicity has been reported as reduced fertility in female mice after dosing at 1 mg/kg on three occasions during gestation.

Teratogenic effects attributable to DDT have not been observed in any of the reproductive toxicity, multigeneration studies in the mouse, rat, or dog. Because of the estrogenic effects of DDT, it was considered as a possible cause of abortion in cattle and humans, but no evidence for a causal link could be established.

Toxicity in Humans

Oral Exposure. Accidental exposure cases have led to the conclusion that 10 mg/kg generally causes illness, but without vomiting. Convulsions have been reported at doses of >16 mg/kg, and 285 mg/kg caused convulsions and vomiting, without fatality.

Oral dosing of human volunteers has resulted in the finding that DDT (in vegetable oil) is tasteless. Longer term studies were conducted by Hayes, in which men were dosed of<0.6 mg/kg/day for 12 or 18 months. No symptoms were found to be treatment-related, and neurological and liver function tests were normal.

A follow-up study included 21 months of dosing and 27 months of observation. The storage of DDT in body fat revealed that, at 0.6 mg/kg/day, 65% of the administered dose was stored as DDE, whereas early in the dosing, only 14% was stored in this way.

The selective storage of this DDT metabolite in fat is a result of its greater lipophilicity than DDT. The ratio of DDE:DDT stored in fat can be used to estimate when an exposure to DDT occurred it increases with time. The o,p′-isomer was excreted more rapidly than was p,p′-DDT. Urinary excretion of DDT in humans is principally as DDA.

Dermal Exposure. Under controlled conditions, dermal exposure to DDT has not been associated with any illness or skin irritation. DDT-impregnated clothing, which was used by U.S. and British troops during World War II to effectively eliminate fleas and lice, was also found to be nonirritating.

Inhalation Exposure. Volunteers were subjected to almost continuous daily exposure to DDT, sufficient to leave a white deposit of DDT on the nasal passages. Except for moderate irritation of the nose, throat, and eyes during dosing, there were no symptoms. Neurologic testing showed no changes resulting from DDT.

Therapeutic Uses. DDT was used extensively for controlling disease-causing insects. For example, malaria was eliminated from Europe and North America by the mid-1950s and it was controlled in many other parts of the world. DDT is still used for human health purposes in parts of the tropics.

In addition, DDT was used medicinally for the successful treatment of a patient with familial jaundice resulting from a deficiency in glucuronyl transferase. Treatment with DDT rapidly reduced plasma bilirubin to the normal range and relieved the symptoms of nausea and general malaise.

The dosage used was 1.5 mg/kg/day, for 6 months, and the plasma DDT level increased from 0.005 ppm to 1.33 ppm. The concentration in body fat reached 203 ppm. This blood level exceeds the highest plasma level found in formulation plant workers of 0.996 ppm.

DDT was also used effectively as a single dose of 5000 mg in three patients who had taken a phenobarbital overdose. The recovery of the patients was at least partly explained by the induction of liver mixed function oxygenases to break down the barbitutrate. It is interesting to note that barbiturate sedatives have been used in humans as antidotes to DDT poisoning.

Accidental & Intentional Poisoning. Ingestion of DDT has been reported to cause hyperesthesia, initially, of the mouth and face followed by paresthesia of the face and tongue, dizziness with loss of equilibrium, tremor, confusion, malaise, headache, fatigue, and delayed vomiting.

Recovery times are variable, but even after severe poisoning, complete recovery has usually occurred within a few days. Liver toxicity has sometimes been reported, such as in the case of three men who accidentally ate 5000 to 6000 mg with a meal. Slight jaundice appeared within 4 to 5 days and lasted 3 to 4 days. Death in humans has rarely been caused by the ingestion of solid DDT, but solutions have sometimes been fatal, but largely attributable to the toxicity of the solvent.

Use Experience. DDT has been used on a massive scale in the past in delousing operations to control typhus. These operations have involved massive dermal and inhalation exposure, especially by workers applying the DDT. However, the only effect that was found in DDT applicators was dermatitis, and this was found principally with liquid formulations where the causative agent was the solvent, usually kerosene.

A study of workers at a DDT production plant found that the 35 most heavily exposed had worked for 11 to 19 years at the plant, making 2722 tonnes per month. The plant had made DDT exclusively and continuously from 1947 to 1966. There we no ill effects at the plant that were attributable to the work.

Similarly, 63 men who had each worked at the plant for up to 5 years were also free of cancer and other diseases that might be linked to the plant. No case of cancer occurred during the 19 years of operation of the plant, in a total workforce of between 110 and 135, even though two of the workers had a prior history of being successfully treated for cancer.

In a follow-up study, analysis of body samples from 31 of these men showed that their exposure to DDT was equivalent to 3.8 mg/man/day for 16 to 25 years. Clinical chemistry and liver function tests were normal except for a slight increase in plasma alkaline phosphatase and SGPT in one man, an elevated alkaline phosphatase in another, and an elevated SGPT in a third worker.

The appearance of these two enzymes in plasma is mainly associated with toxicity to the liver. Morgan and Lin (56) reviewed data on illnesses and exposure from 2620 persons exposed to (a variety of) pesticides along with 1049 persons who were unexposed. Once again, there was an increase in blood serum alkaline phosphatase levels that was associated with an elevated level of DDT plus DDE in serum.

There were also significant increases in SGOT and LDH, both of which are often associated with toxicity to the liver. There was also a reduction in serum bilirubin in exposed workers, perhaps as a result of liver microsome enzyme induction.

Secretion in Milk. There is ample evidence that DDT is excreted in human milk. However, there is no indication in the studies reviewed by Smith of ill effects in babies receiving such contaminated milk. High levels of DDT and metabolites were found in milk of mothers living in home that had been sprayed with DDT for malaria control. Although the children's blood clearly showed elevated DDT levels, there were no reported ill effects.

Compounds Related to DDT

DDD

DDD (3) [72-54-8], 1,1-dichloro-2,2-bis-(4-chlorophenyl) ethane, mp 112°C, also known as TDE, is g metabolite of DDT, which found use in agriculture. As with DDT, the p,p′-isomer is insecticidal and the o,p′-isomer is an impurity.

Toxicology

Animal Studies. DDD is less acutely toxic than DDT, with an oral LD_{50} of 3400 to 4000 mg/kg in the rat and a dermal LD50 *ca.* 1200 mg/kg in the rabbit. In a 2-year chronic dietary study in the rat, a LOEL of 100 ppm (5 mg/kg/day) was quoted for tissue damage, similar to that produced by DDT.

Human Studies

Oral Exposure. Most of the studies conducted in humans have used the o,p′-isomer of DDD. This has been used as a drug, with medical and veterinary uses, for the treatment of Cushing's syndrome and adrenal carcinoma, under the generic name of mitotane. After oral ingestion, o,p′-DDD binds to lung and adrenal tissue.

Its toxicity to the adrenal appears to be species-dependent, however, causing gross atrophy in the dog and reducing corticosteroid production without pathological effects in other mammals, including humans. The site at which corticosteroid synthesis is blocked appears to vary: In some species, it is 11β-hydroxylation that is inhibited, and in others, including humans, 3β-hydroxy-Δ^5-steroid dehydrogenase is inhibited. In adrenal tissues surgically removed from patients who had been treated with o,p′-DDD, there were reduced levels of cortisol, corticosterone, 18-hydroxycorticosterone, and aldosterone, but the synthesis of corticosteroids was unaffected by o,p′-DDD, *in vitro.*

This suggests that a metabolite of DDD might be responsible for the adrenal effects, *in vivo.* Alternatively, it has been shown that o,p′-DDD induces microsomal mixed function oxygenases, leading to oxidation of steroids and drugs, and such an (induction) effect could be related to the suppression of corticosteroids, *in vivo.*

Therapeutic Effects. Dosages of o,p′-DDD used in humans have generally been *ca.* 100 mg/kg/day, lasting for several weeks. A favourable response was obtained in 25% to 50% of patients with inoperable adrenocorticoid carcinoma. Such large doses used clinically have caused symptoms such as general lassitude, anorexia, nausea, vomiting, diarrhea, and dermatitis; these are similar to many other anti-cancer drugs.

There has been no histological damage reported in human adrenal glands, but electron microscopy revealed degeneration of mitochondria in the *zona fasciculata* (of the cortex) in a patient who had received 3000 mg/day for 1 month. In addition to effects on the adrenals, o,p′-DDD has been used to treat spanomenorrhea in humans, with a success rate of almost 90%.

Dicofol (kelthane)

Dicofol (kelthane) **(13)** [115-32-2], 2,2,2-trichloro-1,1-bis(4-chlorophenyl)ethanol (IUPAC), mp 78.5 –79.5°C is an acaricide and has been used against plant-feeding mites.

Toxicology. Dicofol was not carcinogenic in the rat or mouse in chronic dietary studies, except for a dose-related increase in benign liver adenomas in the male mouse. It was not genotoxic in a battery of tests. Liver toxicity in chronic studies was observed in both rodents in the form of necrosis, vacuolation, hyperplasia, and hypertrophy, with a NOEL of 5 ppm (0.25 mg/kg/day) in the rat.

This liver toxicity could be the basis for the liver tumors, mentioned above. In a chronic dog study, liver cell hypertrophy was also reported. Vacuolation of cells in the adrenal cortex was also observed in chronic rodent studies.

There was no indication of developmental toxicity in the rat or rabbit. In a rat reproductive toxicity study, reduced pup survival was noted, but only at dose levels that were maternally toxic. Neurotoxicity was measured using FOB* tests after an oral gavage. Several effects were noted, including ataxia, at 350 mg/kg, with a NOEL of 75 mg/kg.

Methoxychlor. Methoxychlor **(22)** [72-43-5], 1,1,1-trichloro-2,2-bis-(4-methoxyphenyl)ethane (IUPAC), (mp 89°C. tech grade mp 70–85°C), is an insecticide in which methoxy groups replace the ring chlorines of DDT. It is soluble in water to 0.1 mg/L. It can be obtained in good yield by condensation of chloral with anisole in the presence of sulfuric acid. The technical grade contains not less than 88% of the p,p′-isomer and a small amount of the o,p′-isomer.

Methoxychlor is used for control of a wide range of insect pests in field crops, forage crops, fruit, vines, flowers, vegetables, and in forestry. It is also used for control of insect pests in animal houses, dairies, and in household and industrial premises. Methoxychlor is one of the safest insecticides with a rat oral LD_{50} of >6000 mg/kg. It is useful in the home garden, for the control of insect pes of vegetable and fruits, for veterinary hygiene, and for the control of bark beetles that are the vectors of Dutch elm disease. Methoxychlor is a contact insecticide and also has stomach action.

It is much less readily dehydrochlorinated in alkalin solution or in biological systems than is DDT. However, the p,p′-methoxy groups are rapidly attacked by microsnmal oxidase systerns in higher animals to form phenols that are conjugated and eliminated. Pathways of degradation of methoxychlor in microorga-nisms, mammals, mosquito, larvae, algae, fish, and snails, are primarily dechlorinrtion and O-dealkylation. The principal route of degradation in mammals is by O-dealkylation to the corresponting phenol and bisphenol and by dehydrochlorination to 4,4′-dihydroxybenzophenone. Thus, methoxychlor does not bioaccumulate as does DDT and is favoured for general environmental use.

However, it is more expensive than DDT and has little effectiveness toward a number of insects. It is more readily degraded by biota than is the fully chlorinated analog and is less likely to store in the body fat of animals or be excreted in milk.

Methoxychlor is stable to oxidizing agents and to ultraviolet light, but it becomes pink- or tan-coloured on irradiation. It reacts with alkalies, especially in the presence of catalytic metals, with the loss of hydrogen chloride.

The major product of photolysis of methoxychlor in air-saturated water, irradiated at wavelengths >280 nm, was 1,1-dichloro-2,2-bis(4-methoxyphenyl)ethylene **(23)**, whereas1,1-dichloro-2,2-bis(4-methoxyphenyl)ethane **(24)** was formed along with **23** in degassed water-acetonitrile solutions.

Subsequently, **23** was photolyzed to benzaldehyde. Products of photolysis in aqueous alcoholic solutions were 4,4'-dimethoxybenzophenone **(25)**, 4-methoxybenzoic acid, and 4-methoxyphenol **(26) (60).** When **22** was irradiated by ultraviolet light (carbon arc, 220–330 nm) in milk, 4-methylanisole (27), **25, 26,** 1,1,4,4-tetrakis(4-methoxy)-2,3-dichloro-2-butene (2) and 1,1,4,4-tetrakis(4-methoxy)-1,2,3-butatriene **(29)** were identified as products.

Chemical decomposition was slow in water, and at 27°C, DT_{50} at pH 5 to 9 was 100 days. Major products of hydrolysis are anisoin, anisil, and **23 (62).** DT_{50} of methoxychlor in water is about 46 days. Dechlorinated and dehydrochlorinated methoxychlor are major products observed in studies with bacteria. Degradation by *Aerobacter aerogenes* gave **23** and **24**.

Methoxychlor degraded more rapidly in flooded soils than under aerobic conditions. In sediments, DT_{50} was <28 days under nitrogen aeration. DT_{50} under static aerobic conditions was 49–55 days and 115–206 days in air-purged flasks. The major degradation products were dechlorinated methoxychlor and its mono- and dihydroxy(demethylated) derivatives.

In a study with ^{14}C-labele methoxychlor, dihydroxy-derivatives represented 15% of extractable radioactivity in pond sediments and 28% in lake sediments after 448 days of aerobic incubation. Unde, nitrogen aeration, the mono and dihydroxy-derivatives represented 44–66% of extractable ^{14}C after 28 days incubation. When methoxychlor was administered to mice orally and topically to houseflies, both species excreted mono- and dihydroxy derivatives formed by O-demethylation.

When methoxychlor was incubated with mouse liver or housefly microsomes, similar results were obtained. Methoxy-chlor was more effective against DDT-resistant houseflies than was DDT, owing to its lower capacity for dehydrochlorination by the enzyme, DDTase.

The high synergistic ratios with piperonyl butoxide suggested that in the housefly, multifunction oxidases are involved in detoxication of methoxychlor, which was metabolized to 1,1-dichlorobis(4-hydroxyphenyl)ethylene and conjugated derivatives. When methoxychlor was incubated for 2 hours with a preparation of mouse liver microsomes, 97.09% intact methoxychlor, 2.55%, 1,1-dichloro-2-(4-hydroxyphenyl)-2-(4-methoxyphenyl)ethylene, and a trace of a bisphenol-1,1-dichloro-2,2-bis(4-hydroxyphenyl)ethylene **(31)** were recovered, suggesting that O-dealkylation in mammalian liver is a major pathway.

Dehydrochlorination of methoxychlor occurred in both susceptible and resistant strains of the grain weevil (*Sitophilus granarius* L.) to give 1,1-dichloro-2,2-bis(4-methoxyphenyl)ethylene, and some bis(4-methoxy-phenyl)acetic acid **(13)** was also formed.

When (phenyl-^{14}C)methoxychlor was given orally to two lactating goats, 17 metabolites plus methoxychlor were excreted in feces and urine and identified by gc/ms. Recovery of ^{14}C was 95% for the goat given DDT and 99% for the goat given 200-mg methoxychlor. Fecal and some urinary metabolites were formed by O-dealkylation, dechlorination, and dehydrochlorination. Most metabolites from urine were completely demethylated and conjugated with glucuronic acid, and ring hydroxylation occurred in one urinary metabolite.

Toxicology

Methoxychlor did not appear to be either oncogenic or neurotoxic in chronic rodent dietary toxicity tests. The only effect observed was body weight loss, with an approximate NOEL of 200 ppm (10 mg/kg/day) in the rat. Developmental toxicity was observed in the rat in the form of increased resorptions and fetal death, at doses causing only minor maternal toxicity (body weight loss).

The apparent NOEL was 18 mg/kg/day. At maternal doses of 50 to 400 mg/kg/day in the rat during gestation, reduced maternal body weight was accompanied by reduced fetal weight, fetal number, and skeletal variations, without evidence of teratogenicity.

In multigeneration reproductive toxicity studies, a severe reduction in female fertility was reported, with a NOEL of 10 mg/kg/day. The endocrine effects of methoxychlor have been studied in detail. At doses above *ca.* 100 mg/kg, it appears to be estrogenic in the rat, causing reduced fertility largely through preimplantation loss.

Demethylation is considered to be necessary for reproductive toxicity, and this readily occurs *in vivo.* One result is that methoxychlor is generally inactive in *in vitro* assays to study endocrine effects. Methoxychlor stimulates estrogen controlled behaviour in the rat and hamster, but the effects are not identical to those of estrogen.

It has been suggested that in mice, methoxychlor acts as an estrogen agonist in uterus and an antagonist in ovary. This is consistent with the action of other chemicals, such as tamoxifen, at *beta* and *alpha* estrogen receptors; agonists at one of these receptors tend to act as antagonists at the other.

Another effect of methoxychlor and its demethy-lated metabolite has recently been demonstrated in the male rat. Both of these compounds inhibit the biosynthesis of testosterone in Leydig cells at 1 to 1000 nM. Thicinhibition did not involve an interaction at the level of the estrogen or androgen receptor. Instead, it was attributed to the inhibition of P450 mixed function oxygenases responsible for side-chain cleavage of cho esterol, a necessary step in testosterone biosynthesis.

Perthane

Perthane [72-56-0] 1,1-dichloro-2,2-bis-(4-ethylphenyl)ethane (mp 60–61°C) is a rapidly biodegradable insecticide with very low mammalian toxicity, rat oral LD_{50} 8170 mg/kg. It has been used as a household insecticide.

Fluorine Analog of DDT

The fluorine analog of DDT, 1,1,1-trichlcro-2,2-bis-(4-fluorophenyl)ethane [475-26-3(DFDT) (mp 45°C) has very similar properties to the chloro compound, but it is less persistent. It is more expensive to produce than DDT. It was used as a sanitary insecticide by the German army in World War II, but is too phytotoxic for agricultural use. The rat oral LD50 is 900 mg/kg. .

Secion 2

Hexachlorocyclohexane

Hexachlorocyclohexane (HCH) or benzene hexachloride (BHC) is an insecticide obtained as a mixture of isomers by chlorination of benzene in a simple manufacturing process. Derivatives of cyclohexane may exist in a number of conformations. The cyclohexane molecule adopts the chair form and atoms 1, 3, and 5 lie in one plane and atoms 2, 4, and 6 lie in another parallel plane.

Figure 17.21: Cyclohexane: axial and equatorial conformations.

The position of substituents on the cyclohexane ring may be either in the plane of the ring atoms (equatorial, *e*) or perpendicular to the plane of the ring atoms (axial, *a*). Figure elsewhere in this chapter shows the chair form of the cyclohexane ring with all substituents axial (left) and the ring with all substituents equatorial (right). Consideration of the biological activity of *gamma* HCH (lindane or the gamma-isomer of hexachlorocyclohexane) provided an early illustration of the benefits of examining both biological properties and environmental behaviour of stereoisomeric compounds. *Gamma* HCH is a product of catalytic chlorination of benzene.

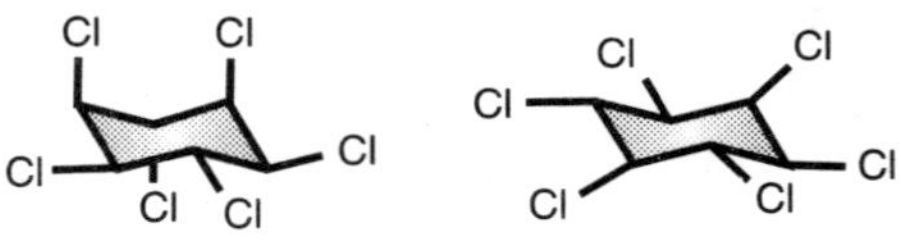

Gamma-HCH Beta-HCH

Figure 17.22: Gamma HCH and beta HCH.

This reaction yields a mixture of eight stereoisomersThese atoms, indicating the respective configuration of chlorine, are named: alpha: α, aaeeee; beta: β, eeeeee; gamma: γ, aaaeee; delta: δ, aeeeee; epsilon: ε, aeeaee; zeta: ζ, aaeaee; eta: η, aeaaee;, and theta: θ, aeaeee. The crude product is a grayish or brownish amorphous solid with a characteristic odor; it begins to melt at 65°C.

It consists of 10-18% of the active gamma isomer, 1α,2α,3α,4β,5β,6β-hexachlorocyclohexane (IUPAC) ((mp 112°C) with at least four other isomers: a-isomer (mp 157°C), 55-70%; 6-isomer (mp 309°C), 5-14%; δ-isomer (mp 138°C), 6-8%); ε-isomer (mp 219°C), 3-4%; and a trace of η-*isomer* (*aeaaee*) (mp 90°C). A heptachlorocyclohexane is present up to 4% as is a trace of an octachlorocyclohexane; both are insecticidally inactive.

The gamma isomer is by far the most toxic of the isomers, being from 500-1000 times as active as the δ isomer and *ca* 5000-10,000 times as active as the β isomer; the β isomer and s isomer are nontoxic. The various isomers differ greatly in their solubilities, and a pure gamma isomer can be prepared by treating the crude product with methyl alcohol or acetic acid, in which the α and β isomers are nearly insoluble (leaving a marketable product containing 30-40% *gamma* isomer), and then fractionally crystallizing the alcohol-soluble fraction from chloroform.

Purification by selective crystallization is essential because most isomers are not effective insecticides. The *gamma* isomer represents about 10-18% above of the product. Isomers can be distinguished by physical characteristics, such as solubility and melting point, and chemical reactivity. The pure gamma isomer has a slight aromatic odor (*d* 1.85 g/cm^3, vp 1.3 mPa at 20°C).

It is very stable to the action of heat, light, and oxidation and can be burned without appreciable decomposition but is readily decomposed by alkaline materials to form, principally, 1,2,4-trichlorobenzene and three moles of hydrogen chloride. The gamma isomer is soluble in water to 7.3 mg/L and is soluble in aromatic solvents.

The rat LD_{50}s are 88, 91 (oral), 900, and 1000 (dermal) mg/kg. The *gamma* isomer of *hexachlorocy-clohexane* has similar insecticidal action to DDT. However, it is more soluble in water than is DDT and may be used for seed treatment.

Gamma-HCH is used to control a wide spectrum of phytophagous and soil-inhabiting insects, public-health pests, and animal ectoparasites. It is used on a wide range of crops, in stored products warehouses and storerooms, public health applications, and in seed treatments. γ-HCH is an antagonist of the GABA-receptor activated chloride channel complex.

Environmental Fate

The technical product HCH contains predominantly four stereoisomers (α, β, γ, and δ), and the product has been used in this form. Each isomer may behave differently in the environment, and it may be difficult to interpret environmental residue analytical data because interconversion of isomers may occur under certain conditions.

Rates of disappearance of individual isomers in soils, water, and food differ significantly. Some investigators found that α- and β-isomers accumulate in soils, rice, straw, milk, and other commodities following extensive use of HCH. However, other workers reported that the rate of degradation of all four HCH isomers (α, β, γ, and δ) was essentially the same. *Gamma-HCH* is very stable to sunlight and is photodegraded in organic solvents by irradiation at 254 nm.

The conversion of γ-HCH to a-HCH in the laboratory and in aquatic sediments by a strain of *Pseudomonas putida* isolated from soils has been reported. Degradation of gamma-HCH and its isomers was more rapid in anaerobic environments.

Metabolism

Metabolic pathways are complex, and more than 80 metabolites of gamma-HCH have been identifieed. In animals, metabolism of gamma-HCH **(32)** generally led to less-chlorinated unsaturated metabolites. Chlorinated phenols may be formed and excreted as glucuronides. For detailed discussion of the biodegradation of *gamma*-HCH, see Kurihara and Nakajima. Oxygenation or glutathione conjugation is an important initial stage in reaction. Key intermediates in metabolic pathways are hexachlorocyclohexene, pentachlorocyclohexene **(34)**, and

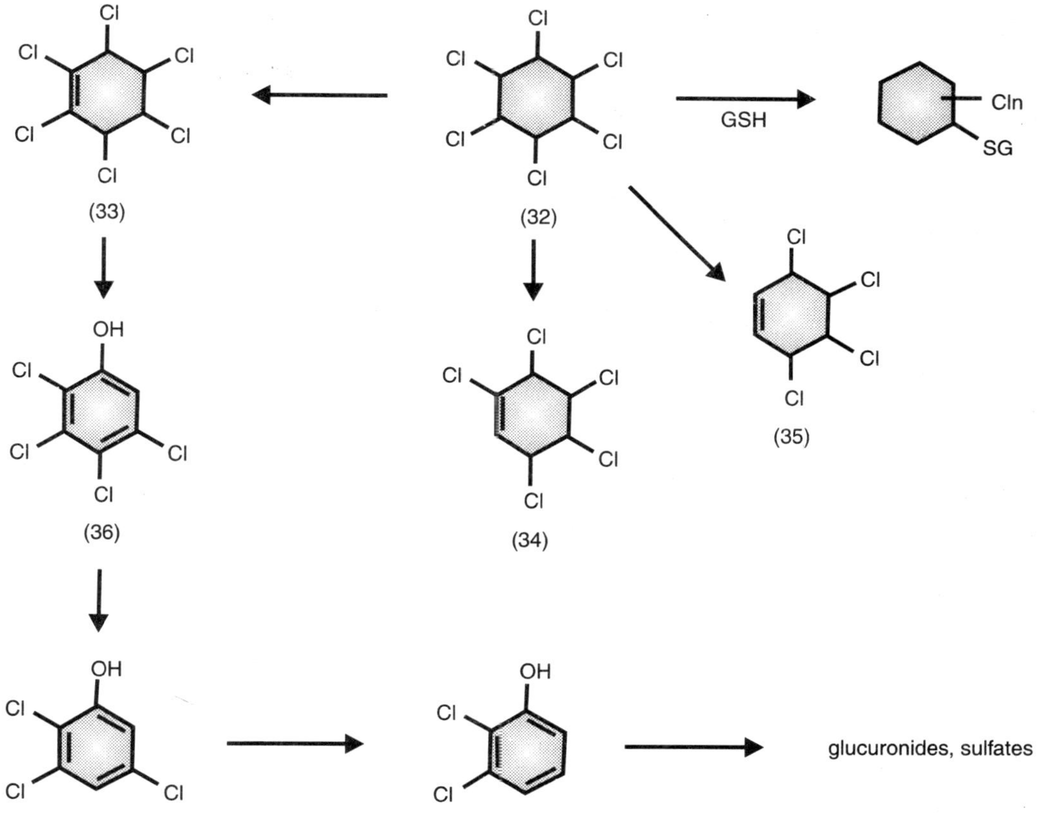

Figure 17.23: Initial stages in metabolism of HCH to chlorobenzenes, chlorophenols, glucuronides, and sulfates.

tetrachlorocyclohexene **(35)**. Interconversion of HCH isomers and conversion of γ-HCH to α-HCH and hexachlorobenzene occurred in plants (and microorganisms).

The metabolism of *gamma*-HCH in higher plants and microorganisms proceeded via pentachlorocyclohexene, whichwasdehydrohalogenated in peas and corn to give polychlorinated benzenes. Oxidative reactions converted these to polychlorinated phenols. Dehydrogenation reactions were important in metabolism of gamma-HCH in the rat. In 6 days, female rats excreted free and conjugated 2,3,4,5 tetrachlorophenol **(36)**, a secondary metabolite derived from **33,** the primary product of dehydrogenation of *gamma*-HCH to the extent of 11% of the administered dose.

Polychlorophenols were obtained, and 2,4,6-trichlorophenol may havLarisen by direct oxygenation of HCH by rat liver microsomes to form an intermediate pentachlorocyclohexanon, chlorohydrin. This gave 2,4,6-TCP after a two-step dehy-drohalogenation via the enol form. Reactions involving glutathione S-transferase gave a number of metabolites containing mercapturic acid moieties or thiophenol groups.

Toxicology

Gamma-HCH (lindane) caused liver tumors in mice, probably as a secondary consequence of toxicity to the liver. In the rat, liver damage (hepatocyte hypertrophy) was reversible on discontinuation of dosing. The principal toxic effect in the rat was kidney damage in the form of hyaline droplets in the proximal tubules, specificcally in the male.

This effect has been shown to be caused by α2μ-globulin formation, a mechanism that is not relevant to humans. A battery of genotoxicity tests was negative. Likewise, developmental and reproductive toxicity tests did not reveal any adverse findings.

Section 3

Cyclodiene Insecticides

Mode of Action

The GABA-activated chloride channel has been established as the nerve target site for three types of insecticide: cyclodienes, lindane, and toxaphene. Each of these structural types acted as potent, competitive inhibitors of the binding of a radioligand [^{35}S-TBPS] to rat brain membranes. The binding inhibition was isomer-specific, and the potency *in vitro* correlated very closely with the toxicity (LD_{50}) to rodents following oral or intraperitoneal administration.

The epoxides of many of the cyclodienes, such as heptachlor and aldrin, were both more potent on the assay and more toxic *in vivo* than were the parent molecules. Among the active compounds using the ^{35}TBPS assay were aldrin, chlordane, dieldrin, α-endosulfan, endrin, heptachlor, isobenzam, isodrin, 12-ketoendrin, lindane, and toxaphene.

Chemically related insecticides that were inactive on this assay were DDT, kepone, and mirex. It was also demonstrated that lindane and heptachlor epoxide acted as inhibitors of GABA-induced chloride permeability increases in cockroach leg muscle. The authors also showed that these two compounds inhibited the specific binding of tritiated picrotoxinin to rat brain membranes.

There is thus little doubt that the critical lesion resulting from dosing both mammals and insects with lindane and related compounds is the same, i.e., antagonism at the GABA-activated chloride channel in nerve/muscle membranes. Bloomquist and colleagues have also recently shown that cyclodienes such as heptachlor cause some of the same effects in the mouse brain as are observed in Parkinson's Disease (PD); however, there are also some differences.

Following intraperitoneal dosing at 6 mg/kg (4% of LD_{50}), heptachlor increased dopamine in the striatum through an increase in the level of the dopamine transporter protein (DAT). Cytotoxicity occurred at higher doses, causing a reduction in dopamine transport. However, *in vitro* heptachlor stimulated dopamine release from preloaded striatal synaptosomes. The significance of these findings in the etiology of idiopathic PD is uncertain, however, because a depletion rather than an increased uptake of striatal dopamine is associated with the disease, *in vivo.*

The manufacture of cyclodiene insecticides was based on the availability of a common synthesis intermediate, hexachlorocyclopentadiene, or "hex". The Diels–Alder diene reaction of this compound

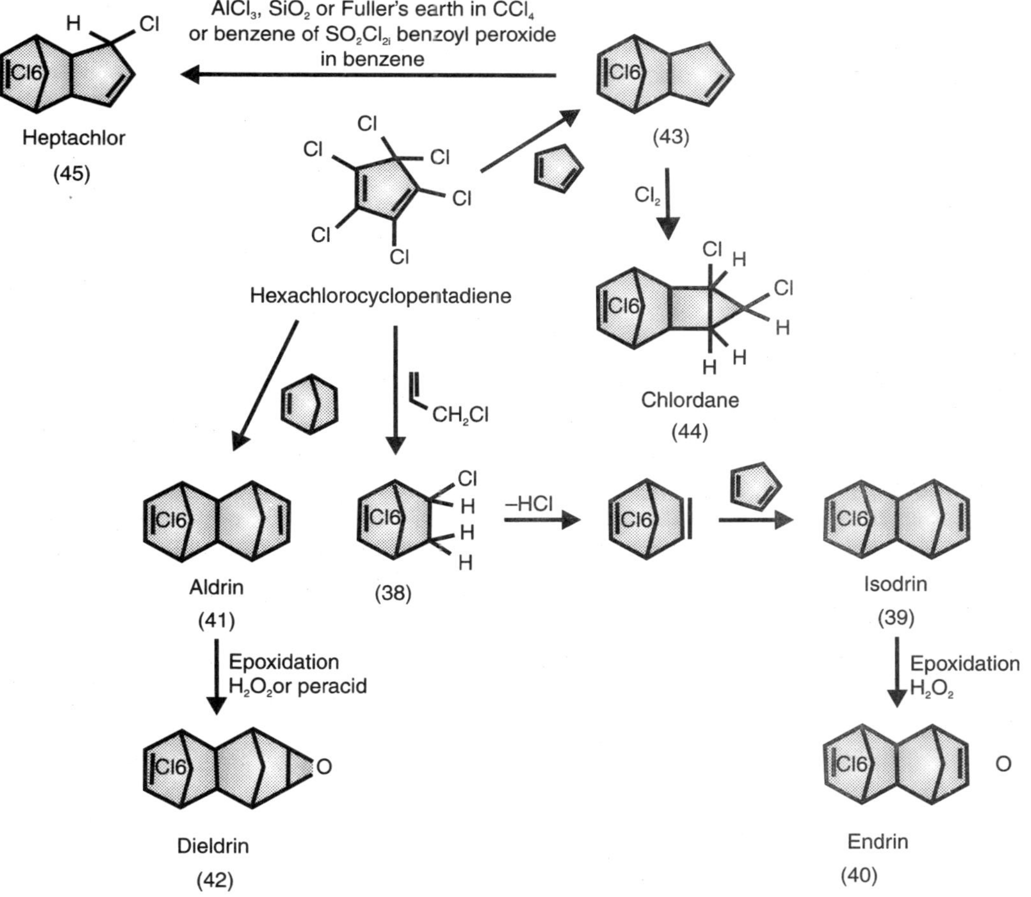

Figure 17.24: Synthesis of aldrin, dieldrin endrin, isodrin, heptachlor, and chlordane.

with suitable dienophiles led to a variety of compounds, including mirex, endrin, aldrin, dieldrin, chlordane, endosulfan, and dienochlor.

The reaction of hex with itself gave dimeric products: dienochlor and mirex. The later was used for fire-ant control until its use was discontinued because its caged highly chlorinated structure was extremely resistant to biological and environmental degradation. In fact, it was also used as a fire retardant. The reaction of hexachlorocyclopentadiene with cyclopentadiene gives chlordane. *Hexachlorocyclopentadiene* may be produced by direct chlorination of cyclopentadiene with chlorine in two stages.

Polychlorocyclopentanes are formed during low temperature chlorination, and then these are subjected to exhaustive chlorination at 350-500°C. Octachlorocyclopentane is formed, and thisbreaksdownto hexachlor-ocyclopentadiene at an elevated temperature.

Hexachlorocyclopentadiene is also produced industrially by chlorination of cyclopentane from petroleum. Cyclopentane is chlorinated to give polychlorocyclopentanes, which are further subjected to high temperature chlorination. Pentanes or amylenes from petroleum may be used as starting materials. Chlorination of aliphatic hydrocarbons under free radical conditions affords polychloropentanes are subjected to high temperature chlorination and cyclization that yield hexachlorocyclo-pentadiene.

Hexachlorocyclopentadiene may be utilized as a component in the Diels-Alder diene synthesis. On prolonged heating, it may dimerize. The action of cuprous chloride in methanol on hexachlorocyclopentadiene gives dienochlor (Pentac, 37).

Dienochlor

Dienochlor (37), [2227-17-0], Perchloro-1,1′-bicyclopenta-2,4-diene (IUPAC), mp 122-123°C has been used as a selective acaricide to control mites on crops. It has predominantly contact action and interferes with oviposition. Dienochlor is stable in storage at 54°C (14 day) and at 42°C (2 years). It undergoes hydrolysis [DT_{50} 30.5 days (pH 9), 93 days (pH 7), and 184 days (pH 5)].

Dienochlor decomposes in simulated sunlight (DT50 1.6 min). Degradation is mainly environmental rather than metabolic, and photochemical breakdown is rapid. It decomposes in soils, DT_{50} 3.1 days and DT50 2-3 days on plants exposed to sunlight.

The major degradation products of dienochlor in plants are perchloroketones. It is rapidly degraded in rats.

Cyclodienes

The cyclodienes are chlorinated cyclic hydrocarbons with endomethylene-bridged structures obtained as products of the Diels-Alder diene reaction using hexachlorocyclopentadiene (or "hex") as a diene component. The major structural groups have a dimethanonaphthane or dimethanoindane skeleton. Isodrin-endrin and aldrin-dieldrin are pairs of compounds that differ in stereochemistry at the ring junction.

Figures elsewhere in this chapter outline the major reaction paths. The reaction of "hex" with vinyl chloride gives *heptachloronorbornene* **(38)**, and this by loss of hydrogen chloride gives hexachloronorbornene nucleus, a basic intermediate in the synthesis of isodrin and endrin. The condensation of hexachloronorbornadiene with cyclopentadiene produces isodrin (39), in which

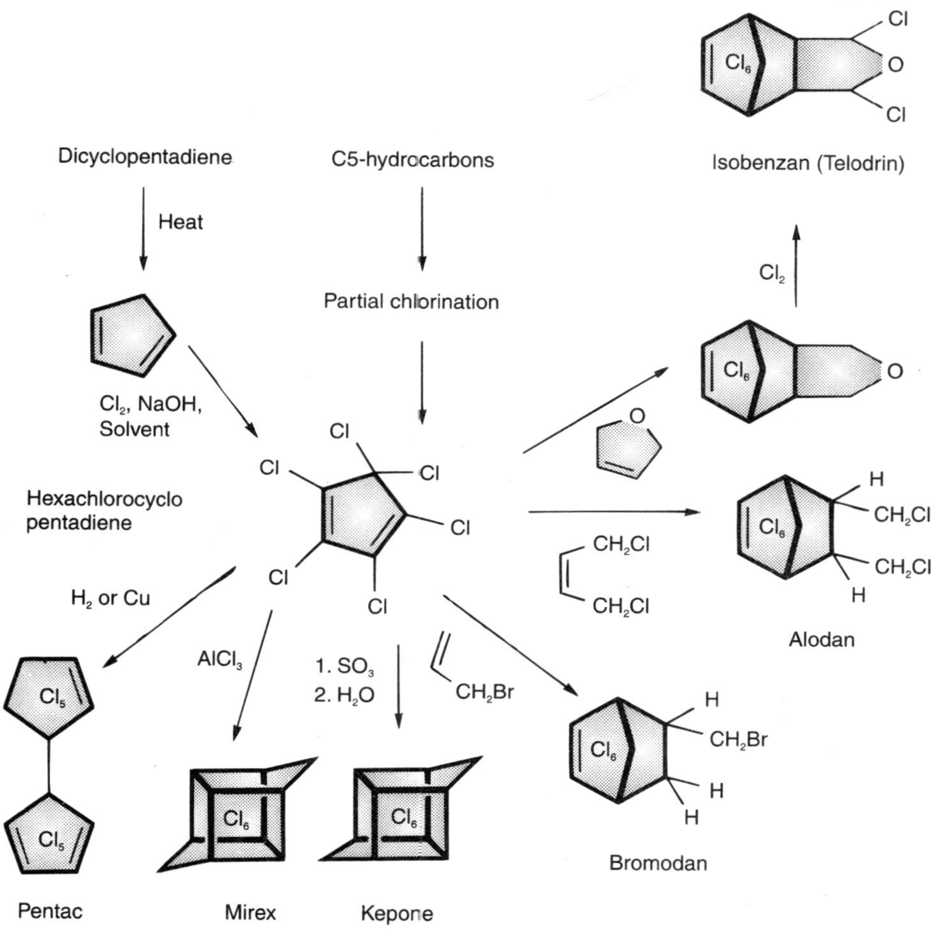

Figure 17.25: Synthesis of mirex, dienochlor, telodrin, and other inseticides from hexachlorocyclopentdiene.

norbornyl ring fusion takes place through its endo positions to the endo positions of the hexachloronorbornene nucleus. Endrin (40) is obtained by epoxidation of isodrin.

Aldrin

Aldrin (41) is formed when "hex" reacts with norbornadiene to give an adduct in which the new norbornene nucleus is fused to another hexachloronorbornene molecule. This is generated concomitantly in the reaction. Ring fusion takes place through the endo positions. The reaction may be brought about by boiling "hex" in an excess of norbornadiene.

In the product, the unchlorinated norbornene nucleus is fused through its exo-positions to hexachloronorbornene. Epoxidation of the product by treating aldrin with perbenzoic acid (or other peroxyacid) in chloroform gives HEOD (42). This is the main constituent of the technical material known as dieldrin (contains not less than 85% HEOD). However, in many publications, the pure product may be referred to as dieldrin.

Chlordene

Chlordene (43) is the product of the reaction of cyclopentadiene with hexachlorocyclopentadiene at 80–90 °C. It is an intermediate in the production of chlordane (44) or heptachlor (45). Heptachlor may be synthesized by reacting chlordene with N-bromosuccinimide to give the intermediate 1-bromochlordene followed by chlorination with hydrogen chloride in nitromethane in the presence of aluminum trichloride. Alternatively, 1 bromochlordene may be hydrolyzed to 1-hydroxychlordene, which is subsequently chlorinated with thionyl chloride in benzene. Heptachlor may also be produced by low temperature catalytic chlorination of chlordane or by chlorination of chlordene with sulfuryl chloride.

Cyclodienes form epoxides readily. These are frequently detected as residues. For example, aldrin forms the more stable epoxide dieldrin. If epoxides are formed in tissues of animals that ingest or absorb residues, they may be concentrated and stored preferentially in fatty $tissue_f$. Epoxides are also formed in plants and soil. Dieldrin and heptachlor epoxides are quite stable, and their lifetimes on foliage or soil are greater than those of the parent compounds, aldrin and heptachlor.

Chlordane

Chlordane (44), [57-74-9] Chlordane; [12789-03-6] chlordane tech. grade; [5103-71-9] cis-isomer (formerly [22212-52-8]); [5103-74-2] trans-isomer; 1,2,4, 6,7,8,8-octachloro-2,3,3a,4,7,7a-hexahydro-4,7-methano-1H-indene (IUPAC) is a persistent nonsystemic ^{co}tact and ingested insecticide with some fumigant action.

Chlordane is prepared by chlorination of chlordene at 50–60 °C. The technical grade contains 60–75% of the *cis* and *trans* isomers. Major components are two stereoisomers: the *alpha* or cis-isomer (1α,2α,3aα 4β,7β,7aα), and the trans-isomer (1α,2β,3aα,4β,7β,7aα), usually known as *gamma,* but occasionally as beta-chlordane.

The nomenclature at C(1) and C(2) has been confused in the literature. The remainder of the technical grade comprises other stereoisomers (each not more than 7%) and heptachlor (84). Technical chlordane is a viscous, amber liquid (bp 175 °C/267 Pa, vp 1.3 mPa at 25 °C) soluble in water to about 9 µg/L. It has rat LD_{50}s of 335, 430 (oral), and 840, 690 (dermal) mg/kg. Technical chlordane contains about 60% of the isomers and 10–20% of heptachlor.

Chlordane, the first of the cyclodiene insecticides to be used in agriculture and industry, was developed by Julius Hyman. According to Brooks (85), the insecticidal properties of chlordane were first described in publication by Kearns et al. (86) in 1945. They were evaluating a product made in the previous year by Velsicol in an attempt to produce an insecticide less costly than DDT.

The addition of two Cl atoms across the double bond of the five-membered ring of chlordene gives the two isomers of chlordane 1,2,4,5,6,7,8,8-octachloro-2,3,3a,4,7,7a-hexahydro-4,7-methano-1H-indene, a-trans. The β-isomer has significantly greater insecticidal activity.

Chlordane was used extensively as a soil insecticide for termite control and as a household insecticide. It was also used as a wood preservative, a protective treatment for underground cables, and to reduce earthworm populations in lawns. It was used against Formicidae, Coleoptera, Noctuidae larva, Saltatoria, subterranean termites, and many other insect pests.

It controls household insects, pes of human and domestic animals, and was applied to soil or directly to foliage or as a seed treatment. The cyclodiene insecticides generally have high mammalian toxicities and are absorbed dermally. Chlordane acts as an antagonist of the GABA receptor chloride chanel complex. Chlordane is decomposed by alkalies with the loss of chlorine. DT_{50} in soil of chlordane is about 1 year (87). Its solubility in water is 0.1 mg L^{-1}(25°C).

Metabolic processes include oxidation, reductive dechlorination, hydrolysis, and epoxidation. These reactions give a variety of products in which the norbornyl moiety generally remains unaltered, except in photochemical reactions when bridged compounds may be formed or dechlorination may occur.

Acetone sensitized photolysis of chlordene **(43)** and cis-chlordane **(46)** giving bridged derivatives with cage-like structures (**47** or **48**). The process is governed by the stereochemistry of the molecule. The isomeric *trans*-chlordane **(49)** and gamma-chlordane failed to yield bridged products as the double bond in these compounds interacts with the endo-chlorine atom.

Later it was reported that with *trans*-chlordane **(49),** bridging did occur and involved carbon-1, which has a noninterfering exochlorine atom. Unsensitized photolysis of trans-chlordane **(49)** in aqueous organic solvents at wavelengths less than 300 nm gave two isomeric monodechlorinated products (**50** and **52**) and ultimately the bis-dechlorination product **(51)**.

When cis-chlordane **(46)** was irradiated as a solid film, up to 70% was converted after 16 to 20 hours irradiation to a mixture of products, more than half of which were bridged isomers (depending on conditions, the dechlorinated compounds may also form bridged compounds). Chlordane, beta-chlordane, and beta-dihydroheptachlor were converted into the isomeric bridged derivatives using the same technique. The isomeric *alpha* chlordane, gamma chlordane, and nonachlordane did not form bridged products as the double bond in these molecules interacts with an endo-directed chlorine atom.

Pure *cis* and *trans* chlordane were degraded to dichlorochlordene **(53)**, oxychlordane **(54)**, heptachlor **(45)**, hectachlor endo-epoxide, heptachlor exo-epoxide **(55)**, chlordene chlorohydrin **(56)**, and 3-hydroxy-trans-chlordane by an actinomycete, *Nocardopsis* sp., from soil. Oxychlordane was slowly degraded to 1-hydroxy-2-chlorochlordene. Incubation with a microbial mat (an algal bacterial consortium) totally dechlorinated chlordane to 4,7-methanoindene.

It was 91% degraded after 21 days, and no parent remained. Organisms evaluated for their potential in removing chlordane from water include a lignin-degrading fungus, *Trametes versicolour,* which removed 75% of added chlordane within 90 days in microcosm experiments. A single oral dose of chlordane (9.7 mg kg^{-1}) was metabolized by the rat via **(53)** 1,2-dichlorochlordene **(57)** and oxychlordane **(54)** to 1-exo-hydroxy-2-chlorochlordene **(58)** and 1-exo-hydroxy-2-endo-chloro-2,3-exo-epoxychlordene **(55)** and other hydroxylated products.

Microsomal preparations of rat liver converted trans-chlordane to *54* via *57. Two* stable products are formed from 53: 56 and 55. Fecal extracts contained hetachlor (0.1% of total radioactivity extracted from feces), 1,2-dichlorochlordene (2.5%), (54) (0.5%), *cis* chlordane (13%), 58 (19%), 55 (7.5%), 1-exo-hydrox-2-endo-chlorodihydrochlordene (chlordene chlorohydrin) **(56)** (3%), monohydroxylated

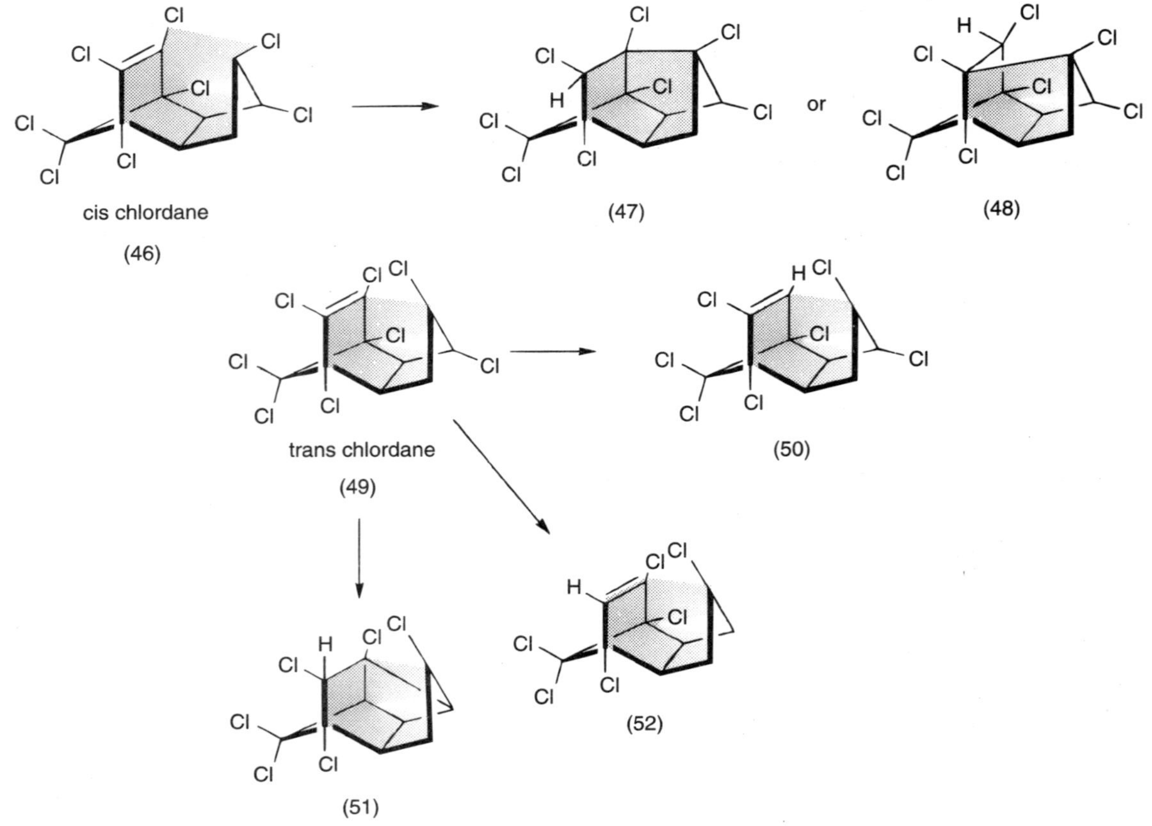

Figure 17.26: Photochemistry of chlordane.

dihydrochlordene (15.5%), 1,2-dihydroxydihydrochlordene (26.5%), trihydroxydihdrochlordene (3%), and three unidentified metablites.

Oxychlordane **(54)** characterized as a metabolite in the fat of rats, dogs, pigs, and cows fed *cis*-chlordane, trans-chlordane, or a 50 : 50 mixture of *cis* and *trans*-chlordane (95) is probably a terminal metabolite. It was suggested that compounds **(56,55)** and **(59)** may not be derived from **(54)** but may be artifacts formed during separation chromatography on alumina G. Base-labile chlorinated cyclodienes are known to form dehydration and dehalogenation products on alumina (96).

Rats treated by a single dose or continuous feeding of radioactively labeled ^{14}C-chlordane excreted labeled products that included cis- and *trans*-chlordane, **(46,49)**, two monochlorodihydroxy derivatives of chlordane, *cis*- or *trans*-dihydroxychlordane, trihydroxylated chlordane, and conjugated derivatives of hydroxylated chlordane metabolites. Similar metabolites to those in feces were found in urine (97).

Oxychlordane (54) was the major residue in all tissues, and when chlordane was removed from the diet after 56 days, it became almost the sole radioactive residue in the tissues. There

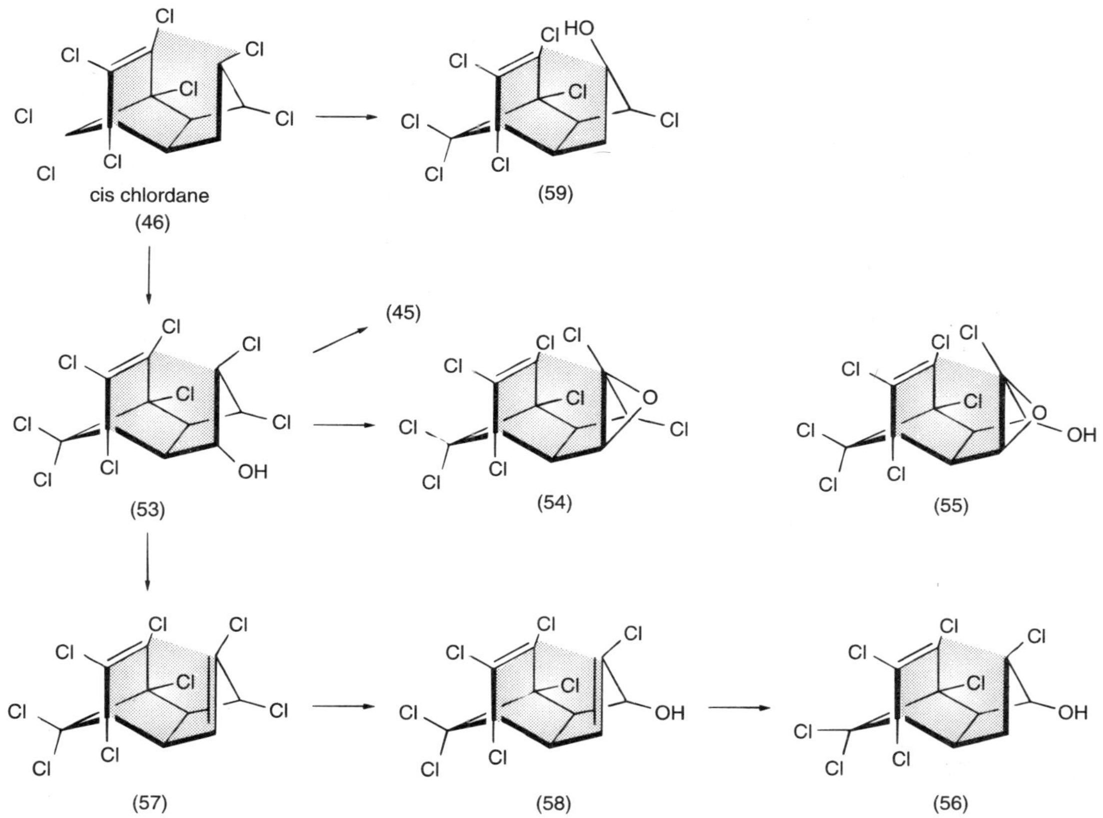
cis chlordane
(46)
(59)
HO
(45)
(53)
OH
(54)
O
(55)
OH
(57)
(58)
OH
(56)
OH

Figure 17.27: Metabolism of chlordane.

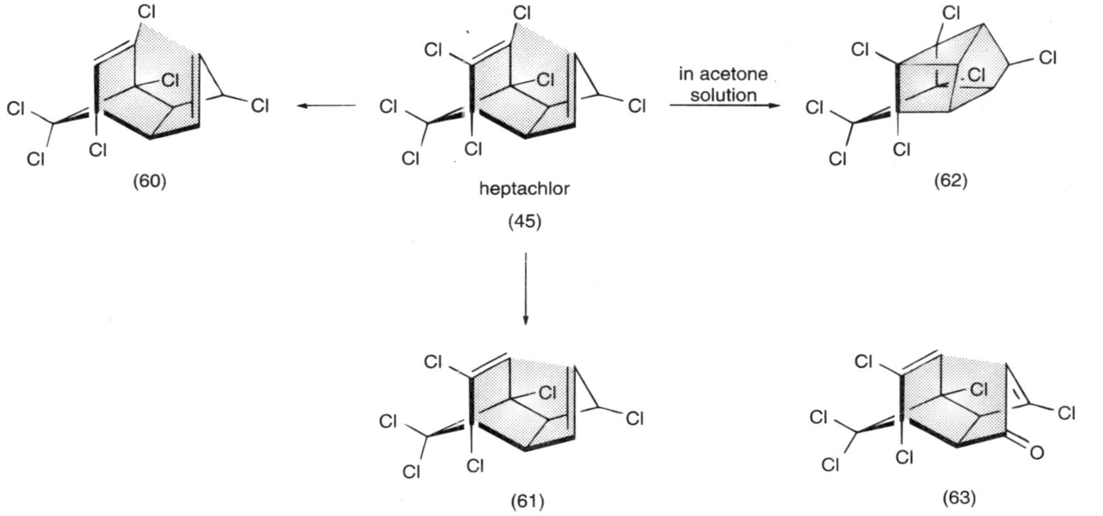
(60)
heptachlor
(45)
in acetone solution
(62)
(61)
(63)
O

Figure 17.28: Heptachlor photochemistry.

was no oxychlordane or dichlorochlordene (57) in feces. More polar metabolites were produced as time after treatment increased and metabolites in urine followed a similar pattern to that in feces.

Toxicology

In both the rat and mouse, there was some evidence of malignant or benign liver tumors following chronic dietary chlordane administration for 2 years. Liver toxicity was reported as either hypertrophy or hyperplasia. A chronic (nononcogenic) NOEL of 1 ppm (0.05 mg/kg/day) was established in the rat for liver toxicity. Genetic toxicity was reported in some studies but not in others: Increased gene mutation and UDS (unscheduled DNA synthesis) were sometimes reported. IARC has designated chlordane as a Class 2B or "possible human carcinogen." Neurotoxicity was reported in some experiments but was not quantified.

Heptachlor

Heptachlor (45) [76-44-8], 1,4,5,6,7,8,8-heptachloro-3a,4,7,7a-tetrahydro-4,7-methano-1H-indene (IUPAC). It is a white crystalline solid mp 95-96 °C vp 0.04 Pa at 25 °C). Technical grade has mp 46-74 °C and contains 65-74% heptachlor, and 28-35% related compounds (including chloroindane, nonachlor, and octachlor), is soluble in water to 56 µg/L. It is about 3-5 times more active than is chlordane as an insecticide. The rat LD_{50}s are 100, 162 (oral), and 195, 250 (dermal) mg/kg.

It was used to control soil-inhabiting pests and other insects, including ants and termites. It is more inert chemically than is chlordane and resistant to water and caustic alkalies. It can be oxidized to the corresponding epoxide. This reaction takes place *in vitro* or under the influence of soil microorganisms, in animals and in insects. Heptachlor epoxide is the major metabolite of heptachlor. It is also more toxic than heptachlor.

In soils, it is extremely persistent Heptachlor epoxide [1024-57-3] (mp 159 °C), rat oral LD_{50} 47 mg/kg is an important and highly persistent environmental pollutant. Hydrogenation of heptachlor produces 6-dihydroheptachlor [14168-01-5] (mp 135 °C), which retains high insecticidal activity with very low mammalian toxicity, rat oral LD_{50} > 5000 mg/kg. Heptachlor may be obtained by treatment of chlordane with N-bromosuccinimide, followed by chlorination with hydrogen chloride in nitromethane in the presence of aluminum trichloride or with monochloro iodide in carbon tetrachloride. Irradiation of heptachlor at 253.7 nm in hexane or cyclohexane solution afforded two isomeric monodechlorination products **(60)** and **(61)**. In acetone at 300 nm, a cage compound **(62)** was the sole product.

Cage formation by heptachlor to give (**62**) may be postulated on the basis of analogous photochemical reactions of endocyclopentadiene derivatives. Photodechlorination in cyclohexane occurred via an excited singlet state (98). In the presence of a photosensitizer, such as acetone, cage formation occurs via a triplet state (99). Under these conditions, heptachlor epoxide **(64)** gave a ketonic product **(63)** in an isomerization reaction in which the epoxide ring was opened.

Metabolism

Soil microorganisms transform heptachlor by epoxidation, hydrolysis, and reduction. Heptachlor incubated with a mixed culture of organisms gave chlordene (43), which was further metabolized

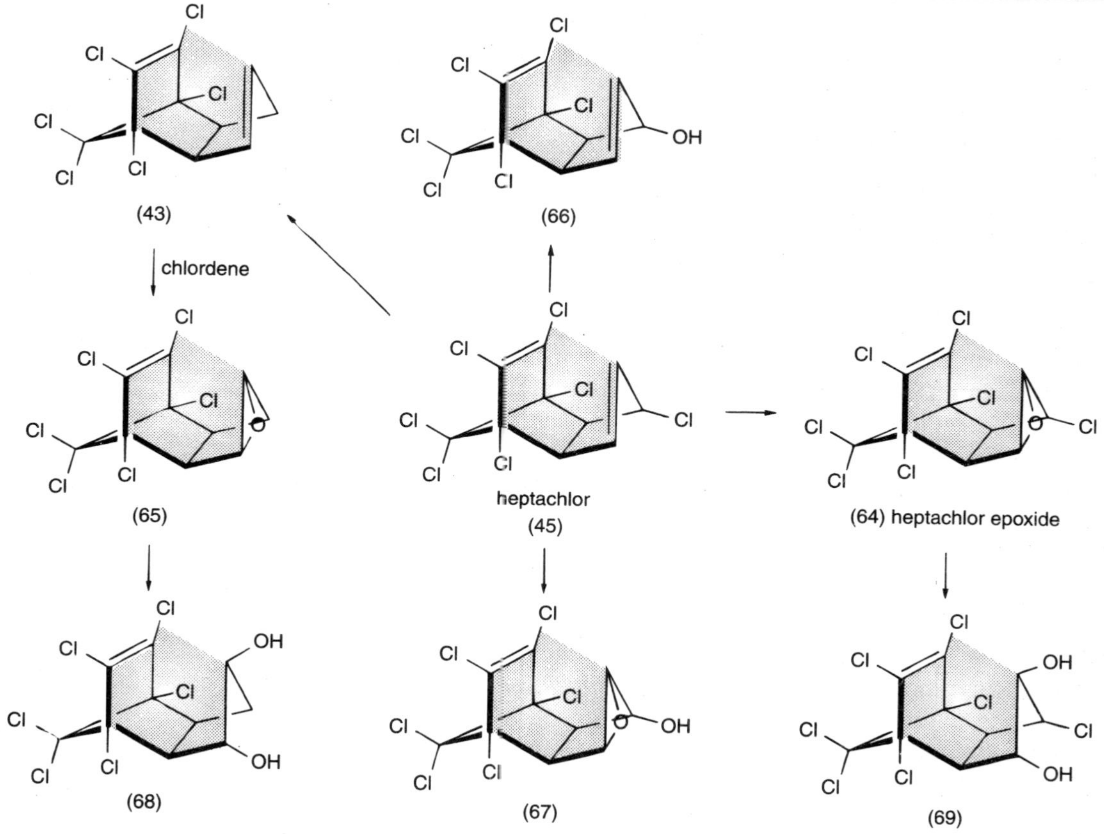

(43)

chlordene

(66)

(65)

heptachlor (45)

(64) heptachlor epoxide

(68)

(67)

(69)

Figure 17.29: Heptachlor metabolism.

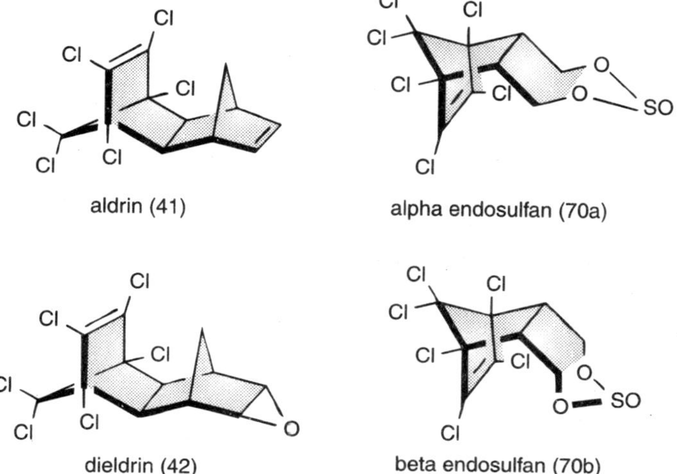

Structures 41 and 42; structures 70a and 70b

to chlordene epoxide (65). Also detected were 1-hydroxychlordene (66), 1-hydroxy-2,3-epoxychlordene (67), and heptachlor epoxide (64). The hydrolysis product of heptachlor (45) was metabolized by soil organisms to a product, which may be a ketochlordene.

Metabolic products identified when rats were fed a diet containing 100 ppm of heptachlor during a 4-week period were the epoxide **(64)** and 1-exo-1-hydroxyheptachlor epoxide (13) and 1,2-dihydroxydihy-drochlordene **(68)** (100,101). Additional metabolites formed from heptachlor epoxide after incubation with microsomal preparations from the liver of pigs and from houseflies were the diol **(69)** and 67.

Toxicology

Heptachlor was toxic to the rodent liver, resulting in hepatic hyperplasia, with a NOEL of 5 ppm (0.25 mg/kg/day) in the rat. Hepatocellular carcinoma in mice was observed, but this may have been secondary to liver damage because the genotoxicity assays we considered negative. IARC has designated heptachlor as a Class 2B or "possible human carcinogen." No adverse effects were noted in developmental or reproductive toxicity tests in the rabbit and rat, respectively.

Aldrin

Aldrin, (41) [309-00-2], 1,2,3,4,10,10-hexachloro-1,4,4a,5,8,8a-hexahydro-1,4-endo,exo-5,8-dime thanonaphthalene. It is a colourless solid, mp 104–104.5°C, vapor pressure 5.2 mPa at 20°C. It is chemically stable and is not degraded by water or caustic alkalies at room tempeature.

This compound is very slightly soluble in water (to 0.027 µg/L) but soluble in petroleum hydrocarbons. Aldrin has rat LD_{50} values of 39, 60 mg/kg (oral) and 98 mg/kg (dermal). It has been widely used as a seed treatment and soil insecticide, where it is gradually oxidized to its epoxide dieldrin.

Toxicology

According to the U.S. Environmental Protection Agency, there is good evidence that aldrin/ dieldrin is oncogenic in the mouse, causing a dose-related increase in benign and malignant liver tumors, along with some evidence of lung tumors.

There is some evidence of oncogenicity in the rat. Liver toxicity was noted in both rodents and the dog, at doses as low as 0.5 ppm (0.025 mg/kg/day) in the rat.

DNA damage and UDS recorded in transformed human cells following dieldrin administration indicated possible genetic toxicity. Nonetheless, IARC has designated aldrin and dieldrin as Class 3 or "not concluded to be a human carcinogen." Signs of neurotoxicity were reported in rodents in dietary studies, in the form of hyperactivity and tremors, at doses as low as 2.5 ppm (0.4mg/ kg/day) in the mouse.

Dieldrin

Dieldrin **(42)** [60-57-1] or 1,2,3,4,10,10-hexa-chloro-1,4,4a,5,8,8a-hexahydro-6,7-epoxy-1,4-endo, exo-5_n 8-dimethanonaphthalene (mp 176°C, vp 0.4 mPa at 20°C is formed from aldrin by epoxidation with peracetic or perbenzoic acids. It is soluble in water to 27 µg/L. Aldrin and dieldrin have had extensive use as soil insecticides and for seed treatments.

Dieldrin, which is very persistent, has had wide use to control migratory locusts, as a residual

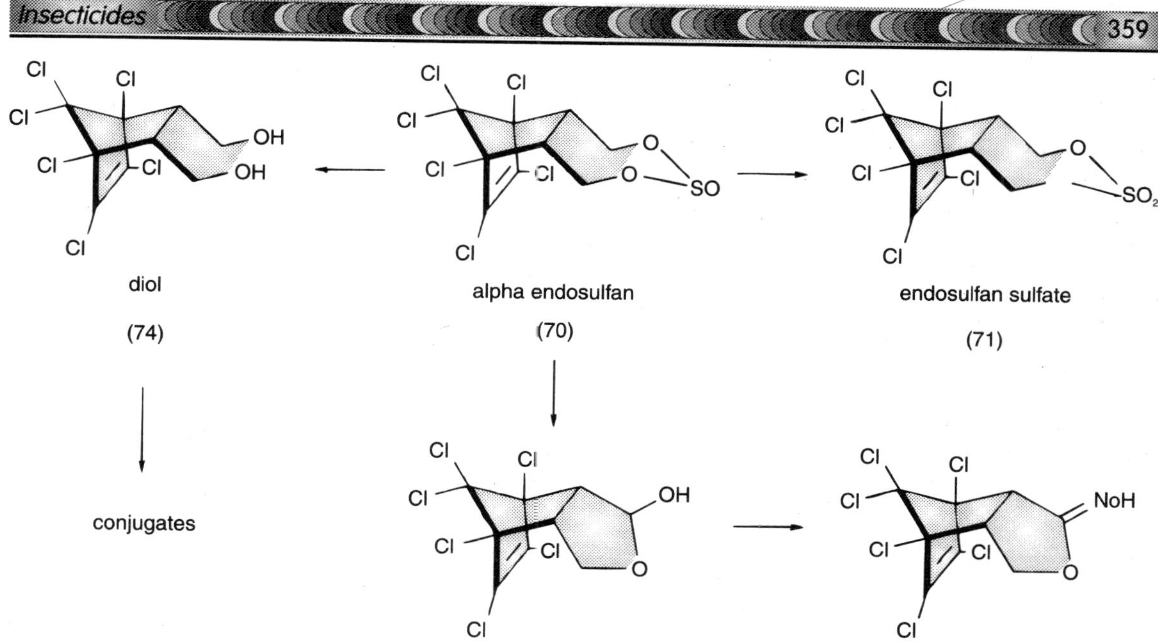

Figure 17.30: Endosulfan transformations.

spray to control the Anopheles vectors of malaria, and to control tsetse flies. Because of its environmental persistence and propensity for bioaccumulation, registrations in the United States were canceled in 1974.

Endrin

Endrin (40) [72-20-8] is 1,2,3,4,10,10-hexachloro-1,4,4a,5,8,8a-hexahydro-6,7-epoxy-1,4-endo,endo5,8-dimethanonaphthalene (mp 245 dec, vp 0.022 mPa at 25°C) and is soluble in water to 23 µg/L.

This compound is the *endo,endo* isomer of dieldrin, which is less stable and more toxic than dieldrin with rat LD_{50} values of 17.8 and 7.5 (oral) and 15 (dermal) mg/kg. It was used as a cotton insecticide, but because of its high toxicity to fish, its use was restricted..

Toxicology

Endrin did not appear to be carcinogenic in rat and mouse chronic dietary studies. It has been designated by IARC as Class 3 or "not considered to be a human carcinogen." Enlarged liver and kidneys were reported in both rodents and the dog. The NOEL was 1 ppm (0.05 mg/kg/day) in the rat. Tissue degeneration was observed in both organs in the dog. All studies were compromised by high mortality, resulting from neurotoxic effects causing convulsions.

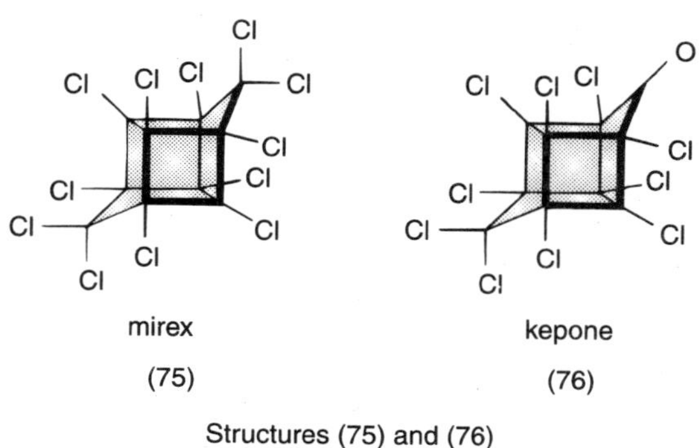

Structures (75) and (76)
mirex and kepone

Endosulfan

Endosulfan (70) [115-29-7], 6,7,8,9,10,10-hexachloro-1,5,5a,6,9,9a-hexahydro-6,9-methano-2,4,3,-benzo-dioxathiepine 3-oxide (IUPAC) [The technical product is a mixture of two isomers: α-endosulfan: 3α,5α,6, 9α,9α,6 (64–67%) and 6-endosulfan 3,5α,6,9,6,9aα, (29–32%) (70b)]; [959-98-8] (formerly [33213-660]) (6-endosulfan);[33213-65-9] (formerly [891-86-1] and [19670-15-6]) (β-endosulfan) is the adduct of hexachloro- cyclopentadiene and 1,4-dihydroxy-2-butene reacted further with $SOCl_2$ to produce 6,7,8,9,10,10-hexachloro- 1,5,5α,6,9,9α-hexahydro-6,9-methano-2,4,3-benzodioxathiepin-3-oxide.

The technical product is a brownish solid, mp 70–100°C, vapor pressure 1.3 mPa at 25 °C, soluble in petroleum solvents but having low solubility in water. In consists of about four parts of a-isomer (mp 108 °C, cis with regard to the sulfite group) and one part of the α-isomer (mp 206 °C, *trans* with regard to the sulfite group).

The α-isomer, which is somewhat more insecticidal, is slowly converted to the more stable α-isomer at high temperature$_s$ and both isomers are oxidized slowly to endosulfan sulfate [1031-07-8] (mp 181 'C). In acid media, both isomers form endosulfan diol [2157-19-9] (mp 203°C).

The rat LD_{50} values are 43, 18 mg/kg (oral) and 13074 mg/kg (dermal). The α-isomer has somewhat greate insecticidal activity and is slowly converted to the more stable β-isomer at a high temperature. Both isomer oxidize slowly in air and in biological systems to endosulfan sulfate [1031-07-8], mp 181–182 °C. In acid media, both isomers form endosulfan diol [2157-19-9], mp 203–205°C.

Endosulfan is a broad-spectrum insecticide used to ^{co}trol pests of vegetables, fruit, field crops, and ornamentals. Unlike other cyclodiene insecticides, it is biodegradable by hydrolysis at the sulfite ester bonds and is more readily metabolized. It is also less persistent on plant surfaces, and 50% of the residues are lost in 3–7 days.

Volatilization may be the major route of loss. Endosulfan is readily hydrolyzed in water to the diol (74), but it is moderately persistent in soil. Endosulfan $^{(c}$and 6-endosulfan) is degraded in soil with DT_{50} 30 to 70 days. The major metabolite is usually endosulfan sulfate (71), which is degraded more slowly. In the field DT50 for total endosulfan (a- and ,6-endosulfan and endosulfan sulfate) is 5 to 8 months.

Metabolism

Endosulfan is metabolized rapidly in mammalian organisms to less toxic metabolites and to polar conjugates. The sulfate is also a major metabolite in plants and occurs as a metabolite in some mammals. Endosulfan is quite toxic to water organisms, and residues were found in runoff water, sediment, infiltration water, and soil following a single application.

Rats dosed orally or intraperitoneally with endosulfan (4–8 mg/Kg) excreted unchanged endosulfan, the hydroxy ether (72), the lactone (73), and unidentified metabolites in urine in the ratio 3 : 1 : 1 : 2. The diol, the hydroxy ether, and the lactone were identified in most samples of urine and feces. The metabolite most frequently recovered from tissues, organs, and feces was endosulfan sulfate.

Transient amounts of endosulfan and endosulfan sulfate were detected in the body fat and liver of mice after they were dosed with ^{14}C labeled endosulfan. The mice excreted endosulfan metabolites. Cows fed 2.5–5 ppm endosulfan for 30 days excreted 0.1–0.2 ppm endosulfan sulfate in milk.

After a single dose of 14 mg/kg of ^{14}C endosulfan, sheep excreted 0.25 ppm in milk in the 6–24-h period following ingestion (102). Residues fell to 0.04 ppm and 0.01 ppm after 3 and 11 days, respectively. The main metabolites in urine were the diol and the hydroxy ether.

Endosulfan sulfate, the ether, the hydroxy ether, the lactone, and one unidentified metabolite were detected on the surface when male migratory locusts (*Pachytilus* migratoides) were exposed to endosulfan by oral, cutaneous, or subcutaneous administration. Similar metabolic pathways were observed in the housefly and the cockroach. Six to seven days after the last dose, neither endosulfan nor its metabolites could be detected in locusts.

Endosulfan is very toxic to fish and caused many fish deaths when the Rhine River became contaminated in June 1969 (concentration was 0.1 ppm). The only residues detected in fish exposed to acute and multiple subchronic concentrations of endosulfan were endosulfan and the diol and the glucuronic acid conjugate of the diol.

Toxicology

Endosulfan was not carcinogenic in the rat or mouse in chronic dietary studies. It was not genotoxic in a variety of tests. Kidney toxicity was observed in the rat at dietary levels of 100 ppm (5 mg/kg/day) and above. Clinical signs of hyperactivity and tremors were reported in many studies. No evidence was found of developmental or reproductive toxicity.

Mirex

Mirex (75) [2385-85-5] is 1,2,3,4,5,5,6,7,8,9, 10,10-dodecachloro-octahydro-1,3,4-metheno-2H-cyclobuta-[c,d]-pentalene. The rat LD_{50}s are 306, 600 (oral) and >2000 (dermal) mg/kg. Mirex was patented in 1954. It is extremely resistant to biodegradation and was once considered the perfect stomach poison insecticide for use in baits to control imported fire ants. However, even at doses of a few milligrams per 10 m^2, it was found to bioaccumulate in birds and fish and its registrations were canceled in the United States in 1976.

Kepone

Kepone (76) or chlordecone [143-50-0] decachloro-5-oxo-pentacyclo-[5.3.0.$0^{2,6}$.$0^{3,9}$,$0^{4,8}$]-decane, mp 350 °C (dec.) is the 2-keto analog of mirex and is soluble in water to 4 g/L by hydration. The rat LD_{50}s are 95, 140 (oral) and > 2000 (dermal) mg/kg. Chlordecone is a stomach poison used in baits for the control of cockroaches and ants and for the control of banana thrips. Because of bioaccumulation, its registrations were canceled in the United States in 1978.

Section 4

Chlorinated Terpenes

A group of incompletely characterized insecticidal compounds has been manufactured by the chlorination of the naturally occurring terpenes. Toxaphene [8001-35-2] is prepared by chlorination of the bicyclic terpene, camphene [79-92-5] to yield a product containing 67–69% chlorine and has the empirical formula $C_{10}H_{10}C_{18}$. The technical product is a yellowish, semicrystalline gum (mp

65–90°C, d 1.64) and is a mixture of 175 polychlorinated derivatives. Toxaphene is unstable in the presence of alkali, upon prolonged exposure to sunlight, and at temperatures above 155°C, liberating hydrogen chloride and losing some of its insecticidal potency. It is very soluble in organic solvents, but only soluble to 0.4 mg/L in water. The oral LD_{50} to the rat is 69 mg/kg.

The most active ingredients in technical toxaphene are

2,2,5-endo-6-exo-8,9,10-heptachlorobornane [51775-36-1 (mouse intraperitoneal LD_{50} 6.6 mg/kg) and 2,2,5-endo-6-exo-8,9,9,10-octachlorobornane [58002-18-9] (mouse ip LD_{50} 3.1 mg/kg). Each constitutes *ca* 2–6% of the technical mixture.

Environmental

Toxaphene is extremely toxic to fish LC_{50} values to trout and bluegill of 0.003–0.006 ppm. At water concentrations as low as 0.00005 ppm, toxaphene-treated fish suffer broken-back syndrome, a crippling

collagen deformity. Bioaccumulation occurs from water to fish at levels up to 100,000-fold. Toxaphene also is highly toxic to birds (oral LD_{50} to pheasant 40 and 71 mg/kg).

The soil persistence of toxaphene is difficult to asses because of the complex mixture, but published estimates for half-life range from 2 months to 10 years. Toxaphene is a broad-spectrum, persistent pesticide that was widely used on cotton and other field crops. Its registration was revoked by the U.S. Environmental Protection Agency in 1983.

Toxicology

Toxaphene has been found to be carcinogenic in rats and mice. It was found to cause an increased incidence of thyroid and pituitary adenomas in the rat and hepatocellular carcinomas in the mouse. Developmental toxicity in the rat was reported as a significant reduction in the number of fetal ossification centers with increasing dose.

Section 5

Hexachlorobenzene

Hexachlorobenzene (77) [118-74-1], HCB, a fungicide, is used as a seed protectant. Hexachlorobenzene acts as a selective fungicide and exerts a fumigant action on fungal spores. It is a white crystalline compound mp 226°C and is almost insoluble in water. Hexachlorobenzene is very stable, unreactive toward acids and bases, and persistent in the environment.

Photolysis is very slow, and in artificial sunlight, solid hexachlorobenzene photodecomposed after 5 months. In sunlight, 20 g of hexachlorobenzene contained in a borosilicate flask gave 64 ppm of pentachlorobiphenyl after 56 days (104).

Sensitized photolysis of HCB at wavelengths greater than 285 nm in acetonitrile/water containing acetone gave dechlorinated products: pentachlorobenzene (78) (71%), 1,2,3,4-tetrachlorobenzene **(79) (0.6%)**, 1,2,3,5-tetrachlorobenzene **(80) (2.2%)**, and 1,2,4,5-tetrachlorobenzene **(81) (3.7%)**.

Without acetone, products included pentachlorobenzene (78) (76.8%), 1,2,3,5-tetrachlorobenzene **(80) (1.2%)**, 1,2,4,5- tetrachlorobenzene **(81) (1.7%)**, and 1,2,4-trichlorobenzene

(82) (0.2%) (105). Irradiation of hexachlorobenzene in methanol solution at wavelengths greater than 260 nm gave a mixture of reductively dechlorinated products (pentachlorobenzene and a tetrachlorobenzene, probably 80) and pentachlorobenzyl alcohol **83,** and also a tetrachlorodi(hydroxymethyl)benzene (106). A similar product mixture was obtained by exposing a methanolic solution of hexachlorobenzene in methanol to sunlight outdoors. After 15 days, only 30% of hexachlorobenzene was recovered

Photolysis rates were enhanced by the addition of sensitizers (diphenylamine, tryptophane, and naturally occurring organic substances), but no products were identified. In an anaerobic sewage sludge, hexachlorobenzene was reductively dechlorinated and the principal product was 1,3,5-trichlorobenzene (84). Pentachlorobenzene, 1,2,3,5-tetrachlorobenzene, and dichlorobenzenes were also identified (107). In activated sludge, 1.5% of hexachlorobenzene was mineralized as carbon dioxide after 5 days.

Metabolism

Metabolism in mammals is slow, and metabolites include polychlorinated phenols and benzenes, and many sulfur derivatives. Enzyme preparations from liver, lung, kidney, and small intestine dechlorinate HCB, and hepatic mixed function oxidases are responsible for the formation of pentachlorophenol and other phenols. The principal metabolites in mammals are pentachlorophenol (85), tetrachlorohydroqu none, and pentachlorothiophenol.

Lesser amounts of pentachlorobenzene, tetrachlorobenzene, 2,3,4,6- and 2,3,5,6-tetrachlorophenols, and 2,4,5-and 2,4,6-trichlorophenols are also produced. Adult male rats excreted less than 1% of a single oral dose of ^{14}C-labeled HCB in urine within 7 days.

Pentachlorophenol, 2,4,5-trichlorophenol, tetrachlorobezene, and pentachlorobenzene were detected. A total of 16% of the dose was excreted in feces, but no metabolites were detected. A total of 70% of the dose remained in the body, mainly in fat, and this was mostly HCB with traces of dechlorinated metabolites (108).

Enzyme studies in vitro indicated that HCB was dechlorinated by enzyme preparations from liver, lung, kidney, and small intestine and that pentachlorophenol was formed by hepatic mixed function oxidases. The identification of sulfur-containing metabolites, including N-acetyl-S-pentachlorophenyl cysteine (pentach-lorophenylmercapturic acid) **(86)** and pentachloromethylthiobenzene (109) indicated that the reaction of glutathione with HCB may be an intermediate step in metabolism.

A key step in the metabolic pathway is the formation of pentachlorophenylmercapturic acid as a polar intermediate. The distribution pattern of metabolites was similar in rat urine and feces, except that tetrachloro-1,hydroquinone was detected only in urine. When rats were treated orally with 50 mg Kg^{-1} HCB every other day for 2 weeks, 21 urinary metabolites could be separated by capillary gas chromatography.

The may have been more than 21 metabolites in the excreta because conjugates were hydrolyzed during the procedure. Of the known HCB metabolites in humans, pentachlorophenol occurred in urine in larger amounts than in feces, and urinary concentrations were dependent on the concentrations of HCB in human adipose tissue.

Metabolites were formed by reductive dechlorination, ring hydroxylation, or replacement of chlorine by sulfur containing moieties (104). Pentachlorophenol, 2,4,5-trichlorophenol, and pentachlorobenzene were detected *in vivo* (103). Enzyme studies *in vitro* indicated that HCB was dechlorinated by enzyme preparations from liver, lung, kidney, and small intestine and that pentachlorophenol was formed by hepatic mixed function oxidases. A variety of sulfur-containing metabolites, including N-acetyl-S-pentachlorophenyl cysteine and pentachloromethylthiobenzene, were identified (110). In higher animals, HCB metabolites are excreted primarily in feces, which contain unchanged HCB and polar metabolites. In lactating animals, HCB may be excreted in milk, but an important route may be passive elimination across the intestinal wall into the contents of the gut.

Toxicology

According to WHO, HCB has been found to be carcinogenic in several animal studies as well as having adverse non-neoplastic effects on a number of organ syst-ems. An accidental human poisoning took place in Turkey in 1955–1959, during which HCB-contaminated wheat flour was used to make bread.

There were over 60 cases of porphyria cutanea tarda, related to disturbances in porphyrin metab-olism, with high mortality. Seve developmental toxicity was also reported: Nursing infants of dosed mothers developed pembe yara or pink sore, and most died within a year.

Although follow-up studies on survivors were conducted over 20 to 30 years, no consistent epidemi-ological evidence was developed for an increased cancer incidence, but other abnormalities persisted.

Pentachlorophenol

Pentachlorophenol [87-86-5] mp. 178 °C (technical), 191 °C (anhydrous) is an insecticide, fungicide, and nonselective contact herbicide. It is used to control termites, and it has been used extensively as a wood preservative to protect timber against rot and marine borers, and in 1972, 38 million pounds of an estimated U.S. production of more than 50 million pounds was used for this purpose.

The sodium salt has been used as a general disinfectant. It was widely used as a wood preservative, but its use in proximity to water led to leaching of pentachlorophenol and its associated impurities. It has now been displaced because of its potential for contamination of many ecosystems. Its metabolism, toxicology, and environmental effects have been investigated intensively, and pentachlorophenol and its associated impurities are the subject of extensive literature (112,113).

Impurities of manufacture included dioxins, such as TCDD, dibenzofurans, and hexachlorobenzene. Pentachlorophenol was manufactured in the United States by direct chlorination of phenol or chlorophenols. Chlorination is performed at atmospheric pressure.

The temperature in the primary reactor is in the range 65–130°C (preferably 105°C) and is held in this range until the melting point of the product reaches 95°C. The temperature is increased to maintain a temperature of about 10°C above the product melting point until the reaction is complete in 5–15 hours. The process gave rise to a number of related impurities.

The commercial product contained about 10% tetrachlorophenol. Dow commercialized a product containing 88% chlorophenol, 2,3,4,6-tetrachlorophenol, less than 30 ppm octachlorodibenzo-p-dioxin, and less than 1 ppm hexachlorodibenzo-p-dioxin. Some commercial products contained up to 2500 ppm octachlorodibenzo-p-dioxin and up to 27 ppm of the hexachloro-dibenzo-p-dioxins. Such compounds are extremely toxic to a variety of organisms, including mammals. Pentachlorophenol (PCP) (85) decomposed on exposure to sunlight. There have been a number of studies. Ultraviolet irradiation in hexane or methanol gave 2,3,5,6-tetrachlorophenol **(86)** by reductive loss of chlorine, whereas irradiation of a suspension of the free phenol in water afforded polymeric substances as the major products with a little tetrachlorophenol, chloranil (90), and chloranilic acid (88).

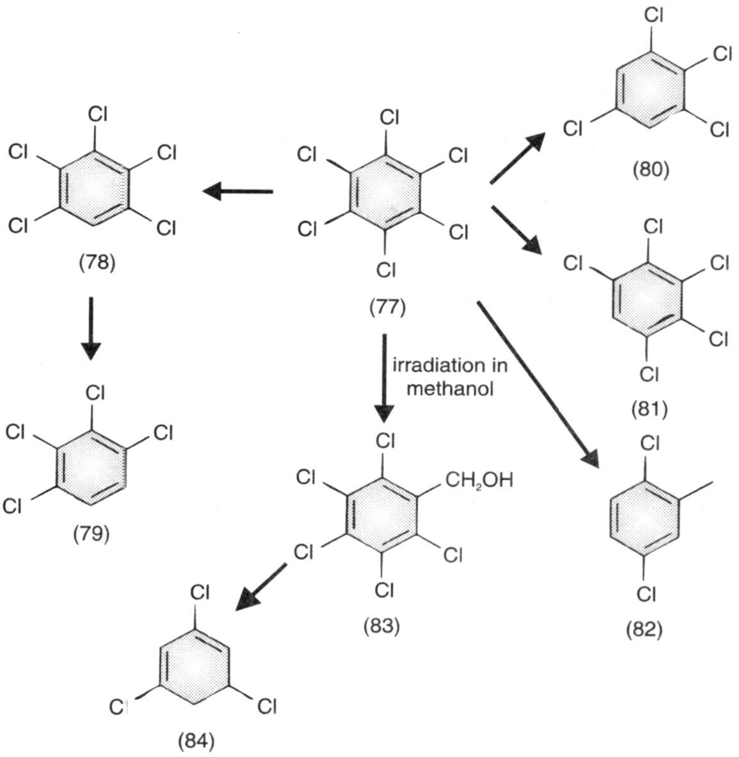

Figure 17.31: Hexachlorobenzene photolysis.

When an aqueous solution was exposed to sunlight for 10 days, the violet-coloured solution contained a number of reaction products (114). The major products were chloranilic acid (88) and 3,4,5-trichloro-6-(2′-hydroxy-3′,4′,5′,6′-tetrachlorophenoxy)-o-benzoquinone. Minor products also identified were tetrachloror-esorcinol (0.10%) (87), 2,5-dichloro-3-hydroxy-6-pentachlorophenoxy-p-benzoquinone **(91)** (0.16%), and 3,5-dichloro-2-hydroxy5- (2′,4′, 5′,6′-tetrachloro-3-hydroxyphenoxy-p-benzoquinone **(93)** (0.08%).

In hexane, 2,3,5,6-tetrachlorophenol **(86)** was the major product (30% after 32-h irradiation) and a 10% of a compound tentatively identified as a tetrachlorophenol (115). Trace amounts of octachlorodibenzo-p-dioxin **(94)** were obtained when sodium pentachlorophenate was irradiated by natural or artificial sunlight (116). In soils, PCP may undergo reductive dechlorination under anaerobic conditions to give tetra-, tri- and dichlorophenols and m-chlorophenol.

In aerobic and anaerobic soils, the major metabolite was pentachloroanisole and lesser chlorinated phenols were also formed (117,118). Microbial conversion in aquatic situations or in activated sludge also gives rise to lesser chlorinated phenols, and pentachloroanisole was also identified among the products when the disappearance of pentachlorophenol was investigated under aerobic or anaerobic conditions in aquaria.

Metabolism

Pentachlorophenol was metabolized in rats by conjugation with glucuronic acid and eliminated

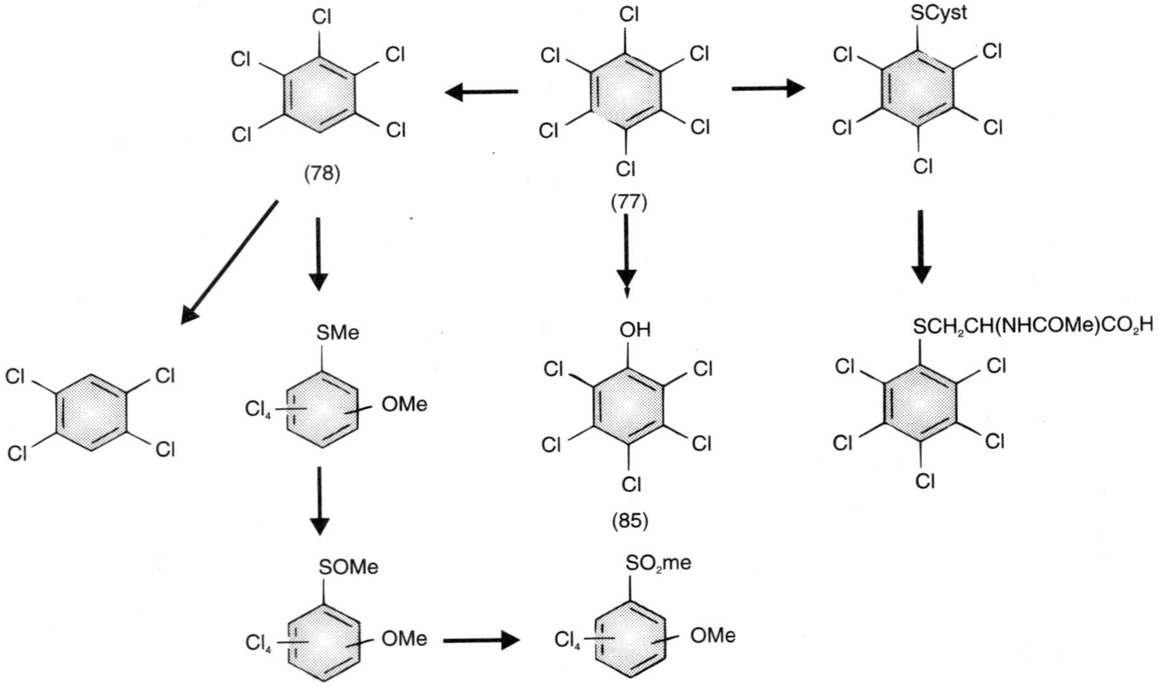

Figure 17.32: HCB metabolism—initial steps in pathways showing formation of sulfur- and oxygen-containing compounds.

as the glucuronide. P_{450} catalyzed oxidative dechlorination alo occurred to form tetrachlorohydroquinone, and this was conjugated to form a monoglucuronide representing 27% of the dose administered. Other metabolites have been reported, including isomeric tetrachlorophenols, tetrachlorocatechol and tetrachlororesorcinol. Trace amounts of benzoquinones were also noted.

Metabolites in female rats were tetrachloromonophenols, diphenols, and hydroquinones.

Mode of Action

PCP acts as a biocide through its ability to uncouple mitochondrial oxidative phosphorylation.

Toxicology

The toxicology has been addressed in a recent risk assessment. Acutely, pentachlorophenol was reported to have LD_{50} values in the rat of 12 mg/kg (inhalation) and 146 mg/kg (M)–175 mg/kg (F) by oral gavage. More detailed studies of the toxicology of pentachlorophenol have been compromised by the toxicity of impurities present in most of the earlier samples used in the evaluation process.

These include dioxins, such as TCDD, dibenzofurans, and hexachlorobenzene, each causing a range of toxicological effects, some of which may have overlapped those caused by PCP itself. In addition, the principal rodent metabolite of PCP is genotoxic, tetrachloro-1,4-hydroquinone (TCHQ) (89). Although it is apparently not formed in humans *in vivo*, its formation has nevertheless been

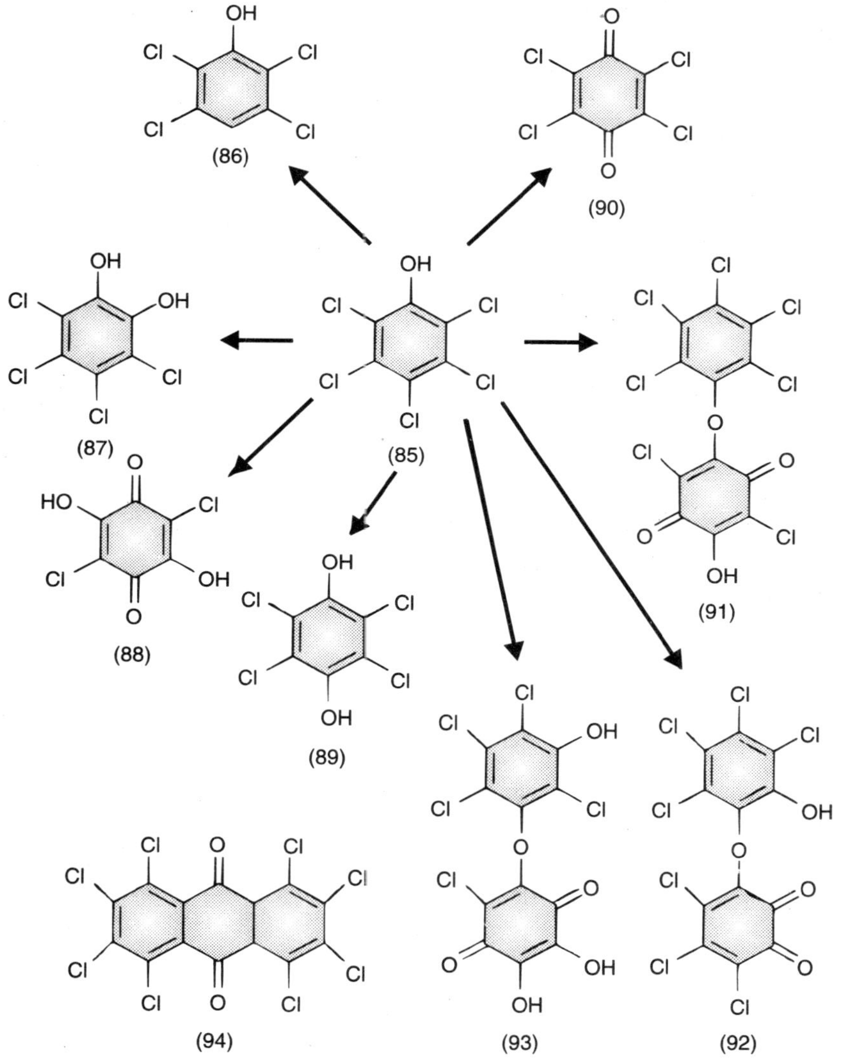

Figure 17.33: Photolysis of pentachlorophenol.

demonstrated *in vitro* using human microsomal mixed function oxygenases. Although a number of toxicity studies have been conducted with both known impurities and TCHQ, it is often difficult to know whether animal experiments are valid for human health risk assessment. Nevertheless, it appears that the main target organ of purified TCP in animals is the liver. This toxicity was manifested as liver inflammation, increased relative weight, and increased serum alkaline phosphatase.

The estimated chronic NOEL in the dog for these effects was 0.15 mg/kg/day, from a 1-year study, based on a LOEL of 1.5 mg/kg/day. In the rat, a significantly increased incidence of mesotheliomas ($p<0.05$) and nasal carcinomas in males was reported at the highest dose tested,

60 mg/kg/day; there were no tumor increases in females. In the male mouse, an elevation of benign and malignant liver tumors was recorded; in females, benign liver tumors along with adrenal and blood vessel tumors had increased incidences. It is possible that one or more of these (liver) tumors arose as a secondary consequence of liver toxicity. Genotoxicity tests on PCP yielded mixed results, generally being negative in bacterial gene mutation assays but positive in mammalian cells, in the presence but not in the absence of rodent S9 liver microsomes.

PCP has been shown not to bind to DNA. The rodent metabolite, TCHQ, is mutagenic and clastogenic in mammalian cells, in each study in which it has been tested. It has also been shown to bind to DNA. Developmental and reproductive toxicity was not detected at doses that were not maternally toxic.

Other toxic effects include acute effects on the thyroid gland, manifested as decreased levels of plasma T3 and T4 hormone levels in the rat, after intraperitoneal injection, with NOEL values of 0.6 mg/kg (PCP) and 6.5 mg/kg (TCHQ). PCP has been classified by U.S. Environmental Protection Agency as B2, a probable human carcinogen.

INDEX

A

B

C

D

E

F

G

H

I

J

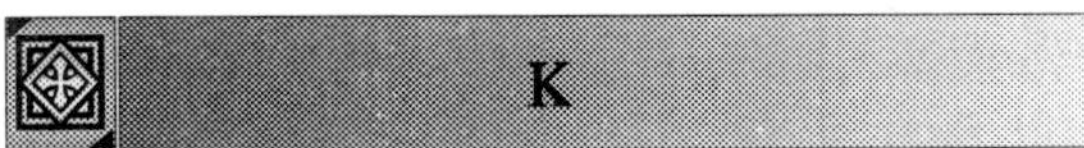

K

L

M

N

O

P

Q

R

S

T

U

V

W

X

Y

Z